$$\sum f = 0$$

$$\sum M = 0$$

APPLIED MECHANICS
STATICS

APPLIED MECHANICS
STATICS

CHARLES E. SMITH

OREGON STATE UNIVERSITY

SECOND EDITION

JOHN WILEY & SONS

NEW YORK CHICHESTER BRISBANE TORONTO SINGAPORE

Library of Congress Cataloging in Publication Data:

Smith, Charles Edward, 1932-
Applied mechanics, statics

Includes bibliographical references and indexes.
1. Mechanics, Applied—Collected works.
I. Title.

TA350.S572 1981b 620.1′03 81-4732
ISBN 0-471-02965-3 AACR2

Printed in the United States of America

10 9 8 7 6 5 4 3 2 1

PREFACE

The objectives of these two volumes are essentially the same as those of the three volumes of the first edition. They have been designed primarily for use by students in all branches of engineering, who usually complete their first courses in engineering mechanics during the sophomore year. The latter portion of *Dynamics* provides further instruction for those students who are specializing in mechanical or aerospace engineering.

The organization of *Statics* is the same as that of the first edition; modifications consist of much more detailed explanations, with additional examples and exercise problems.

Dynamics contains the material on basic kinematics and kinetics of particles and rigid bodies from the second volume of the first edition, also explained in much more detail and with many more examples. This is supplemented with introductory sections on mechanical vibration and orbital motion and material from the third volume of the first edition, on motion relative to rotating reference frames and general rigid body dynamics.

Although strong logical interdependence exists among the sections that are essential to every basic course in mechanics, there is flexibility in the choice of other material. An understanding of all the material in Chapters 1 through 4 and Section 6-1 is essential, and the applications in Chapter 5 provide important practice so that this chapter should also be considered essential.

Chapters 1 and 2 cover concepts that may well be familiar to students from previous courses in physics and mathematics; thus, these may be studied at an appropriate pace, so long as the material is mastered. Experienced teachers of engineering mechanics all recognize that the concepts and methodology presented in Chapter 4 are the key to any analysis involving forces, so that mastery of this part of a statics course is absolutely essential. One who cannot draw a proper free-body diagram and correctly write the corresponding equations of equilibrium is simply not ready to advance. Except for Section 6-1, the topics within Chapters 6 and 7 can be chosen to suit personal preferences and the needs of particular curricula—none of these topics is essential background for *Dynamics*. Of course, concepts and methods presented in Section 6-3 are essential to a part of the study of mechanics of deformable bodies ("Strength of materials"), so there is an obvious advantage to studying this in the statics course instead of postponing it for the later course.

The comments made in the preface to the first edition, on necessary background, reasons for emphasis of geometric concepts, and suggestions for successful study, apply to this edition as well.

Corvallis, Oregon **CHARLES E. SMITH**

PREFACE TO THE FIRST EDITION

These three volumes on applied mechanics, *Statics, Dynamics*, and *More Dynamics* are designed to meet two needs. The first is for a textbook for the course in statics and dynamics, which is usually taught for students in several branches of engineering at the sophomore level. For students majoring in civil, chemical, electrical, and some other branches of engineering, instruction in these subjects usually ends after the equivalent of about six or eight quarter hours. The second need is for instruction in dynamics at the upper division level, for students who are specializing in mechanical, aerospace, and some types of "systems" engineering.

Although some present text books are satisfactory for the first course alone, transition into the second level, usually starting somewhere in the middle of another book, is awkward. Also, those books that are satisfactory for students who plan further study of dynamics are not written with the students of the first group in mind.

The first two volumes of this set, *Statics* and *Dynamics*, are written for a course that teaches students to apply the rudiments of mechanics and, simultaneously, provides the basis for study of the topics covered in the third volume. This volume, *More Dynamics*, covers subjects that are generally within the domain of mechanical, aerospace, and some "systems" engineers.

Concern for the needs of the student who will not study these latter topics dictates that the material offered in these volumes be organized somewhat differently than for a course that is designed only for mechanical and aerospace engineers. Specifically, treatments of kinematics and rigid body dynamics are each split, the more involved aspects, rotating reference frames and rigid body motions in which the direction of the angular velocity vector varies, are discussed in the third volume. Although this may not be the most efficient organization from a logical point of view, experienced teachers know that few students can handle this level of generality without some prior experience with the simpler special cases.

Trigonometry and first-year calculus are the required background for *Statics*. In Chapter 1, the rudiments of Newtonian mechanics are covered in sufficient detail that a course can be given without a physics prerequisite. If a physics prerequisite is used, students can move more quickly through this material. Similarly, a class of students who have covered vector algebra can progress quickly through Chapter 3. As a minimum, a course in *Statics* should cover Chapters 1 through 4, Section 6-1, and enough of the remainder of the book to insure that students are adequately skilled in drawing free-body diagrams and in writing and solving equations of equilibrium. Other sections may be chosen nearly independently of one another, the choices depending on the objectives of the curriculum.

The heavy emphasis on the geometric interpretation of vectors reflects my belief that students cannot learn mechanics by replacing a visual process with formal manipulation of equations. Too often, attention is given to the formal algebra of vectors in terms of sets of components, at the expense of attention to the *meaning* of vector relationships basic to mechanics. Instruction should encourage students to seek a spatial understanding first and to refer to this constantly as analysis proceeds. For example, the symbol $\mathbf{A} \times \mathbf{B}$ should bring to mind immediately the sketch on page 62, but should suggest the formula (3-25) only when a computation is required.

Very few people can learn mechanics solely by observing the analysis carried out by someone else. At some point (the sooner the better) the student must discard the observer's role and, making and correcting the inevitable mistakes, attempt to work problems and carry out derivations with increasing independence. For this reason, these books will be relatively ineffective in the lap of someone sitting in an easy chair. They must be studied at a desk with an ample supply of scratch paper at hand. The paper will be needed for attempting solutions of example problems before given solutions are read, for carrying out steps of analysis that are omitted from the books, and for making supplemental sketches. Students who work in this way, always attempting to relate solution procedures to basic ideas, will achieve an understanding that is satisfying as well as professionally useful. In this endeavor I wish them well.

Corvallis, Oregon **Charles E. Smith**
July 1975

ACKNOWLEDGMENTS

There were many contributors to this project, and I am grateful to all of them. Valuable suggestions came from many students, especially Brad Whiting and John Gale. Dr. Hans J. Dahlke pointed out a number of errors and shortcomings in *Statics*. Dr. William E. Holley, Dr. Robert W. Thresher, and Dr. Robert E. Wilson made many helpful suggestions for Dynamics. A number of the exercises that illustrate realistic engineering applications were ideas from Dr. Robert J. Zaworski. Suggestions from Professor James M. Gere of Stanford University were most helpful in planning the second edition. Thurman R. Poston of Wiley Interscience has been especially helpful in seeing the first edition through production and the second edition into production. I thank Carol Beasley and the rest of the staff of Wiley for the editing and production. Important sources of inspiration have been Emeritus Professor Kenneth E. Bisshopp of Rensselaer Polytechnic Institute and Emeritus Professor Donovan H. Young of Stanford University. Most important of all has been the time made available by the countless sacrifices of Marian, Brian, and Susie.

C.E.S.

CONTENTS

APPLIED MECHANICS
STATICS

1

INTRODUCTORY CONCEPTS

Exciting advancements with significant effects on methods of modern engineering are appearing at an ever-increasing rate. Improvements in electronic computers, new instruments for making measurements and analyzing the information that these instruments produce, and many other innovations provide an effective set of tools with which engineers can carry out their tasks.

Yet, there is hardly a design that does not rely heavily on the relatively old science of *classical mechanics*. This view of the mechanical interactions among objects was first formalized by Sir Isaac Newton in his *Principia* in 1686. Important extensions of Newton's laws of motion were developed by D'Alembert (about 1743), Lagrange (about 1788), and Hamilton (about 1827). The subsequent mathematical developments of vector and tensor analysis, just prior to 1900, further facilitate our understanding and ability to use the principles of mechanics. The knowledge we convey here stems directly from these developments, so it is by no means new.

However, new and interesting *applications* for this science arise as rapidly as ideas for new devices, so that a working understanding of Newton's laws and their extensions is as indispensable to the engineer as modern electronic devices and computer systems.

1-1 NEWTON'S LAWS OF MOTION

Newton's three laws were originally stated as follows.

1. Every body continues in its state of rest, or of uniform motion in a straight line, unless it is compelled to change that state by forces impressed on it.
2. The change of motion is proportional to the motive force impressed and is made in the direction in which that force is impressed.
3. To every action there is always opposed an equal reaction; or, the mutual actions of two bodies on each other are always equal and directed to contrary parts.

Before these notions can be used, the terms "uniform motion in a straight line," "change of motion," "motive force impressed," and "action and reaction" must be understood fairly precisely.

Toward such an understanding let us consider the rigid block depicted in Figure 1-1. The block—in this case the "body" referred to in Newton's laws—is free to slide on the perfectly smooth surface and is under the influence of the attached spring. If the spring is in its relaxed state, it will exert no influence on the block; then, if the block is at rest, it will remain at rest, and if it is moving with some speed it will continue to move at this same speed. If the spring is held in an extended state, the rightward speed v of the block is observed to increase, the time rate of increase dv/dt depending explicitly and only on the extension in the spring. The influence that the spring exerts on the block is called the impressed *force*, and is directly associated with the observed time rate of change of speed, or *acceleration* of the block.

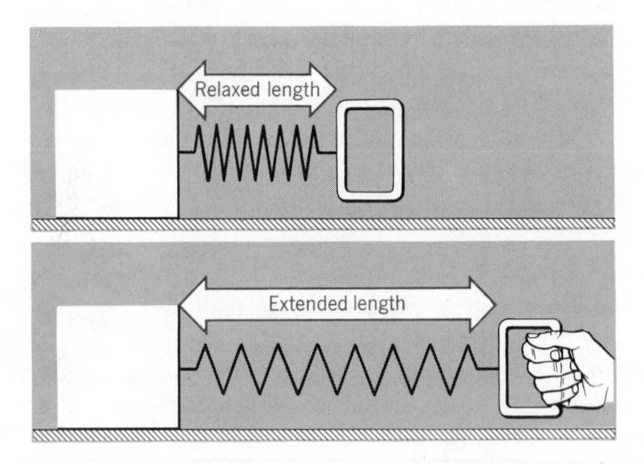

FIGURE 1-1

A quantitative definition of *force* is given operationally in terms of the observed acceleration of a standard object, a platinum cylinder located at the headquarters of the International Bureau of Weights and Measures in Paris. The force that will produce in this cylinder an acceleration dv/dt equal to A meters per second per second, is defined as having a magnitude of A *newtons*. The earth attracts a medium-small apple with a force of about one newton.

Now consider an extension of the above experiment. Suppose that for each of a number of different amounts of spring extension, we measure and record the acceleration of the standard object. We can now determine the force (originally defined in terms of the acceleration of the standard object) in terms of an observable quantity independent of the object, namely, the extension of the spring. Next, let us attach our newly calibrated spring to several different objects and measure the correspondence between force (now determined in terms of spring extension) and corresponding acceleration. The following fact emerges: for each object, the force f is directly proportional to the acceleration a; that is,

$$f = ma$$

The constant of proportionality m is a characteristic property of each object, called its *mass*, and is different for different objects. For example, the force required to impart a given acceleration to a hockey puck is considerably less than that required to impart the same acceleration to a goalie, which means that the mass of the puck is much smaller than that of the goalie. Or, a given force will impart a much greater acceleration to the puck than it will to the goalie.

Quantitative definition of mass is given in terms of that of the standard object discussed above. The *kilogram* is the unit defined as the mass of this standard object, which is called the international prototype of the kilogram. To determine the mass m of an object, we could compare its acceleration with that of the international prototype, both objects being subjected to the same force f. By using the subscript 0 to refer to the prototype, we can write

$$f = m_0 a_0 = ma$$

or

$$m = \frac{a_0}{a} m_0$$

In this way, we can determine the mass m of any object, in kilograms, as the ratio of the acceleration of the prototype to that of the other object. The mass of a hockey puck is approximately 0.16 kg, and that of a typical, suited-up goalie would be approximately 100 kg.

In addition to magnitude, the concept of force must include consideration of direction. (Note that this word appears in Newton's statement of his second law.) We will tackle a precise meaning of direction as applied to velocity and acceleration in later chapters; for now, the direction of a constant force can be taken as the direction in which a rigid object, initially at rest and acted on by this and no other force, is observed to move.

As a means of depicting both magnitude and direction, we use an arrow, its length proportional to the *magnitude* of the force (i.e., the number of newtons measuring it), and the direction indicated by its orientation with respect to the physical objects under analysis. Symbols used in equations and discussions to represent forces and other entities that have both magnitude and direction will appear throughout this book in boldface type, for example, **f**, **a**, and so forth. Thus newton's second law, including its directional aspect, is written in the form:

$$\mathbf{f} = m\,\mathbf{a} \tag{1-1}$$

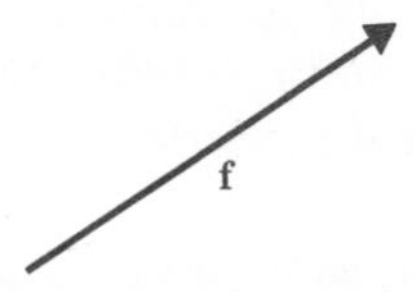

Newton's third law can be explained in terms of Figure 1-2. There, the block and spring have each been isolated from the objects with which they interact (i.e., the block isolated from the spring, and the spring isolated from the block and the hand), and the forces acting on each object are depicted by arrows. These diagrams are called *free-body diagrams*, and they are indispensable in the analysis of forces in mechanical systems. The force **f** from the spring acting on the block is directed to the right, causing the block to accelerate in that direction. Newton's third law states that the block must exert a force of equal magnitude and opposite direction on the spring. The spring is under the influence of this force $-\mathbf{f}$ from the block and the force exerted by the hand at the opposite end.

The force interaction in the example shown in Figures 1-1 and 1-2 occurs at the surface where the spring and block make contact. Force interactions can also occur between bodies that are not in contact but are merely in proximity with one another. Examples of this situation are electromagnetically induced forces and gravitational forces.

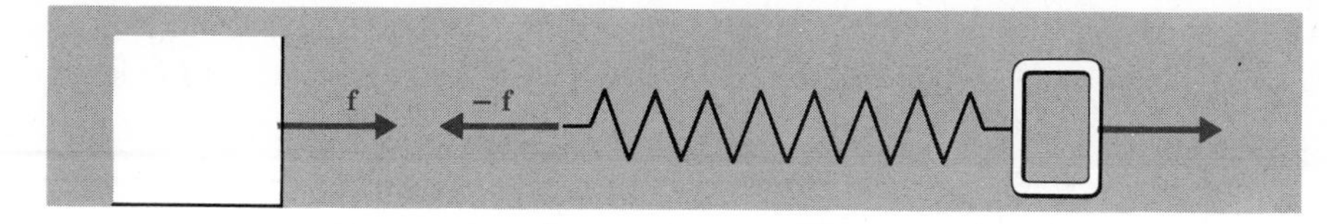

FIGURE 1-2

PROBLEMS

1-1 Explain how Newton's first *two* laws can be expressed by the single Equation (1-1).

1-2 A 0.05-g feather is falling through still air at a constant speed of 0.2 m/s. What is the resultant force acting on the feather?

1-3 What magnitude of acceleration will be imparted to a 0.12-kg fig by a 1.8-newton force? *Ans.* 15 m/s^2.

1-4 What magnitude of force will be required to accelerate a 1500-kg automobile at 2.5 m/s^2?

1-5 A 10-g bullet goes from rest to a speed of 900 m/s in 0.0015 seconds. What is the average force exerted on the bullet during this time? *Ans.* 6000 newtons.

1-6 An extended spring with negligible mass pulls two blocks toward one another, with the spring exerting the only significant force. At the instant that the magnitude of this force is 15 newtons, what is the acceleration of each block?

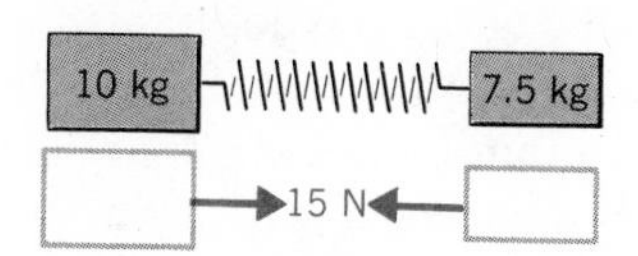

1-7 An extended spring with negligible mass pulls two blocks toward one another, and the spring exerts the only significant force. At the instant the 10-kg block is observed to accelerate at 25 m/s^2, the other block is observed to accelerate at 40 m/s^2. What is the mass of the second block?

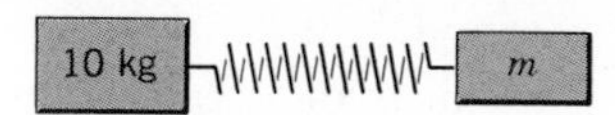

1-8 Two teams are engaged in a tug-of-war. According to Newton's third law, the magnitude of the force that each team exerts on the other is equal to the tension force in the section of rope separating the two teams, so that the magnitude of the force exerted by team A on team B is equal to the magnitude of the force exerted by team B on team A. In view of this, how can either team win?

1-2 NEWTON'S LAW OF GRAVITATION

Another of Newton's great achievements was his deduction of the law that gives the magnitude of gravitational attraction between bodies. Together with his laws of motion, this law must be consistent with observed motions of the bodies in the solar system. It states that the magnitude of the gravitational force is given by

$$f_g = \frac{\gamma m_1 m_2}{r^2} \tag{1-2}$$

in which m_1 and m_2 are the masses of the two objects attracting one another, r is the distance between the centers* of the objects, and γ is the *universal gravitational constant*, having the same value for all pairs of objects (see Appendix A-1, p. 302). When one of the two objects is the earth and the other object is near the surface of the earth (where r is about 6400 km), the group of factors $\gamma m_{\text{earth}}/r_{\text{earth}}^2$ is essentially constant, and we abbreviate it as g. Then the attractive law becomes

$$f_g = mg,$$

in which m is the mass of the object near the earth's surface. Precise measurements show a variation in g of about 0.5% among various locations on the earth's surface. For most estimates of gravitational forces near the surface of the earth, the value $g = 9.81$ newtons per kilogram (N/kg) is used. Thus a person having a mass of 82 kg will be attracted toward the earth by a force of magnitude

$$f_g = (82 \text{ kg})(9.81 \text{ N/kg}) = 804 \text{ N}$$

Consider now an object near the surface of the earth, that has no significant forces acting on it other than the gravitational attraction exerted by the earth. Newton's second law of motion and law of gravitation give us the following value of the earthward acceleration of the body in this free-fall condition:

*The precise meaning of "center" in this case requires some fairly detailed analysis, which we postpone for now.

$$a_g = \frac{f_g}{m} = \frac{mg}{m} = g$$

This tells us that the acceleration is independent of the mass of the falling object. That is, every object will be accelerated at the same rate, regardless of its mass, if the earth's gravitational attraction is the only significant force. The fact that this agrees with observed behavior of free-falling objects provides one of the verifications of Newton's laws of motion and gravitation.

There are, of course, many other implications of Newton's laws of motion and gravitation. The exposition of some of them, and the development of skills that will enable the student to determine other implications that bear on the design of various mechanical devices, are the primary concerns of this book.

Certain observed physical behavior cannot be satisfactorily predicted by Newtonian mechanics, most notably the behavior of particles of atomic size and objects traveling at speeds near that of light. Nevertheless, it is an amazing fact that, 300 years after their formulation, Newton's laws form the basis for successful prediction of the mechanical behavior of the vast majority of the systems that are employed by technology.

PROBLEMS

1-9 From the values of physical constants given in Appendix A-1, estimate the radius of the earth.

1-10 The radius of the moon is approximately 0.27 times that of the earth, and its mass is approximately 0.012 times that of the earth. At what rate will an object dropped near the surface of the moon be accelerated?

1-11 The mean distance between the earth and the moon is 384 400 km and the mass of the moon is approximately 0.012 times that of the earth. Estimate the attractive force between them. *Ans.* 1.94×10^{20} newtons.

1-12 A space capsule is at a location on a direct line between the earth and the moon. The gravitational forces acting on the capsule exactly cancel each other. What is the altitude of the capsule?

1-13 A 45-Mg spaceship is at an altitude above the earth's surface of 200 km. Estimate the gravitational force. Estimate the acceleration of the *earth* caused by the force that the spaceship exerts on the earth.

1-14 A feather is observed to fall more slowly than a rock. This seems at odds with the statement that the free-fall acceleration of an object is independent of its mass. Explain.

1-15 What is the magnitude of the aerodynamic force acting on a 2-kg object that is falling with a downward acceleration of 9.3 m/s^2 near the surface of the earth? *Ans.* 1 newton.

1-16 The gravitational acceleration near the surface of Mars is about 0.38 times that near the surface of the earth, and the mass of Mars is about 0.08 times that of the earth. Estimate the radius of Mars.

1-17 At what altitude will the force of the earth's gravity be half that at zero altitude?

1-18 What will be the acceleration of an orbiting spaceship that is at an altitude of 700 km? *Ans.* 7.96 m/s^2.

1-19 Referring to the previous problem, will the value of the acceleration be affected by whether the spaceship is in a circular orbit, in an elliptical orbit, or falling directly toward the earth?

1-20 With the information given in Problem 1-10 and Appendix A-1, estimate the magnitude of the acceleration of the earth caused by the moon.

1-21 A 60-Mg rocket is launched vertically upward. What is the thrust of the engines

(a) When the rocket is hovering over the launch pad before accelerating?

(b) When the rocket has an upward acceleration of 6 m/s^2?

1-3 DIMENSIONS AND UNITS OF MEASUREMENTS

Science and engineering are largely *quantitative* endeavors, so that they depend heavily on logical and consistent schemes for measuring physical entities and expressing interrelationships among measured values of them. We have just encountered the entities velocity, acceleration, force, and mass and have discussed ways of quantifying the last two. Any scheme of measurement is, ultimately, a set of comparisons with an agreed-on *standard*, such as the platinum cylinder discussed above. Standards for mass, length, and time, developed by the International Organization for Standardization (ISO) and accepted worldwide, are:

Mass The *kilogram* is the unit of mass; it is equal to the mass of the international prototype of the kilogram. (An average man has a mass of about 75 kg.)

Length The *meter* is the length equal to 1 650 763.73 wavelengths in vacuum of the radiation corresponding to the transition between levels $2p_{10}$ and $5d_5$ of the krypton-86 atom. (An average man is about 1.8 m tall.)

Time The *second* is the duration of 9 192 631 770 periods of the radiation corresponding to the transition between the two hyperfine levels of the ground state of the cesium-133 atom. (This is the familiar one-sixtieth part of a minute.)

All other quantities in mechanics can be determined in terms of the quantities of mass, length, and time. For example, velocity is defined as an increment of distance traveled divided by the corresponding increment of time and, thus, can be measured in meters per second. The quantities mass, length, and time are said to be *fundamental*, and those of other mechanical quantities, such as velocity, are said to be *derived*.*

The *dimensions* of any physical quantity give the relationship that that entity bears with the quantities taken as fundamental. These dimensions are expressed in terms of M (indicating mass), L (indicating length), and T (indicating time). The dimensions of velocity, for example, are length divided by time, indicated as L/T. Acceleration is a change in velocity (L/T) divided by an increment of time (T), so that it has dimensions $(L/T)/T = L/T^2$. Force is defined as the product of mass (M) and the acceleration it produces (L/T^2), so that it has dimensions ML/T^2.

Dimensional consistency must be satisfied by every equation used to describe physical phenomena. An equation is dimensionally consistent if every term has the same dimension. Use of this requirement is one of the most helpful ways of finding sources of errors in analysis of physical systems; therefore, the practice of making frequent dimensional checks is a sound habit. This checking can be accomplished by writing the dimensions of each quantity entering the equation, carrying out the multiplications and divisions indicated, and verifying consistency. For illustration, consider a formula that is to give the speed of fall of an object as it is pulled through a resistive medium by gravity. The analysis contains the assumption that the resistive force f_d is related to the speed v of the object as $f_d = cv$, where c is a constant depending on the viscosity of the medium and the size and shape of the object. The velocity of the object is supposed to be given as

$$v = \frac{mc}{g}[1 - e^{-(ct/m)}]$$

in which m is the mass of the object, g is the constant of gravity discussed previously, and t is time. To begin with, c is equal to f_d/v, so that it must have

*In the U.S. engineering system of units, force (measured in pounds-force), length, and time are considered as fundamental quantities and the relationship $f = ma$ is used to derive mass. Further details are presented in following paragraphs.

the dimension $(ML/T^2)/(L/T) = M/T$. Then, the argument ct/m of the exponential function is dimensionless $\{[(M/T)(T)]/(M) = 1\}$. This is as it must be, since the function $[1 - e^{-(ct/m)}]$ is expressible in terms of the series

$$(1 - e^{-\theta}) = \theta - \frac{\theta^2}{2!} + \frac{\theta^3}{3!} - \cdots$$

and the terms within this expression cannot be dimensionally consistent unless $\theta = ct/m$ is dimensionless. Next, the factor mc/g has dimensions $[(M)(M/T)]/(L/T^2) = M^2T/L$. But v has dimensions L/T, and so the equation cannot possibly be correct. The correct equation is

$$v = \frac{mg}{c}[1 - e^{-(ct/m)}]$$

which does pass dimensional inspection. Observe that satisfying dimensional consistency alone does not imply the validity of the relationship, but is necessary.

A bewildering number of systems of units are used in various disciplines and in different parts of the world. For many of them, the word system is overly flattering. In one office one might observe rates of energy transfer being calculated in watts by an electrical engineer, in horsepower by a mechanical designer, in British thermal units per hour by a heat transfer specialist, and in "tons of refrigeration" by a sales agent of air conditioning units. The man-hours consumed by designers of refrigeration systems in making these units compatible would be appalling if known. Fortunately, a world-organized effort is underway to encourage those in all fields to adopt a single, coherent system of measurement units called "Le Systeme International d'Unites," or SI units. Conversion to the SI is not likely to be completed for another decade or two, so that engineers will be expected to be familiar with some of the more awkward units as well as the SI for some time to come.

Units most commonly encountered in mechanics are listed in Table 1-1.* In the SI, the units of mass, length, and time are defined as *basic* (see p. 9), and the unit of force is *derived* from these. The newton (abbreviated N) is defined as the magnitude of that force that will impart, to an object with a mass of one kilogram, an acceleration with a magnitude of one meter per second per second. Thus, the newton is a special name for the kilogram-meter per second squared. This equivalence is written as $N = kg \cdot m/s^2$. In the system listed in the next column, the units of length, time, and force are treated as basic, and the unit of mass is derived from them. The pound-force is defined as

*More extensive lists may be found on the inside back cover and in Appendix A-3, p. 303.

TABLE 1-1

Quantity	SI Unit	Coherent U.S. Engineering Unit	Common Non-coherent Unit
Mass	kilogram (kg)	slug = $\mathrm{lbf \cdot s^2/ft}$ $\approx$ 14.594 kg	pound-mass (lbm) $\approx \frac{1}{32.174}$ slug $\approx$ 0.4536 kg
Length	meter (m)	foot (ft) = 0.3048 m	inch (in) = 25.4 mm
Time	second (s)	second (s)	minute (min) = 60 s
Force	newton (N) $\mathrm{N = kg \cdot m/s^2}$	pound-force (lbf) $\approx$ 4.448 N	kilogram-force (kgf) or kilopond (kp) = 9.806 65 N

4.448 221 615 260 5 N* and the foot as 0.3048 m. The *slug* is defined as that mass possessed by a body to which a one-pound force will impart an acceleration of one foot per second per second. Thus, the slug is a special name for the pound-force divided by foot per second squared. This equivalence is written as

$$\mathrm{slug} = \frac{\mathrm{lbf}}{\mathrm{ft/s^2}} = \mathrm{lbf \cdot s^2/ft}$$

The unit systems shown in the first two columns of the table are said to be *coherent*. This means that when numbers representing physical quantities in one of these systems are used in formulas, there is no need to introduce factors that depend on the units used. For example, the force required to accelerate a 45 000-kg spaceship at 5 $\mathrm{m/s^2}$ can be calculated by substituting $m = 45\ 000$ kg and $a = 5\ \mathrm{m/s^2}$ into the formula $f = ma$, and the result will be in newtons:

$$f = (45\ 000\ \mathrm{kg})(5\ \mathrm{m/s^2}) = 225\ 000\ \mathrm{N}$$

The same formula will give, in pounds-force, the force required to accelerate a

*The 14 digits are given here merely to emphasize that this is a definition. Rarely is precision beyond four significant digits required in an engineering computation.

3000-slug spaceship at 16 $\mathrm{ft/s^2}$, with the substitutions $m = 3000$ slugs and $a = 16\ \mathrm{ft/s^2}$:

$$f = (3000\ \mathrm{slugs})(16\ \mathrm{ft/s^2}) = 48\,000\ \mathrm{lbf}$$

Unfortunately we must deal with units that are not coherent with the kg, m, and s, or with the ft, s, and lbf. Four common ones are shown in the third column of Table 1-1. These are defined as indicated by the factors in the table. When dealing with values that are given in noncoherent units, proper conversions into a coherent system of units should be carried out as a first step.

The best practice for handling the sometimes confusing task of converting units is to write the associated units (in abbreviated form) alongside each numerical value and then to cancel like unit symbols exactly as we do with algebraic symbols. To illustrate, let us suppose that the fuel economy of a new automobile is given as 32.9 miles per gallon, and that we need to know the equivalent in kilometers per liter. Reference to a table reveals that there are 5280 feet in one mile, 0.3048 meters in one foot, and 3.7854 liters in one gallon. Then, the given value of fuel economy is multiplied by 5280 ft/mi and by 0.3048 m/ft, and divided by 1000 m/km and by 3.7854 L/gal:

$$\text{fuel economy} = (32.9\ \mathrm{mi/gal})\frac{(5280\ \mathrm{ft/mi})(0.3048\ \mathrm{m/ft})}{(1000\ \mathrm{m/km})(3.7854\ \mathrm{L/gal})}$$

Observe that each of these factors is equivalent to a dimensionless value of unity because it represents a ratio of two equal quantities; for example, 5280 ft/1 mi = 1. Next, in order to determine the resulting units, the unit symbols are cancelled according to the rules of ordinary algebra:

$$\text{fuel economy} = (32.9\ \cancel{\mathrm{mi}}/\cancel{\mathrm{gal}})\frac{(5280\ \cancel{\mathrm{ft}}/\cancel{\mathrm{mi}})(0.3048\ \cancel{\mathrm{m}}/\cancel{\mathrm{ft}})}{(1000\ \cancel{\mathrm{m}}/\mathrm{km})(3.7854\ \mathrm{L}/\cancel{\mathrm{gal}})}$$

This done, we note that the resulting unit is

$$\frac{1}{\mathrm{L/km}} = \mathrm{km/L}$$

as desired. Thus, the procedure provides a reliable means of verifying which conversion factors belong in the numerator and which belong in the denominator. Had we multiplied by 3.7854 L/gal instead of dividing by this factor, we would find the result to be fuel economy in $\mathrm{km \cdot L/gal^2}$, a valid but not very useful unit. Finally, the arithmetic is done, yielding

$$\text{fuel economy} = 14.0\ \text{km/L}$$

Of course, if one expects to do this conversion repeatedly, it would be wise to calculate and record the factor

$$\frac{(5280\ \text{ft/mi})(0.3048\ \text{m/ft})}{(1000\ \text{m/km})(3.7854\ \text{L/gal})} = 0.4251\ \text{km}\cdot\text{gal/L}\cdot\text{mi}$$

The same algebraic procedure should be used to determine units when numerical values are substituted into formulas. When checking the units, it is frequently necessary to observe equivalences such as $\text{N} = \text{kg}\cdot\text{m/s}^2$ or $\text{slug} = \text{lbf}\cdot\text{s}^2/\text{ft}$.

EXAMPLE

Calculate the acceleration imparted to a 100 000-lb spaceship by a thrust of 50 000 lb.

Although neither "pound" is specified as force or mass (a common occurrence), we can tell from the context that the mass of the spaceship is 100 000 lbm and the thrust is 50 000 lbf. We first convert the mass to slugs:

$$m = \frac{100\,000\ \text{lbm}}{32.174\ \text{lbm/slug}} = 3108\ \text{slugs}\ (\text{lbf}\cdot\text{s}^2/\text{ft})$$

Then we use Newton's second law of motion to compute the desired acceleration, as

$$a = f/m = \frac{50\,000\ \text{lbf}}{3108\ \text{lbf}\cdot\text{s}^2/\text{ft}} = 16.1\ \text{ft/s}^2$$

EXAMPLE

Calculate the acceleration imparted to a 45 000-kg spaceship by a thrust of 22 500 kg.

As in the previous example, we determine from context which "kg" means mass and which means force.* We first convert the thrust to its value in newtons:

*The kilogram-force, a unit of older metric systems, is *not* a unit of the SI. In the SI, the kilogram is used *exclusively* as the unit of mass.

$$f = (22\,500 \text{ kgf})(9.806\,65 \text{ N/kgf}) = 220\,650 \text{ N } (\text{kg}\cdot\text{m/s}^2)$$

Then we use Newton's second law to calculate the desired acceleration as

$$a = f/m = \frac{220\,650 \text{ kg}\cdot\text{m/s}^2}{45\,000 \text{ kg}} = 4.90 \text{ m/s}^2$$

The conversion factors that define the kilogram-force (kilopond) in terms of the newton, and the pound-mass in terms of the slug, are sometimes called the "standard acceleration of gravity." This terminology stems from considering a gravitational field with an intensity of exactly 9.806 65 N/kg (m/s^2), a value that closely approximates that of the earth at sea level and at 45° latitude. The gravity force on an object placed in this field would be, in kgf, numerically equal to the mass of the object in kg. The more direct meaning of all of this is that the kilogram-force is *defined* in terms of the newton as

$$1 \text{ kgf} = 9.806\,65 \text{ N}$$

Expressed in U.S. engineering units the "standard acceleration of gravity" is 32.174 048 556 ··· lbf/slug (ft/s^2). The gravity force on an object placed in this field would be, in lbf, numerically equal to the mass of the object in lbm. Again, there is a simpler meaning: The pound-mass is *defined* in terms of the slug as

$$1 \text{ slug} \approx 32.174 \text{ lbm}$$

Every numerical value written in an engineering computation should be accompanied by its unit, unless the number represents a dimensionless quantity. Experience has demonstrated many times over that the minor amount of time and space saved by neglecting this practice is ultimately lost in tracing unit-related errors, or in puzzling over computations resurrected from old files.

PROBLEMS

1-22 What is the displacement, in liters, of a 325-cubic inch automobile engine? *Ans.* 5.33 L.

1-23 At a place where the atmospheric pressure is 14.7 lbf per square inch, what is this pressure in SI units?

1-24 What is the efficiency (ratio of power output to power input) of an electric motor that draws 2.4 kW of electric power as it delivers 2.6 horsepower?

1-25 What are the mass and diameter, in SI units, of a 180-grain, 0.30-caliber bullet? (The diameter is 0.308 in.) *Ans.* 11.66 g; 7.82 mm.

1-26 Water flows into a container at 300 gallons per hour. How long will it take to collect 0.2 m^3?

1-27 A 10-lb force accelerates a body at the rate of 50 ft/s^2. What is the mass of the body
(a) In slugs?
(b) In lbm?
(c) In kg?

1-28 A 1.25-ton automobile accelerates from rest to 50 mi/h in 6.8 seconds. What is the average resultant force on the car during this time?

1-29 A force of 15 N accelerates a body at the rate of 20 ft/s^2. What is the mass of the body
(a) In kg?
(b) In lbm?

1-30 Before opening his parachute, a 180-lb man is observed to accelerate downward at 24 ft/s^2. What is the magnitude of the aerodynamic force acting on the man?

1-31 A 1.2-ton automobile rounds a curve of radius $\rho_c = 100$ m at a constant speed of 50 mi/h. Use the equation of Problem 1-61 to calculate the centrifugal force on the automobile.

1-32 A 10-lb object is suspended on a spring of stiffness $k = 50$ lbf/in. and oscillates according to the equations given in Problem 1-59. Use the second of these equations to determine the time required to complete one cycle of oscillation. *Ans.* 0.143 s.

1-33 What is the fuel consumption, in liters per kilometer, of an automobile that gets 19 miles to the gallon?

1-34 A person with a mass of 180 lbm is standing on the surface of the moon, where the gravitational field strength is 1.57 N/kg. What is the force of the moon's gravity acting on the person?

For Problems 35 through 40, use the values given in Appendix A to determine the required quantity.

1-35 How many gallons are in one ft^3?

1-36 What is the density of water in lbm/ft^3?

1-37 What is the speed of light in mi/hr?

1-38 How many feet are there in one nautical mile?

1-39 What is the gravitational acceleration near the earth's surface in ft/s^2?

1-40 What is the density of water in g/mL?

For Problems 41 through 57, give the value of the stated quantity in SI units.

1-41 One light-year. *Ans.* 9.46 Pm.

1-42 A spring stiffness of 85 pounds-force per inch of displacement.

1-43 The tension force in a cable from which a 10-ton block is suspended.

1-44 The tension force in a cable from which a 2-Mg block is suspended.

1-45 One kilowatt-hour.

1-46 10 000 horsepower-hours.

1-47 3000 feet per second.

1-48 1700 r/min.

1-49 4000 barrels per day.

1-50 100 bushels per acre of wheat. *Ans.* 871 μm.

1-51 1000 board-feet of lumber.

1-52 48 000 acre-feet of water.

1-53 Thermal conductivity of 127 Btu/hr·ft·°F.

1-54 8729 furlongs per fortnight.

1-55 "Speed 55 miles."

1-56 "60 pounds of water pressure."

1-57 "Six yards of gravel" *Ans.* 4.59 m^3.

For Problems 58 through 67, check the given equations for dimensional consistency.

1-58 $E=mc^2$ (Einstein's equation)

1-59 $m\dfrac{d^2x}{dt^2}+kx=0$

$x=x_0\cos\sqrt{k/m}\,t$

m = mass
x = displacement
k = spring stiffness (force per unit displacement)

1-60 $T=2\pi\sqrt{a/g}$

T = period of oscillation of pendulum
a = pendulum length

1-61 $f_c=mv^2/\rho_c$

f_c = centrifugal force
m = mass
v = speed
ρ_c = radius of curvature

1-62 $p=\rho gz$

p = fluid pressure
ρ = density
z = depth

1-63 $y = a\cosh x/a$

y = vertical distance to a point on a suspended cable
x = horizontal distance

$a = S_0/\rho A g$

S_0 = tension force at apex
ρ = density
A = cross-sectional area

1-64 $A = \int_a^b y(x)\,dx$

A = area under curve
y = distance to curve
x = distance along x-axis

1-65 $a_n = \dfrac{v^2}{\rho_c}$

a_n = normal component of acceleration
v = speed
ρ_c = radius of curvature

1-66 $c = \sqrt{\dfrac{S}{\mu}}$

c = speed of travel of transverse wave in string
S = tension in the string
μ = lineal density (mass per unit length) of the string

1-67 $t = \dfrac{m}{c}\log\dfrac{mg + cv_0}{mg + cv}$

t = time for a particle of mass m to reach upward velocity v, traveling under the influence of gravity through a medium that provides a drag force equal to c times the speed.

1-4 WEIGHT AND MASS

We now understand the mass of an object in terms of its resistance to acceleration, an "absolute" property, independent of the contact or proximity with other bodies. The technically correct meaning of the term *weight* is the force of gravity acting on an object, this force depending on the proximity of the object with a body such as the earth. However, the term is also commonly used to denote different, closely related forces, and to denote mass. Because of this, the term *weight* is often a source of serious confusion in technical communication where precision of meaning is essential. Today, careful technical writers avoid this confusion by using the term *force of gravity* rather than *weight* when force of gravity is meant. Older books and articles that deal with forces of gravity must be read with the traditional definition of *weight* kept in mind.

PROBLEMS

1-68 A spring scale and a balance scale, both of high precision, are calibrated at Gaithersburg, Maryland, to measure mass. If they are taken to various parts of the world, what error can be expected from each?

1-69 A 70-kg astronaut is "floating" inside a spaceship that is in a circular orbit at an altitude of 207 km above the earth. At this altitude, the gravitational field intensity is 9.2 N/kg. What is the magnitude of the force of gravity on the astronaut?

1-70 In discussing a picture of the astronaut of the previous problem, a television commentator might say that the astronaut is weightless. Would you agree? State a definition of "weight" that would be consistent with this usage.

1-71 When the tank is empty, the spring scale reads 16 lb. After the tank is filled with water, the scale reads 13.9 lb. What is the magnitude of the gravity force on the ball?

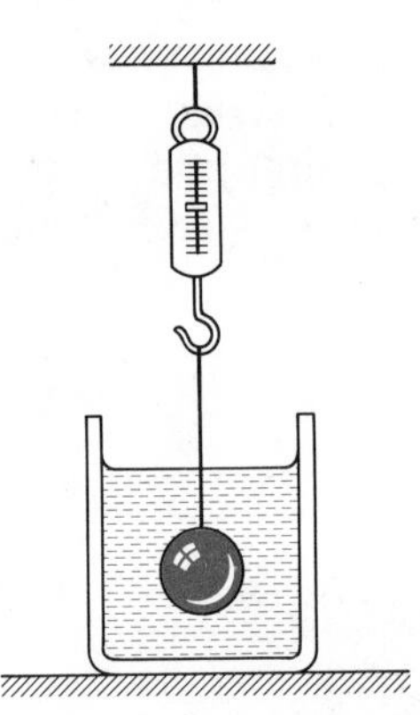

1-72 A marine scientist says that the submerged weight of the ball of the previous problem is 13.9 lb. Would you agree? What does she mean by "submerged weight"?

FORCE SYSTEMS

2-1 COMPOSITION AND RESOLUTION OF FORCES

In analyzing forces it is necessary to account for both magnitude and direction. These aspects are depicted by arrows that have lengths proportional to the magnitudes represented. The magnitude and direction of an individual force is defined in terms of the acceleration that it will produce in a body in the absence of all forces. When two forces act simultaneously, it is an observed fact that the acceleration produced, and hence by definition the resultant force, is determined by the *parallelogram law of addition* (Figure 2-1*a*). An equivalent procedure is the formation of triangles by the tail-to-head placements shown in Figure 2-1*b* and *c*. When any number of forces act, the resultant can be determined by repeated application of this law.

EXAMPLE

What must be the tension in the line of tugboat *B* (Figure 2-2) so that the resultant towing force on the ship will be straight ahead? Also, what will be the magnitude of the resultant force?

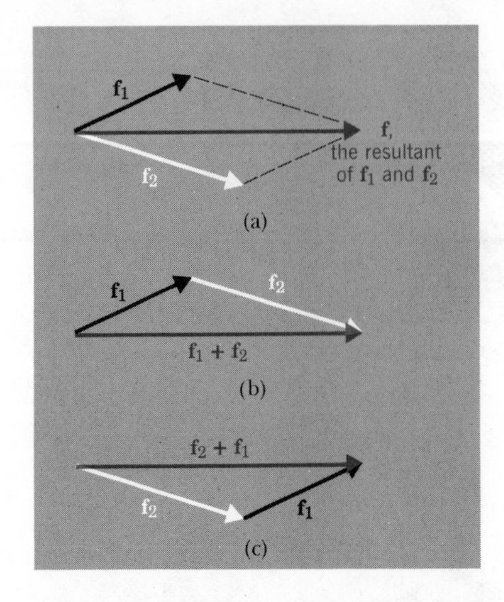

FIGURE 2-1

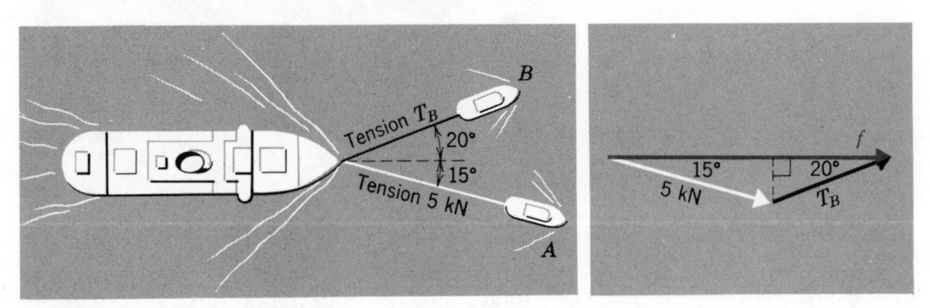

FIGURE 2-2

Referring to the vector triangle we note that if the resultant is to be straight ahead,

$$T_B \sin 20° = (5 \text{ kN}) \sin 15°$$

Or,

$$T_B = \frac{(5 \text{ kN}) \sin 15°}{\sin 20°} = 3.78 \text{ kN}$$

Also from the vector triangle, we can determine the magnitude of the resultant as

$$f = 5 \text{ kN} \cos 15° + T_B \cos 20°$$
$$= 8.39 \text{ kN}$$

EXAMPLE

The 70-kg mountain climber is pulling on the short line with a force of 400 N, while the other line has a tension of 500 N. What is the resultant of the forces from the two ropes and gravity acting on the climber?

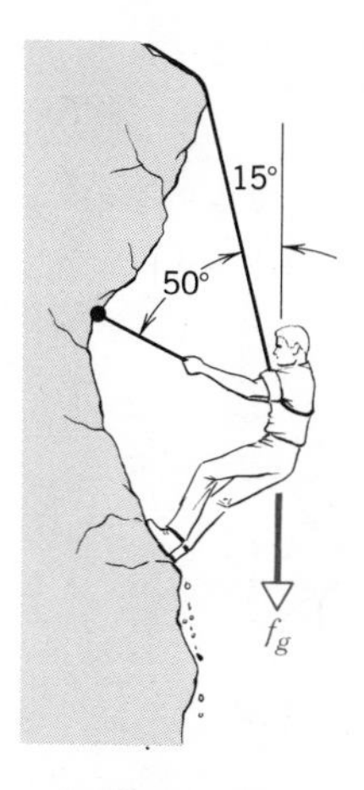

The resultant of these three forces can be determined by applying the parallelogram law to any two of the forces, and again to the third force and the resultant from the first calculation. With reference to the vector diagram, we use the trigonometric rule of cosines to calculate the magnitude of the resultant from the two ropes as

$$T = \sqrt{(500 \text{ N})^2 + (400 \text{ N})^2 + 2(500 \text{ N})(400 \text{ N}) \cos 50°}$$
$$= 816.8 \text{ N}$$

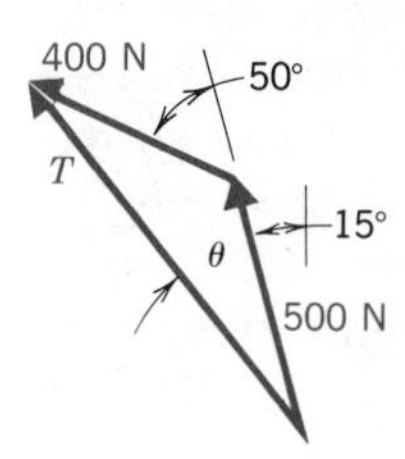

The angle between the upper line and T can then be determined by applying the law of sines to the vector triangle:

$$\frac{\sin\theta}{400\text{ N}} = \frac{\sin 50^\circ}{816.8\text{ N}}$$

which yields

$$\theta = 22.0^\circ$$

By repeating this procedure we can find the resultant of the gravity force and the forces from the lines:

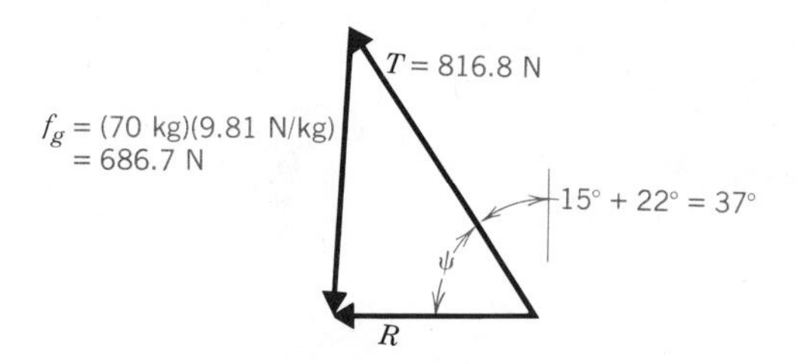

$$R = \sqrt{(816.8\text{ N})^2 + (686.7\text{ N})^2 - 2(816.8\text{ N})(686.7\text{ N})\cos 37^\circ}$$

$$= 493\text{ N}$$

$$\frac{\sin\psi}{686.7\text{ N}} = \frac{\sin 37^\circ}{493\text{ N}}; \qquad \psi = 57^\circ$$

Thus, the resultant of the three forces is a force with magnitude 493 N, acting in a direction 94° from the vertical.

The process illustrated in this example—determining the single force equal to two or more forces— is called *composition* of forces. The opposite procedure—that of determining several forces that sum to yield a given force —is also useful. This is called *resolution* of the given force. The several forces that make up the given force are called *components* of the force. We will see that there are an unlimited number of ways in which a force may be resolved; the particular resolution that will prove advantageous is often suggested by the problem at hand, and will be more readily recognized with experience. Usually, two components will be most useful for planar problems and three components for general problems.

EXAMPLE

Evaluate the two components of the 1000-lb vertical force in directions parallel to the boom OB and parallel to the support cable OA.

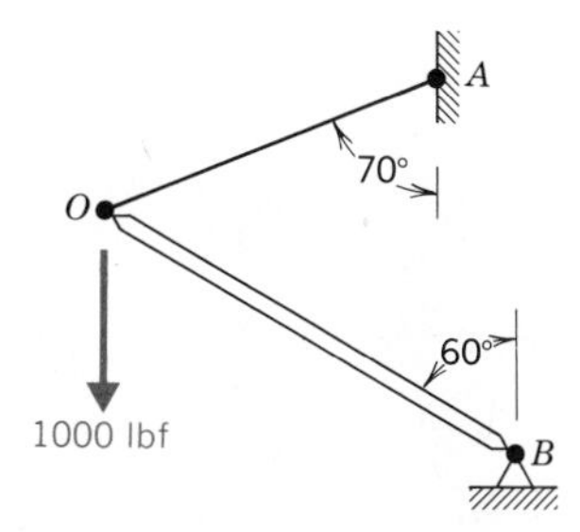

We note first that the angle between the boom and the cable is 180° − 70° − 60° = 50°. Then, applying the law of sines to the vector diagram,

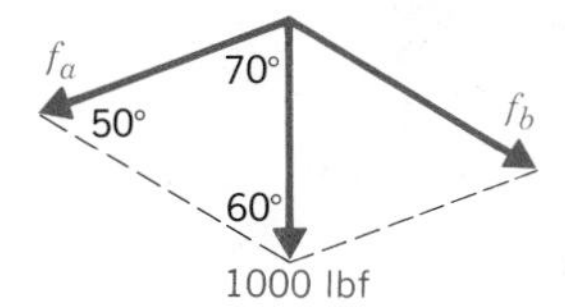

$$\frac{f_a}{\sin 60°} = \frac{f_b}{\sin 70°} = \frac{1000 \text{ lbf}}{\sin 50°}$$

we obtain in the desired magnitudes of these components:

$$f_a = 1131 \text{ lbf} \qquad f_b = 1227 \text{ lbf}$$

EXAMPLE

Resolve each of the three forces on the mountain climber dealt with in the example on page 21 into vertical and horizontal components. Sum the vertical components and sum the horizontal components. Then evaluate the resultant of these vertical and horizontal sums.

Referring to the diagram, the sum of the vertical components (with positive values defined as those acting upward and negative values defined as those acting downward) is

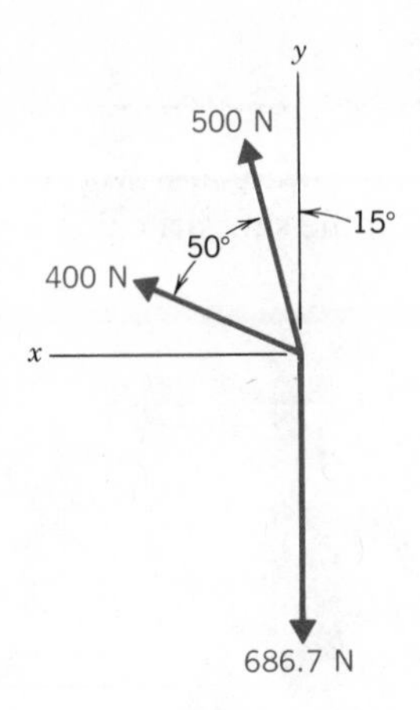

$$f_y = (500\text{ N})\cos 15° + (400\text{ N})\cos 65° - 686.7\text{ N}$$

$$= -34.7\text{ N}$$

Similarly, the sum of the horizontal components acting to the left is

$$f_x = (500\text{ N})\sin 15° + (400\text{ N})\sin 65°$$

$$= 491.9\text{ N}$$

Now the resultant of these two components has the magnitude

491.9 N
34.7 N
R

$$R = \sqrt{(34.7\text{ N})^2 + (491.9\text{ N})^2}$$

$$= 493\text{ N}$$

and acts at an angle from the horizontal

$$\tan^{-1}\left(\frac{34.7}{491.9}\right) = 4°$$

Comparison with the results on page 22 reveals that this is the same result obtained by repeated use of the parallelogram law. Can you show why this should be?

PROBLEMS

For Problems 2-1 through 2-12, determine the resultant force (magnitude and direction) of the given force systems.

2-1

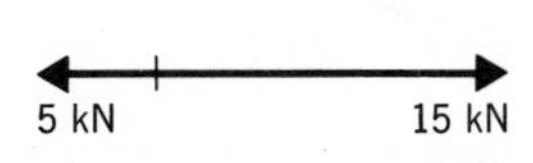

2-2

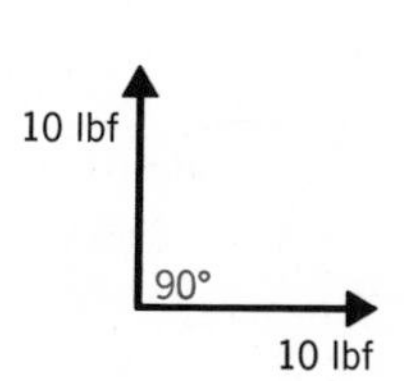

2-3

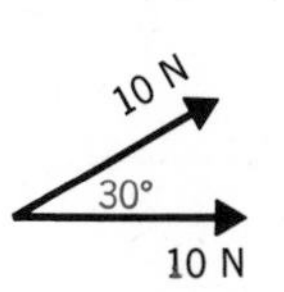

2-4

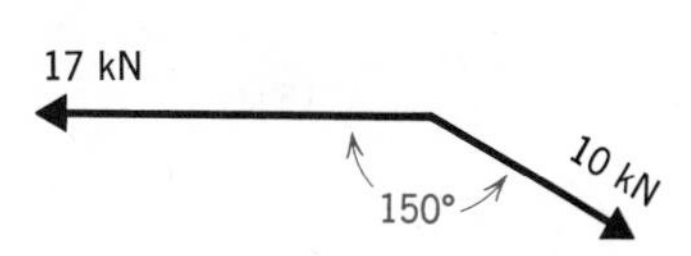

2-5

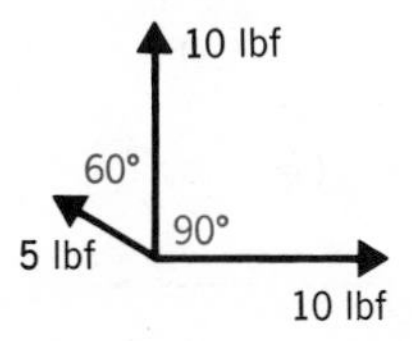

2-6

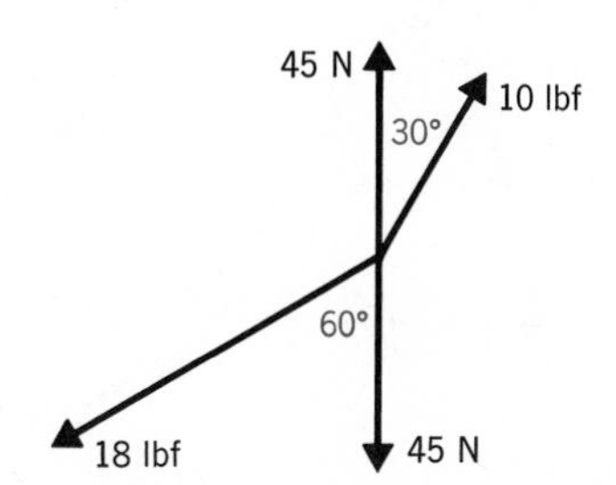

2-7

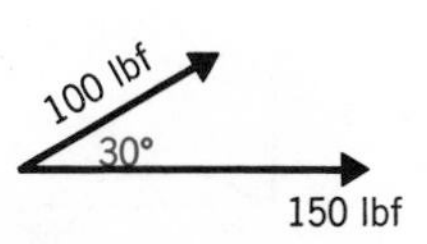

2-8

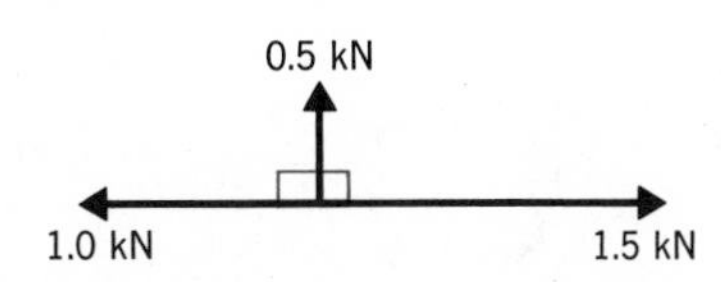

2-9

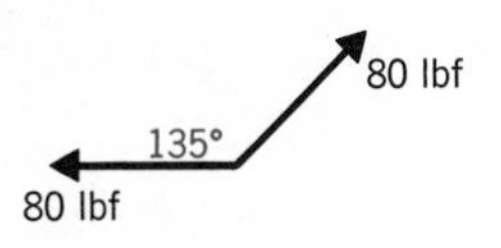

2-10

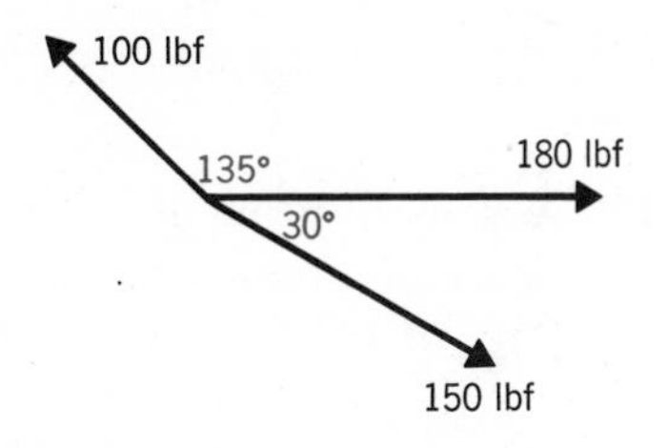

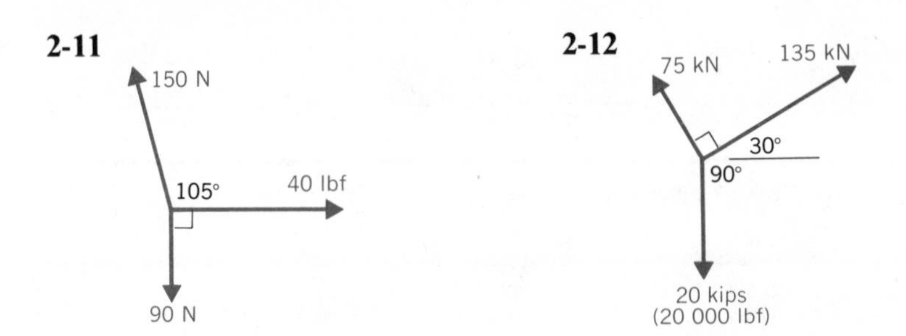

For Problems 2-13 through 2-18, determine the direction and magnitude of an unknown force **F** which, when added to the given force **G**, will yield the given resultant force **R**.

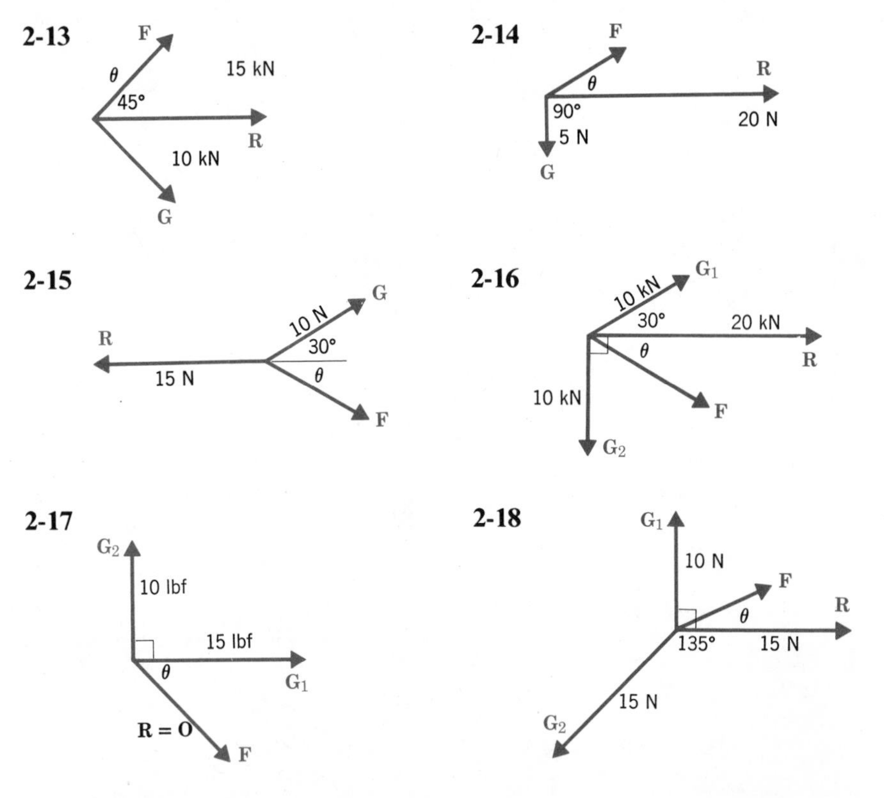

2-19 A ship is being towed by three tugboats as shown. Determine the resultant force exerted by the towlines on the ship. Each towline exerts a force of 5 kN. *Ans.* 12.35 kN; 5.53°

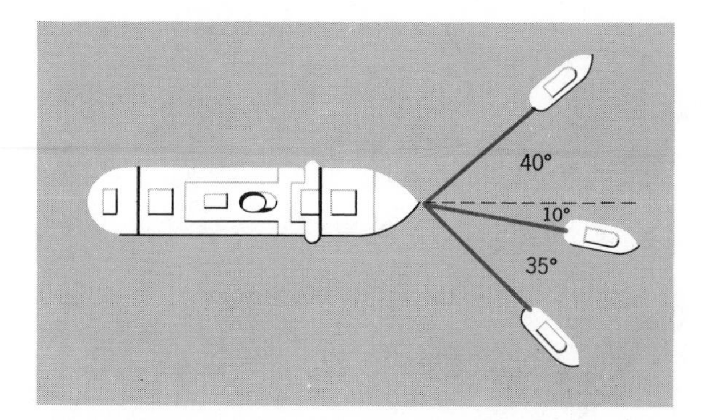

2-20 Three tugboats are towing the ship as shown, each exerting a force of 25 kN. What is the value of θ so that the direction of the resultant force is along the axis of the ship? What is the magnitude of the resultant force?

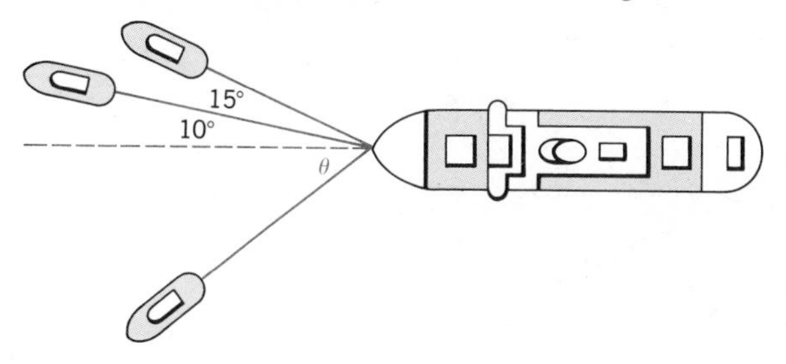

2-21 A drag force f_d is exerted by the surrounding water on the ship of Figure 2-2. What must be the magnitude and direction of this force if the ship is not accelerating?

2-22 At a certain instant, the tension in the cable on which the destruction ball is suspended is 9.3 kN. What is the acceleration of the ball at this instant?

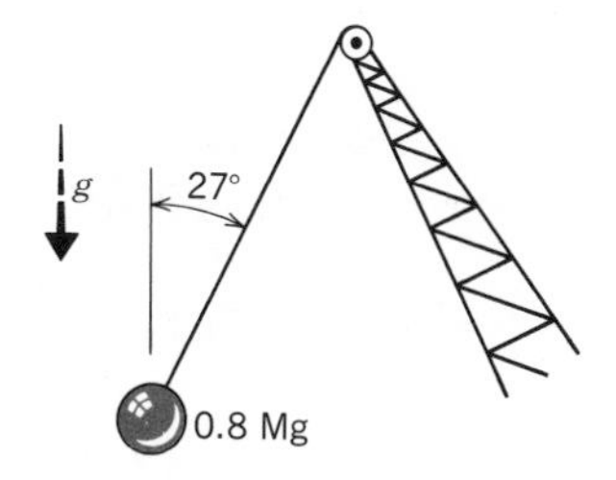

2-23 Determine the magnitude and direction of **f** so that the resultant of this and the 200-lb force is a force of 420 lb acting vertically and downward.

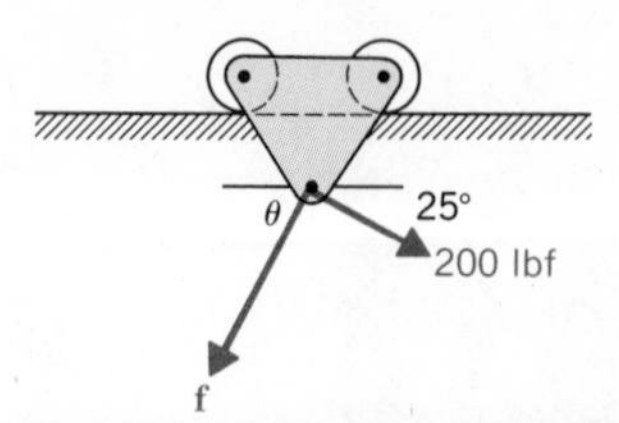

2-24 Determine the resultant force on the pillow bearing.

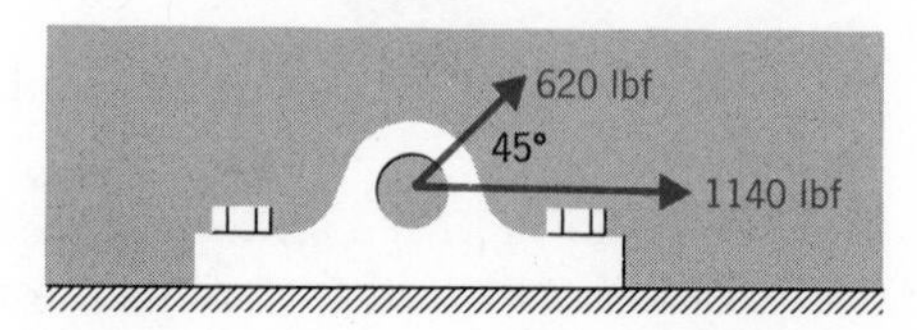

2-25 Determine the resultant force on the eyebolt.

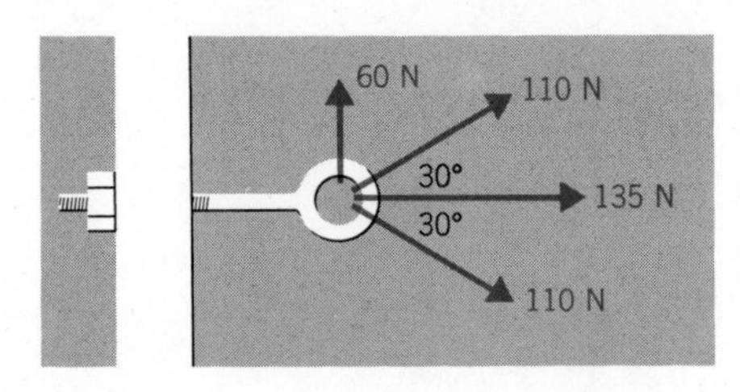

2-26 Determine the resultant force on the end of the diving board shown below.

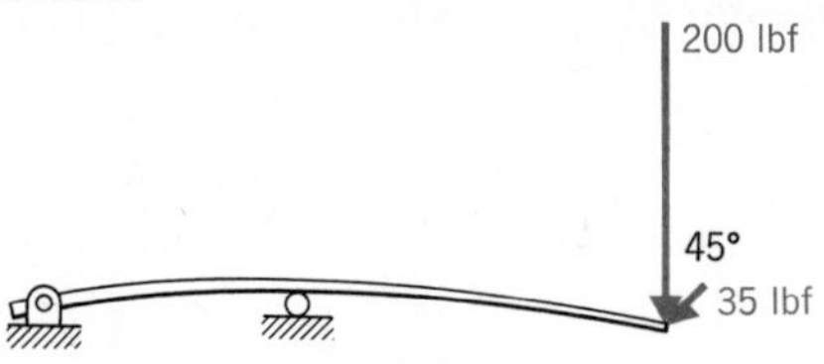

2-27 If the resultant of the 1000-lb vertical force and the tensile force T in the cable is to be in the direction of the boom OB, what must be the magnitude of T?

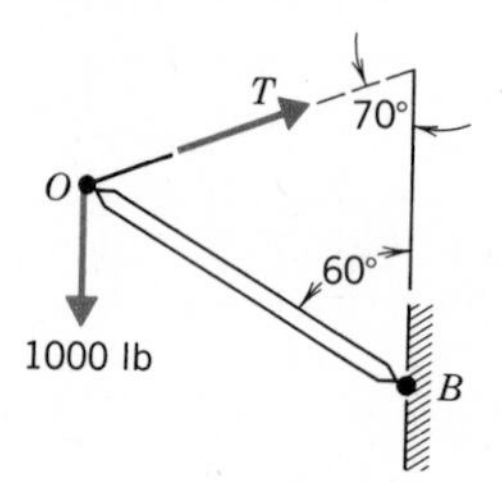

2-28 Determine the resultant force on the edge of the cover of the textbook.

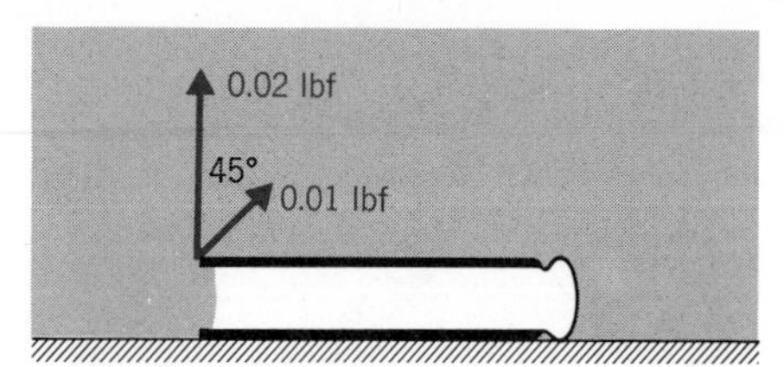

2-29 Determine the resultant force on the bracket.

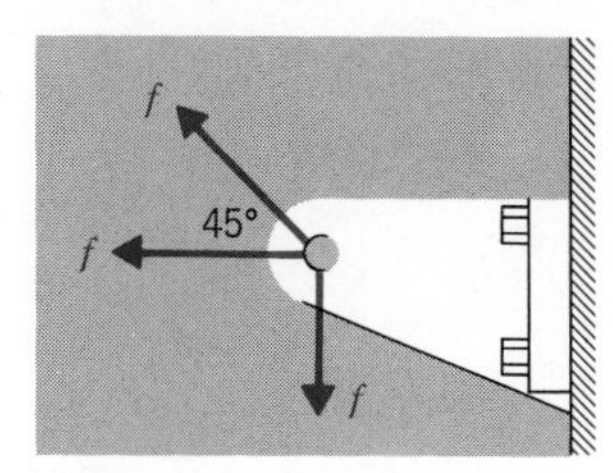

2-30 Spreader bars and stays are often used on vertical masts to increase the rigidity of the mast. If the tension in the upper portion of the cable is 600 N, what must be the tension in the lower portion so that the vertical components of the forces acting on the spreader bar at *A* cancel one another? What is the resultant force transmitted to the spreader bar by the cables? *Ans.* 465 N; 304 N.

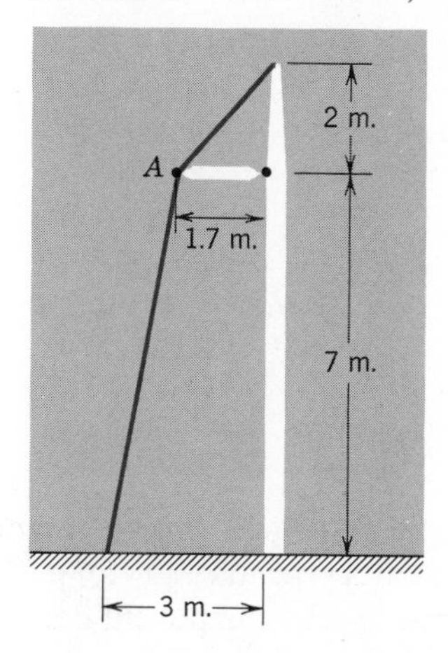

2-31 A rocket with 20 000-lb thrust is fired vertically. A horizontal wind blowing on the rocket tends to push the rocket sideways with a 2000-lb force. What is the resultant of these two forces on the rocket?

2-32 The contact point on a gear tooth has forces on it as shown below. Determine the resultant force on the gear tooth.

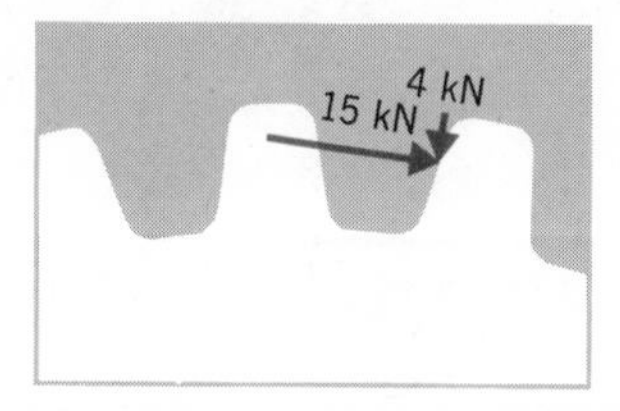

2-33 Determine the resultant of the three mutually orthogonal forces shown.

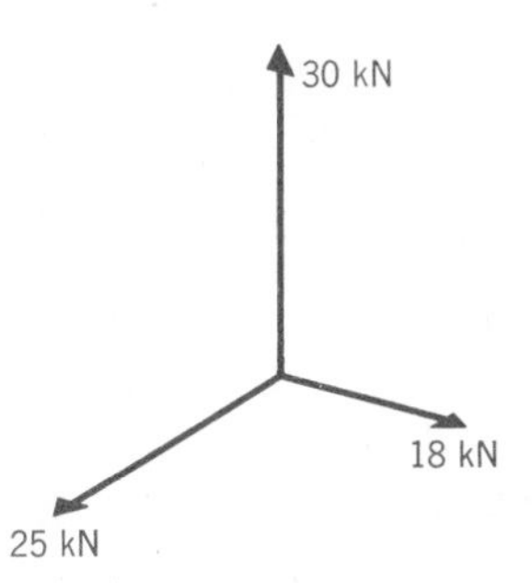

2-34 Determine the resultant force on the doorknob. The three components shown are mutually orthogonal.

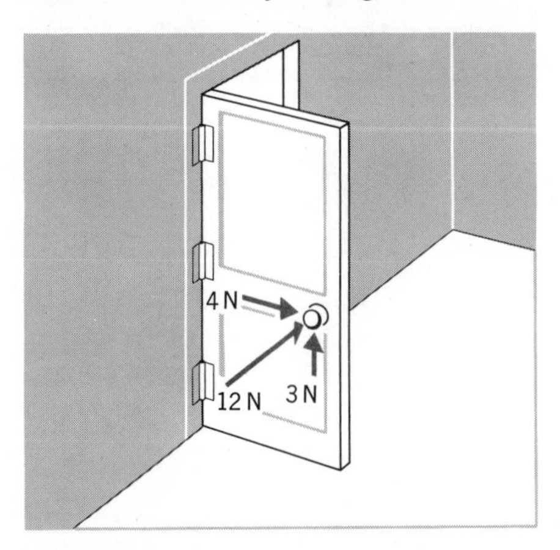

2-35 Resolve the 1000-lb force of the example on page 23 into a component parallel to the boom and a component perpendicular to it. Is the component parallel to the boom the same as that component determined in the example?

2-36 What are the components of the 1000-lb force that are perpendicular and parallel to the support to which the eyebolt is attached?

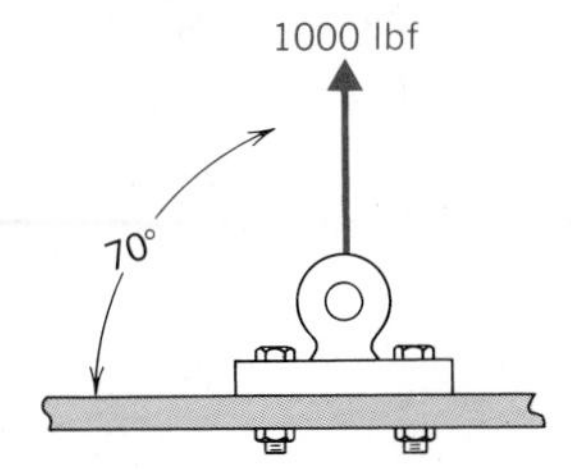

2-37 Resolve the force into components along the given axes

(a)

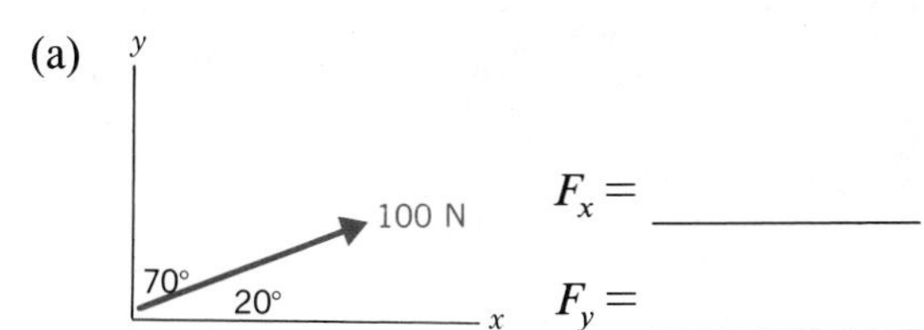

(b)

η

100 N

50°

20°

ξ

$F_\xi =$ ________

$F_\eta =$ ________

(c)

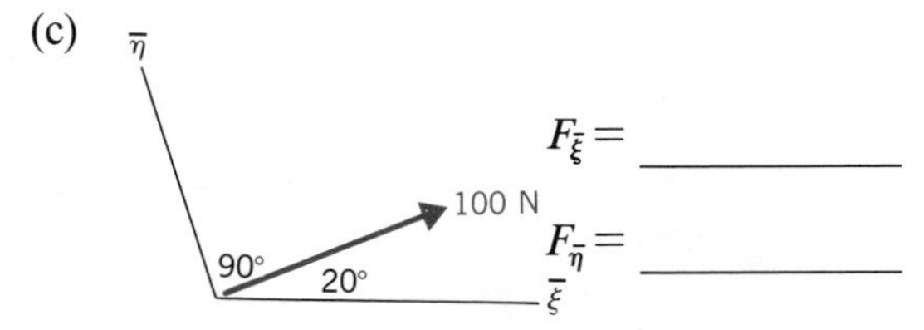

2-38 Resolve the 100-N force into components in the ξ- and η-directions.

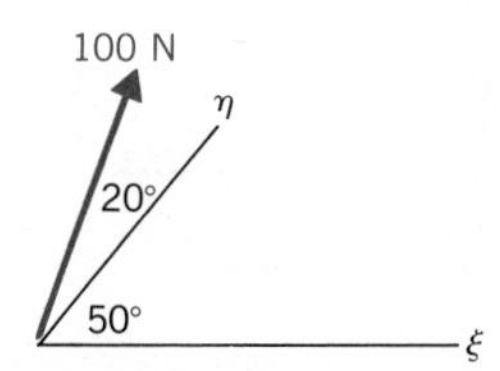

2-39 Referring to Problem 2-24, resolve the force on the pillow bearing into vertical and horizontal components.

2-40 Referring to Problem 2-25, resolve the force on the eyebolt into vertical and horizontal components.

2-41 Referring to Problem 2-26, resolve the force on the end of the diving board into vertical and horizontal components.

2-42 To raise the load, the hydraulic cylinder exerts a force of 50 kN in the direction of its axis *AB*. For the position shown, resolve this force into components parallel and perpendicular to *OB*. *Ans.* 43.76 N; 24.18 N.

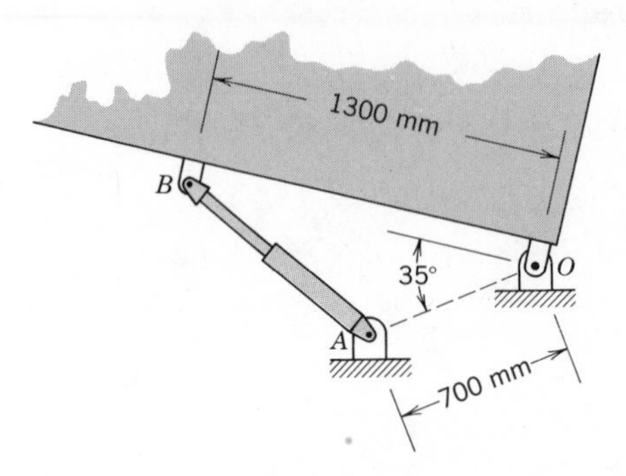

2-43 To support the load, the resultant of the forces in *OA* and *OB* must be an upward vertical force of 3.2 kN. Evaluate the tensions in *OA* and *OB*.

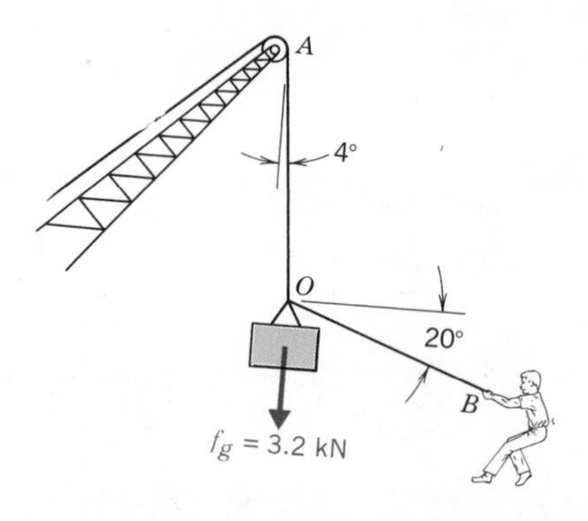

2-2 POINT OF APPLICATION AND LINE OF ACTION

The point of application of a force will in general have an effect on the way the body responds. Attaching a tow line to the front bumper of a stuck automobile might lead to removing the bumper from the automobile, whereas attaching it to a point on the frame might lead to the desired result. For many applications, however, the object may be idealized as *rigid*; that is, the deformation that the forces produce in the body can be assumed to be negligible. (The removal of the front bumper is not a negligible deformation of an automobile.) Under the idealization that the body is rigid, the effect of a force is the same for any point of application along a line parallel to the direction of the force. The line oriented in the direction of a force and passing through the point of application of the force is called the *line of action* of the force. The fact that the effect of a force on a rigid body is the same for any point of application along the line of action of the force is known as the *principle of transmissibility* of forces. For example, a tugboat pushing on the stern of a steamship has the same overall effect on the steamship as one towing from the bow along the same line of action. However, the *internal* forces induced *within* the body *are* affected by the point of application of the force.

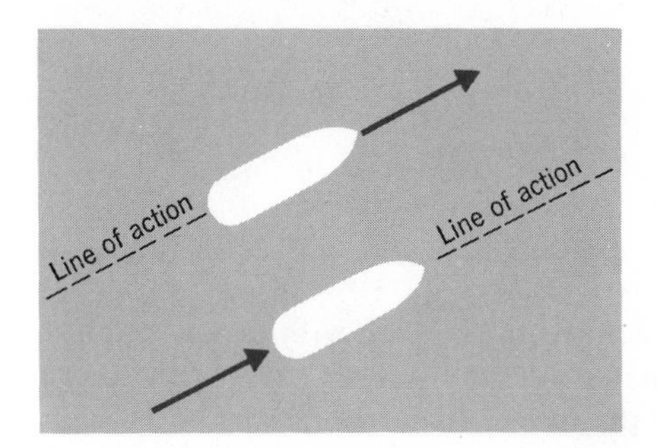

2-3 MOMENT OF FORCE

Anyone who has used a wrench is familiar with the "turning effect" that a force can have on an object. Furthermore, this "effect" is greater as the line of action is placed further from the point about which "turning" is considered. A more precise meaning of this intuitive notion is provided in the definition of *moment of force*.

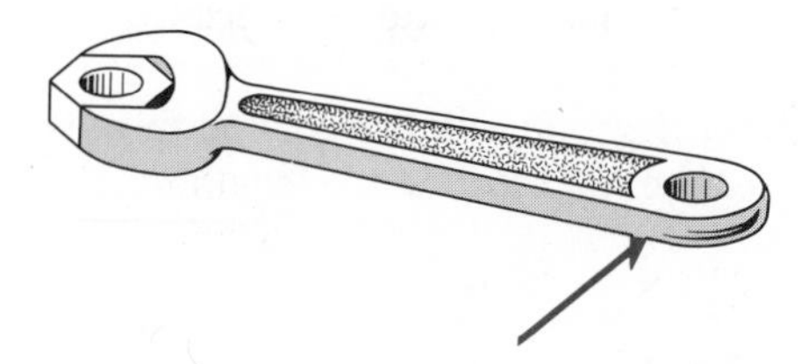

The moment about point O of the force $\mathbf{f}$ is a vector, defined in terms of the quantities shown in Figure 2-3. The magnitude of the moment is the product of the magnitude of the force and the perpendicular distance from O to the line of action of the force,

$$M_O = f\,\overline{OP}$$

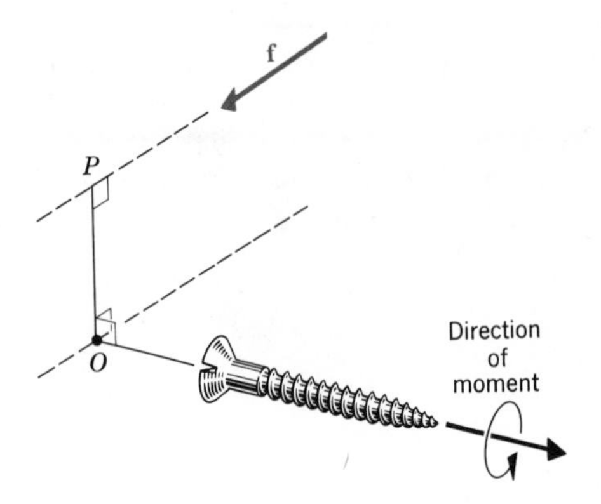

FIGURE 2-3

The *direction* of the moment vector is defined to coincide with the direction of advancement of a right-hand screw rotated by the force, the axis of the screw oriented perpendicular to the plane of the force and the point O.

Observe from the above relationship that the dimensions of moment of force are $(ML/T^2)(L) = ML^2/T^2$, and that it can be measured in newton-meters (N·m).

Keep in mind that the moment of a given force depends on the location of the reference point O. Thus it is ambiguous to refer to a moment unless the force, its line of action, *and* the reference about which the moment is reckoned are all specified.

EXAMPLE

The friction in a pipe joint requires that a force of 220 N be applied at the end of a 360-mm wrench handle to loosen it. What length handle is necessary so that the joint can be loosened by a 150-N force?

Assuming that the same moment about the joint will be required in both cases, we write this relationship in terms of the required handle length d:

$$M_O = (150\text{ N})d = (220\text{ N})(0.36\text{ m})$$

$$(150\text{ N})d = 79.2\text{ N}\cdot\text{m}$$

Then,

$$d = \frac{79.2\ \text{N}\cdot\text{m}}{150\ \text{N}}$$

$$= 528\ \text{mm}$$

This calculation was done with assumption that the wrench handle is perpendicular to the pipe axis and that the force is applied perpendicular to both the pipe axis and the wrench handle. How do you suppose the required force would change if the wrench were moved to the other end of a 45° elbow, as is indicated in Figure 2-4?

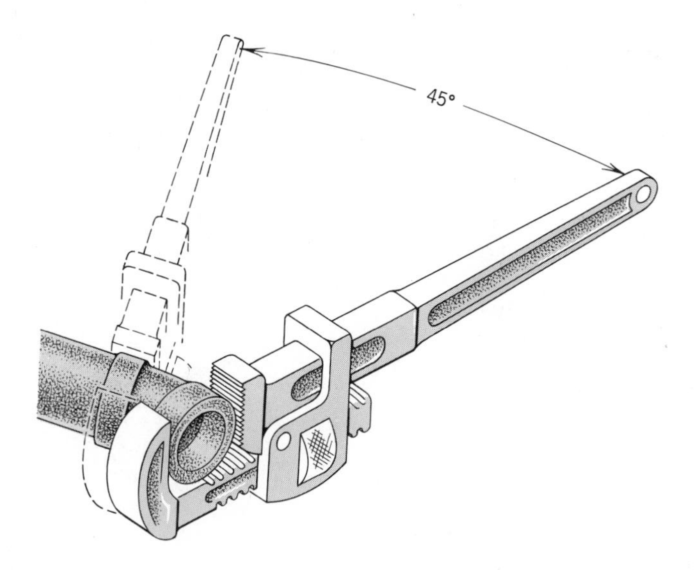

FIGURE 2-4

Consider next the moments about point O of the two mutually perpendicular forces $\mathbf{f}_1$ and $\mathbf{f}_2$ shown in Figure 2-5*a*, where the point O and the lines of action of the forces are all in the same plane. The directions of the moments are perpendicular to this plane, that of $\mathbf{f}_1$ directed into the paper (clockwise), and that of $\mathbf{f}_2$ directed out of the paper (counterclockwise). Let us define the sum of the moments with the convention that outward-directed (counterclockwise-acting) moments are positive and inward-acting (clockwise-acting) moments are negative. Then we see from Figure 2-5*a* that the sum of the moments has the value

$$M_{O2} + M_{O1} = r_1 f_2 - r_2 f_1 \qquad \textbf{(2-1)}$$

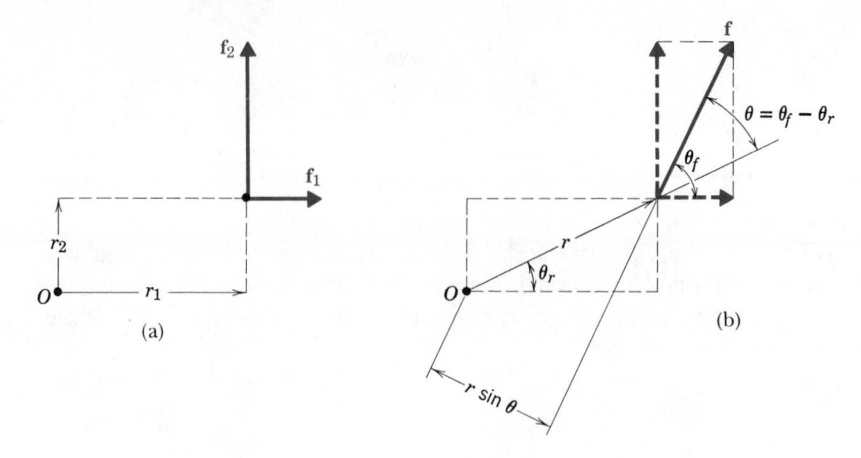

FIGURE 2-5

Now consider the moment about O of the resultant force $\mathbf{f} = \mathbf{f}_1 + \mathbf{f}_2$ shown in Figure 2-5*b*.

$$\begin{aligned} M_O &= fr \sin \theta \\ &= fr \sin(\theta_f - \theta_r) \\ &= fr(\sin \theta_f \cos \theta_r - \cos \theta_f \sin \theta_r) \\ &= (r \cos \theta_r)(f \sin \theta_f) - (r \sin \theta_r)(f \cos \theta_f) \\ &= r_1 f_2 - r_2 f_1 \end{aligned} \tag{2-2}$$

Thus the sum of the moments of $\mathbf{f}_1$ and $\mathbf{f}_2$ is equal to the moment of the resultant $\mathbf{f}_1 + \mathbf{f}_2$! This is a special case of *Varignon's theorem*. Its generalization, to include any number of nonorthogonal forces in three dimensions, is easily established by using results from vector analysis, which is subject of the next chapter.

EXAMPLE

Evaluate the moment about O of the force acting on the end of the bracket (Figure 2-6).

The given force is equivalent to a downward force of 40 lbf, and a leftward force of 30 lbf. With the counterclockwise direction taken as positive, the moment is

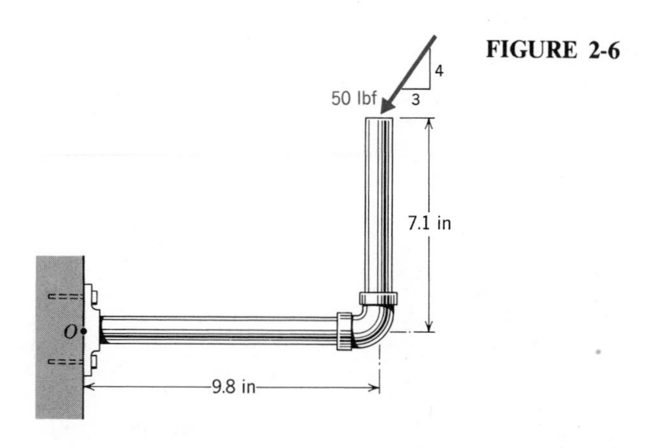

FIGURE 2-6

$$M_O = -(40\ \text{lbf})(9.8\ \text{in}) + (30\ \text{lbf})(7.1\ \text{in})$$

$$= -179\ \text{lbf}\cdot\text{in}$$

The minus sign indicates that the resultant moment about O is directed inward (clockwise).

It is instructive to determine, from the geometry given in Figure 2-6, the perpendicular distance from O to the line of action of the force, and compute the moment according to the definition on page 34. Varignon's theorem tells us that the result must be the same as determined above. As is often the case, it is easier to calculate the sum of the moments of the two force components than to determine the perpendicular distance from the reference point to the line of action of the resultant force.

PROBLEMS

2-44 Referring to Figure 2-6, determine the perpendicular distance from point O to the line of action of the force, and from this evaluate M_O.

2-45 Referring to Figure 2-6, determine the distance from point O to the point where the line of action of the force intersects the axis of the horizontal section of pipe. If the force were applied at this point, what would be the moment about O of the force?

2-46 Determine the sum of moments of the forces about the point C. *Ans.* 2000 lbf·ft.

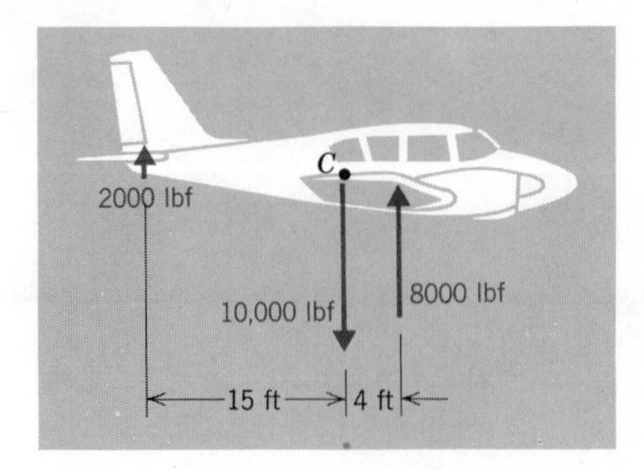

2-47 The moment about A of the vertical force exerted on the springboard by the diver is 3.5 kN·m. What is the magnitude of this force?

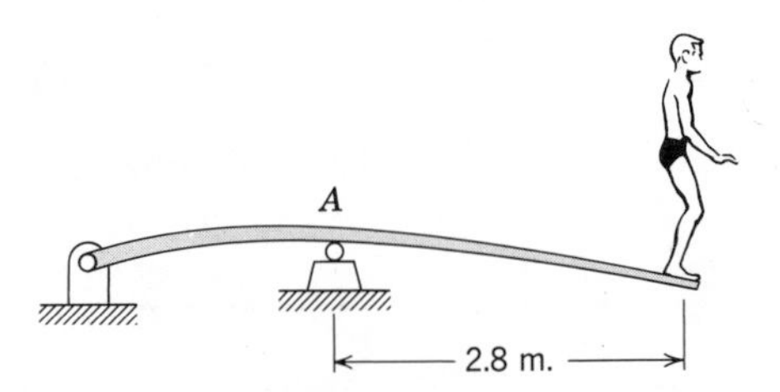

2-48 The connecting rod exerts a force of 1000 lbf on the crank as shown. Evaluate the moment about O of this force.

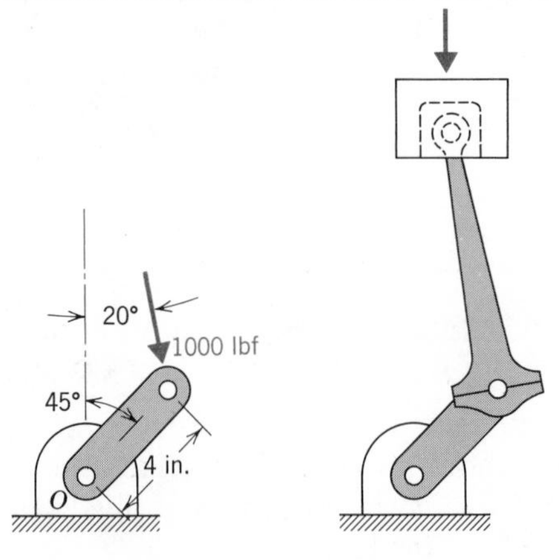

2-49 The signpost will break off at the ground if the sum of moments of forces reaches 1800 ft·lbf. Estimate the magnitude of wind force that the signpost can withstand. *Ans.* 976 N.

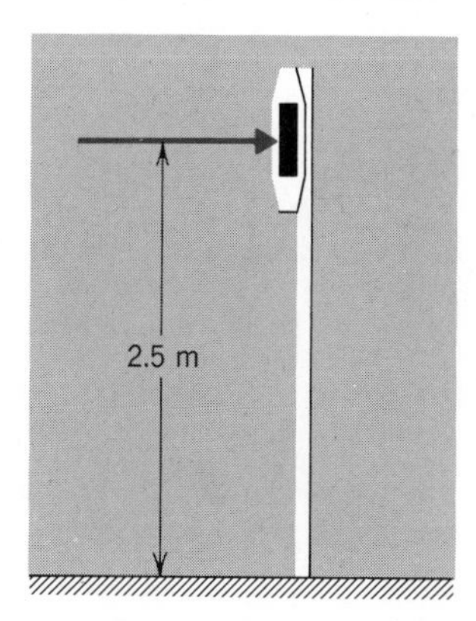

2-50 From the given dimensions in Problem 2-42 (p. 32), compute the perpendicular distance from O to the axis of the cylinder. What is the moment about O of the 50-kN force on the load at B?

2-51 Referring to Problem 2-42 (p. 32), compute the moment about O of the 50-kN force by using the components of this force parallel and perpendicular to OB. From this, what must be the perpendicular distance from O to the line of action of the force?

2-52 Evaluate the moments about O' and about O of the 400-lb force.

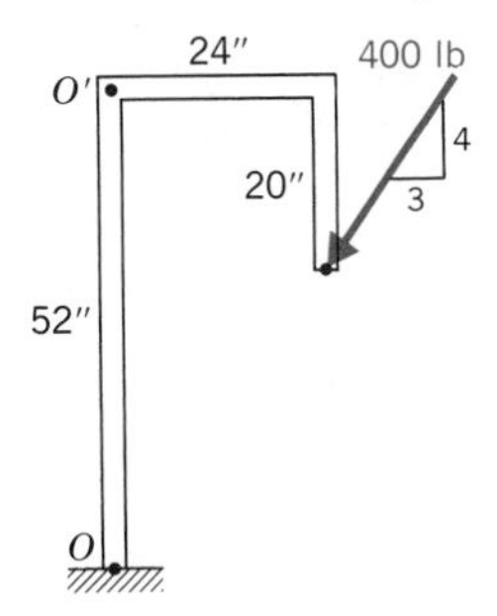

2-53 If the sum of the moments abut O, of the 4-ton gravity force and the force the cable exerts on the log at B, is zero, what is the tension in the cable AB?

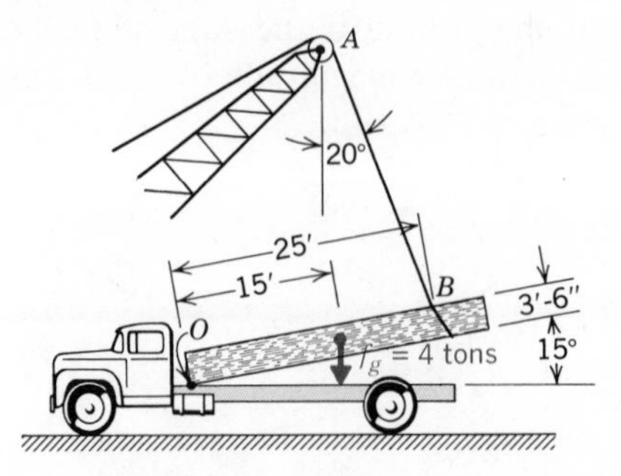

2-54 The tension in the line AB is 3.5 kN.

(a) What is the moment about O of the force the cable AB exerts on the spreader bar OB at B?

(b) What must be the tension in BC if the moment of the force that the cable BC exerts on the spreader bar is equal and opposite to that computed in (a)?

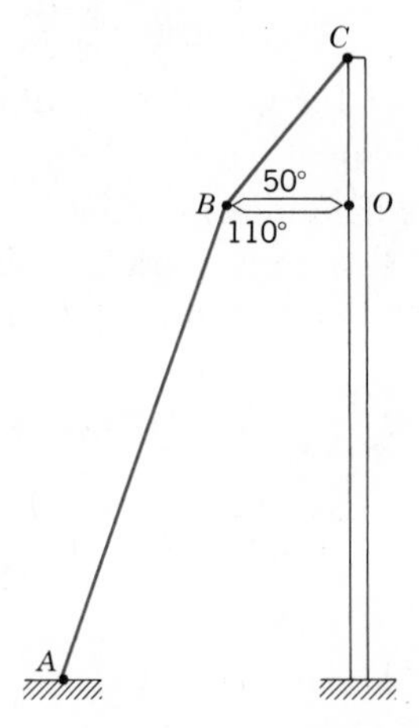

2-55 In terms of the parameters shown in the sketch, what is the moment about O of the gravity force on the oscillating pendulum?

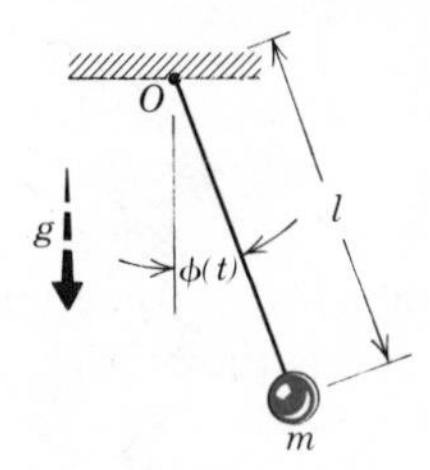

2-56 What is the moment about A of the gravity force acting on the 8.8-Mg falling tree? *Ans.* $(1.73\ \mathrm{MN \cdot m}) \sin \phi$.

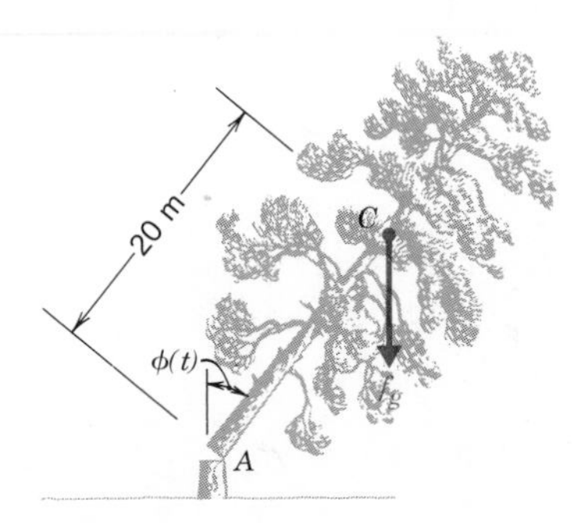

2-57 What will be the tension in the cable AB if the sum of moments about O, of the force this transmits to the boom and the 1000-lb force, is zero?

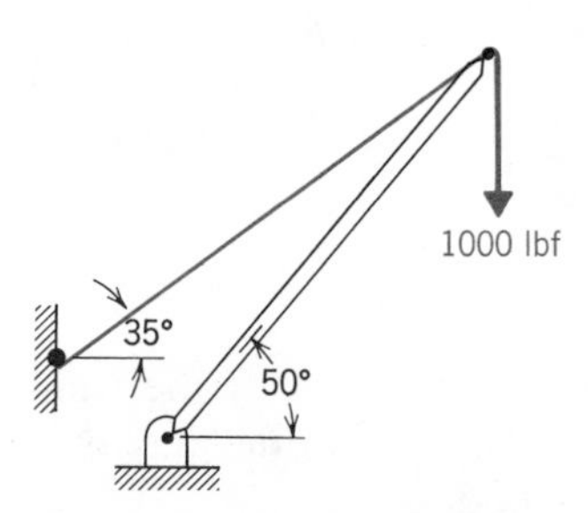

2-58 What will be the sum of the moments about O of the wind forces acting on the flag pole? *Ans.* 1450 lbf · ft.

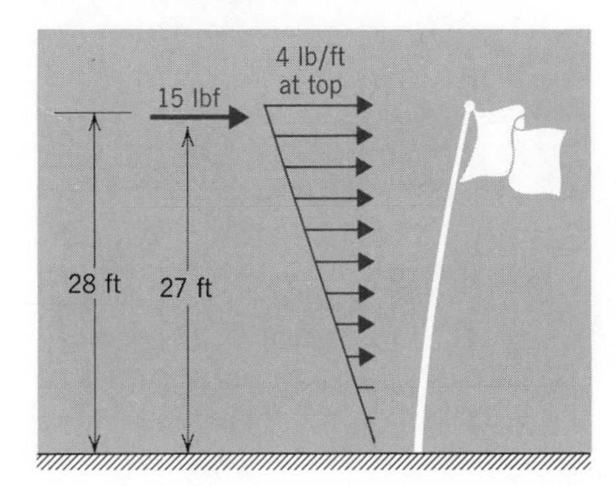

2-59 If the force from the nail is 1.5 kN, what must be the force P in order that the sum of moments about O of the forces acting on the crowbar is zero?

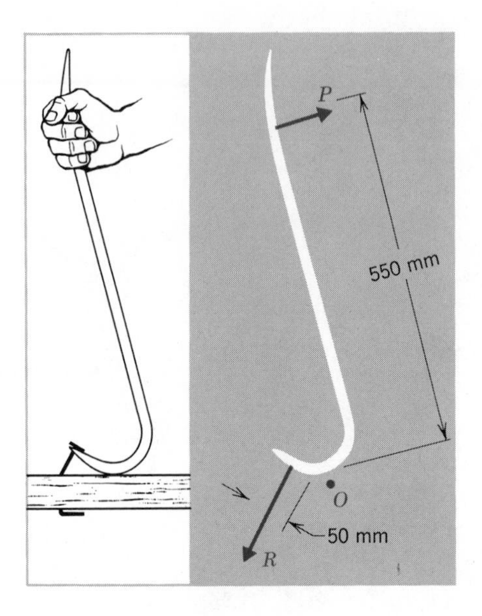

2-60 The crank is being used to raise the 200-kg load. If the sum of the moments about O of the force f on the handle and the gravity force on the load is zero, what must be the magnitude of f?

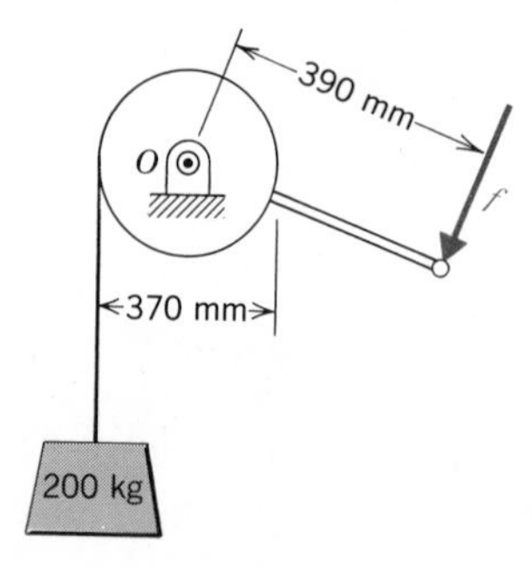

2-61 Determine the moment about O of the gravity force acting on the two-ton load.

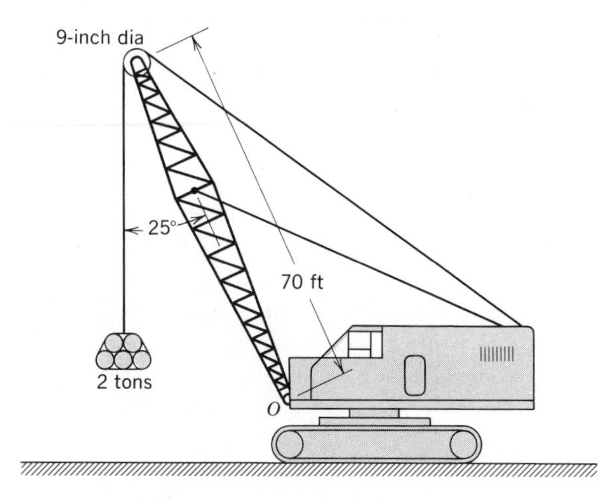

2-62 The plumber exerts a force of 240 lb on the wrench handle. Determine the moment of this force

(a) About point *A*.
(b) About point *B*.
(c) About point *C*.

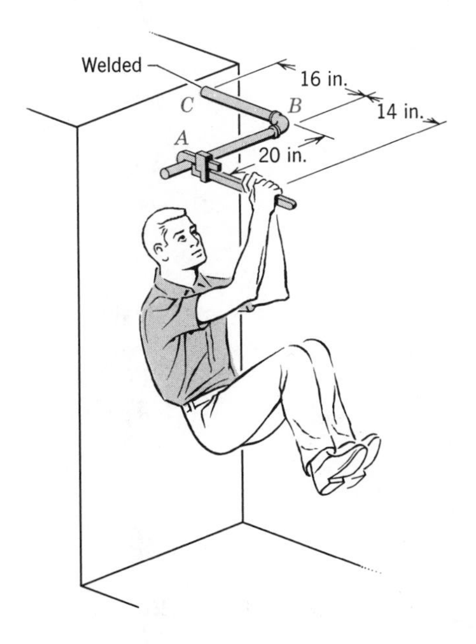

2-4 COUPLES

Two forces, of equal magnitude and opposite direction, and with different lines of action, constitute a force system called a *couple*. It may be depicted by two arrows representing equal-magnitude, oppositely directed forces, or by a curved arrow in the plane of the two forces. A couple has the properties that *its resultant force is zero, and its moment, about any point, has a magnitude equal to the product of the magnitude of either force and the perpendicular distance between the lines of action*. The first property is obvious; the second will be

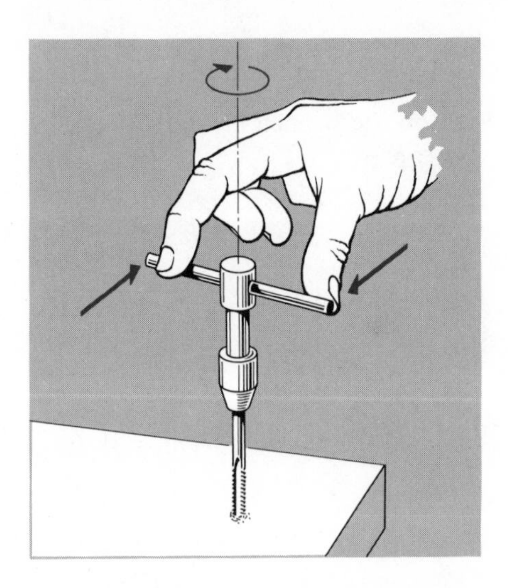

verified after we examine some helpful results from vector analysis. For now, let us examine the planar case where the point about which the resultant moment is evaluated lies in the plane of the lines of action (Figure 2-7). With the clockwise direction denoted by positive values, the sum of moments about O of the two forces has the magnitude

$$f(d + a) - fa = fd$$

which is independent of a. Thus, the moment about any point in this plane will be the same, namely fd.

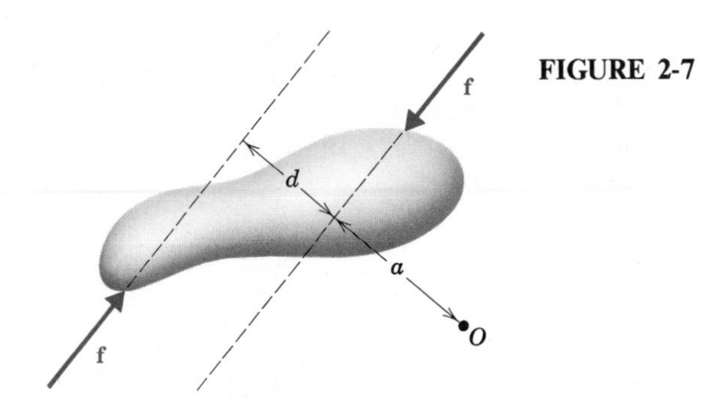

FIGURE 2-7

PROBLEMS

2-63 Determine the direction, magnitude, and line of action of the force which, when acting together with the given force, forms a couple with a moment of 800 lbf·in directed toward the viewer of the sketch.

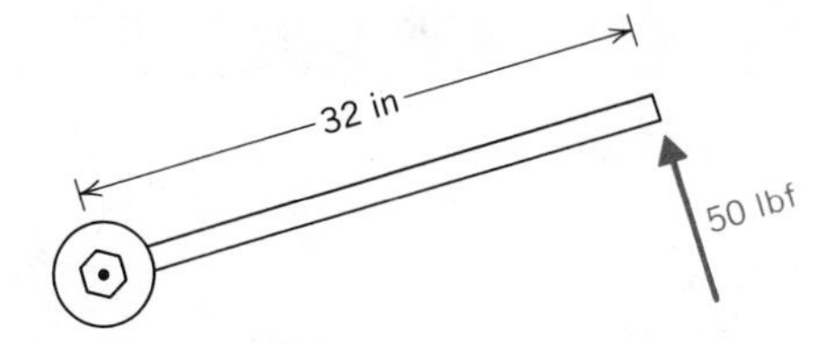

2-64 Each of the ship's two propellers exerts a force of 250 kN on the ship. How much pull must each tugboat exert in order to prevent the ship from rotating?

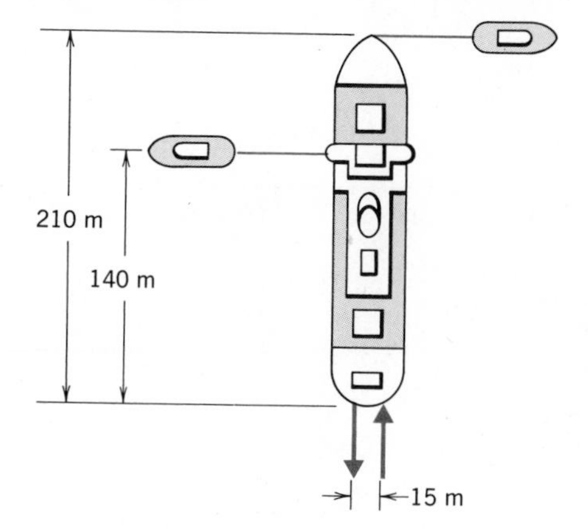

2-65 The brake is set on the wheel, and it will not slip until the moment about the center of the wheel of forces acting on the lug wrench reaches 150 N·m. Will the brake slip? *Ans.* 147.2 N·m; no.

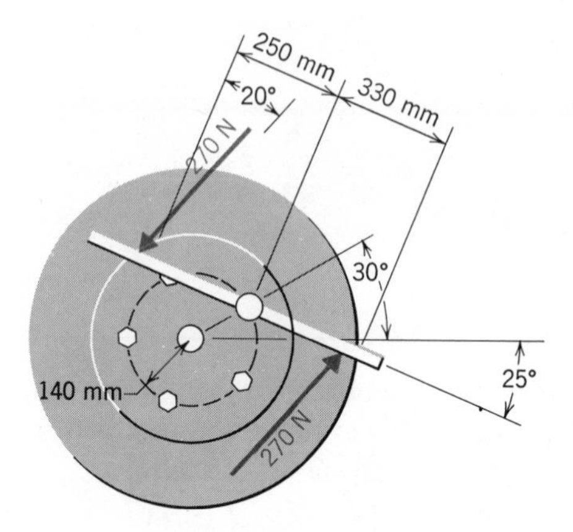

2-66 Evaluate the sum of the moments about A of the two 30-N forces.

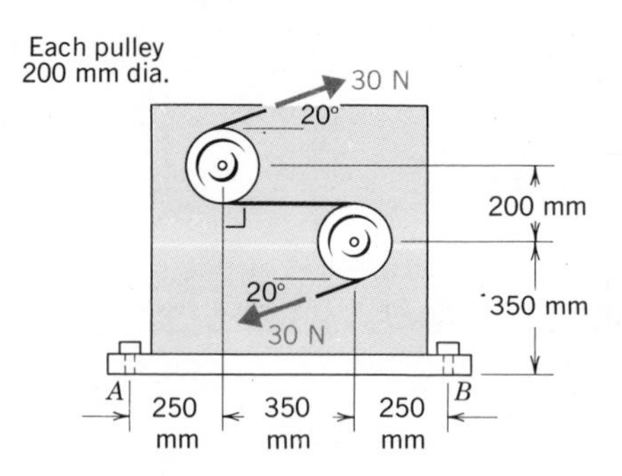

2-67 Evaluate the sum of the moments about O_1 of the four forces. Evaluate the sum of the moments about O_2 of the four forces.

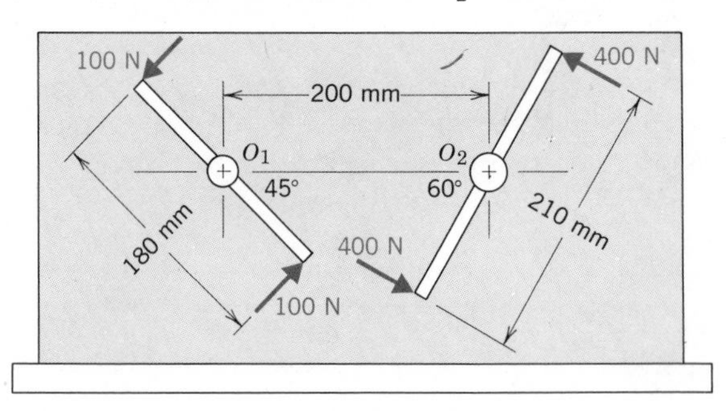

2-68 Evaluate the moment about O of each of the two 36-N forces, and add these two vectors. Compare the sum with the moment about the point of application of one of the forces on the handle.

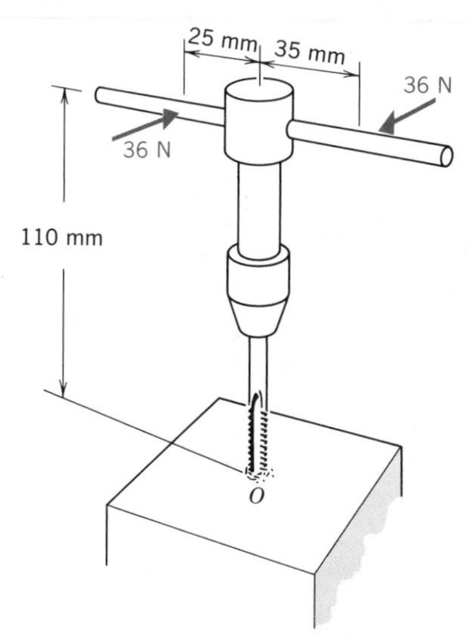

2-69 Should the operator of the tap wrench depicted in the sketch on page 44 be sure that his thumb and forefinger are at equal distances from the vertical axis? Explain.

2-70 Repeat Problem 2-68, but give the force on the right a magnitude of 40 N and the force on the left a magnitude of 36 N.

2-5 EQUIVALENT FORCE SYSTEMS

Two force systems are *equivalent* if they produce the same response in a rigid body. As we will learn later, this is the case provided that (1) the force resultants of the two systems are equal, and (2) the resultant moment about some point is the same for each system. For example, consider the four systems of forces applied to the angle bracket in Figure 2-8. The resultant force in each case is 100 N downward, and the resultant moment about the corner of the bracket is 60 N·m directed outward from the wall, so that each system is equivalent to the others.

Observe that the moment about the unattached end of the bracket, of each of the systems of forces, is also the same. In the next chapter we will show that if, for two systems of forces, the resultant force and the resultant moment

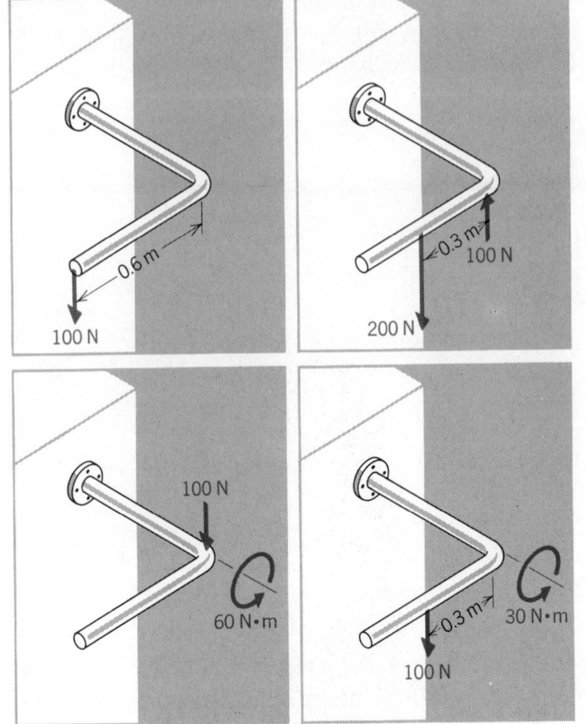

FIGURE 2-8

about one point are equal, then the resultant moment about every point must be equal.

In applications of the ideas introduced here, we often encounter geometric aspects of problems that are much more complex than we have met so far. A most useful tool for handling these complexities is the algebra-geometry of vectors, which will be examined in the next chapter.

PROBLEMS

2-71 For each of the four systems of forces in Figure 2-8, evaluate the resultant moment

(a) About the point midway between the corner of the bracket and the free end.

(b) About the free end of the bracket.

2-72 A vertical force is to be added to the given force f, so that the force system is equivalent to a single force of magnitude P with its line of action passing through point A. Determine the magnitude and line of action of this additional force.

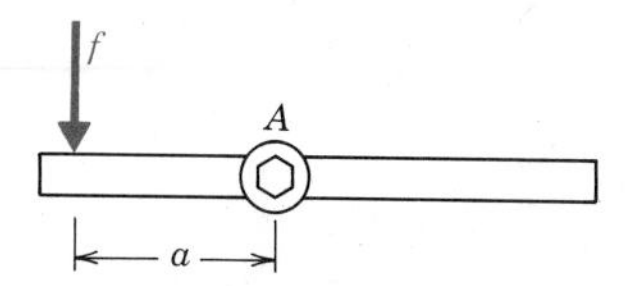

2-73 Determine the line of action of a force that will be equivalent to the given force and couple acting on the wheel. *Ans.* 5 in above center.

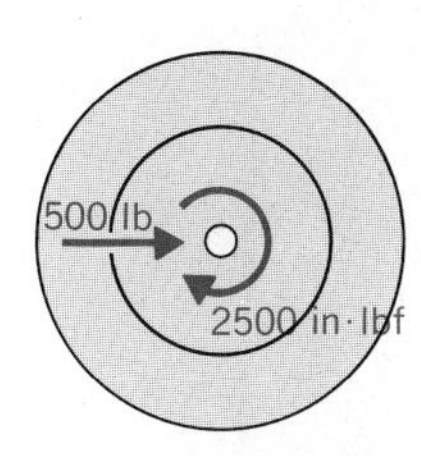

2-74 Indicate the equivalent of the force on the gear at A, as a force at O together with a couple. What is the moment of the couple?

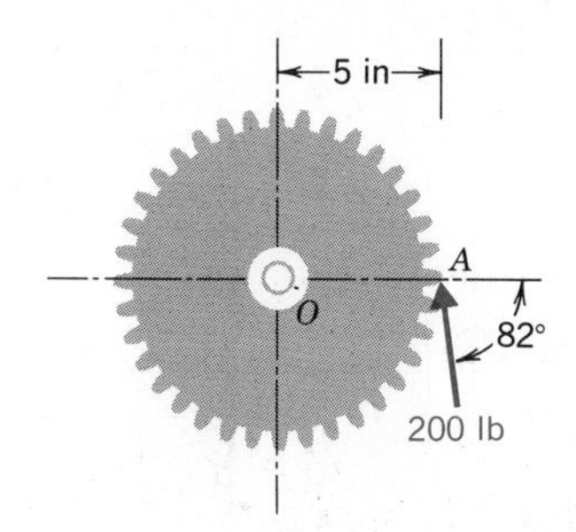

2-75 Specify the equivalent of the given forces as a force through A together with a couple.

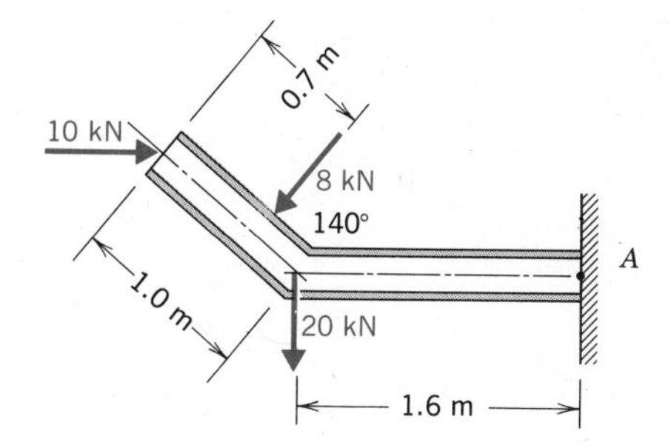

2-76–78 Tugboats are busily maneuvering steamships about the harbor. For each steamship, determine a single force and its line of action, such that it is equivalent to the given set of forces. *Suggestion*: After determining the magnitude and direction of the force, consider moments about the bow of the ship.

2-76

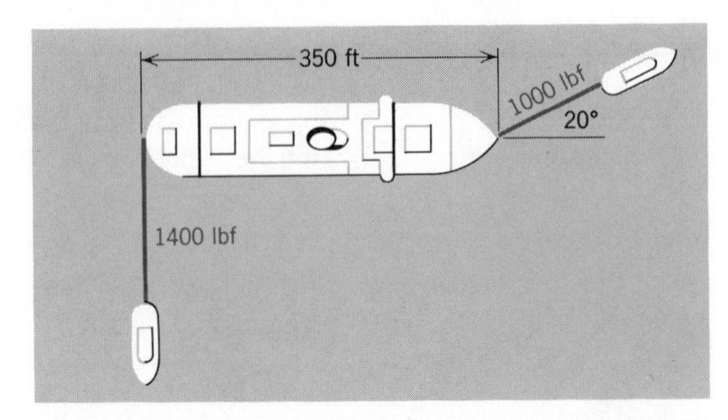

2-77

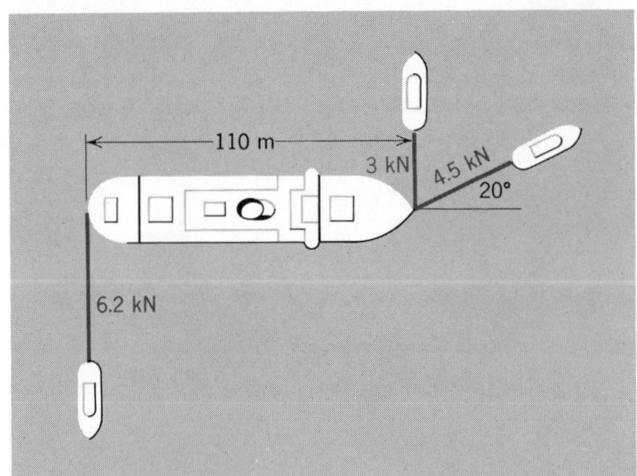

2-78

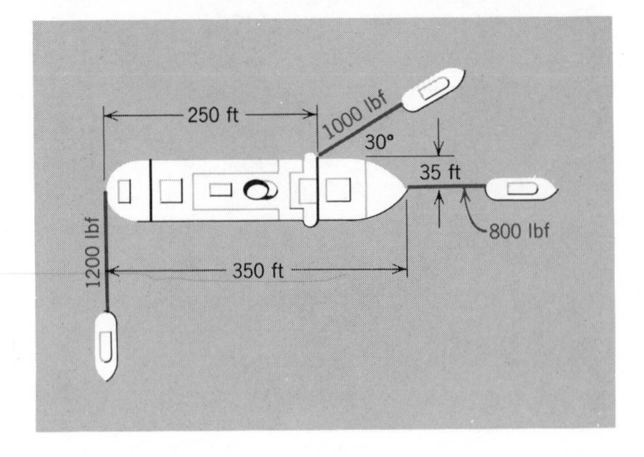

2-79 The bracket in Figure 2-8 has a force of 100 N downward, applied at a point 0.2 m toward the wall from the corner. Evaluate the additional couple that must be applied so that the force system is equivalent to the others shown.

2-80 The bracket in Figure 2-8 has a force of 100 N upward, applied at a point 0.2 m toward the wall from the corner. Determine a set of additional forces that, when added to this upward force, complete a force system equivalent to those shown.

2-81 Specify the equivalent of the given forces as a force through B together with a couple, and as a force through A together with a couple.

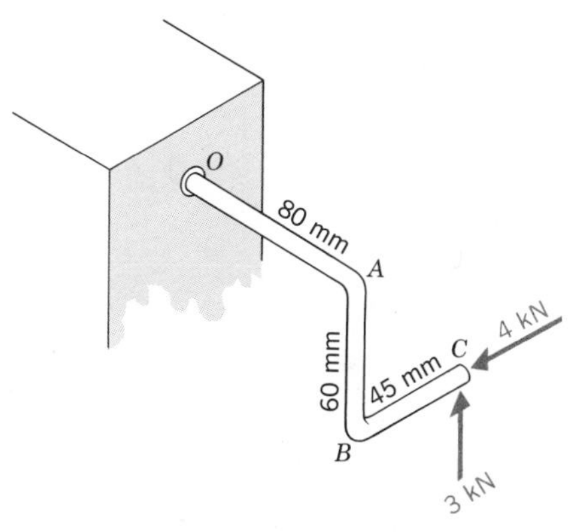

2-82 Determine a force system, equivalent to the 250-N force shown, consisting of a force passing through point O and a couple.

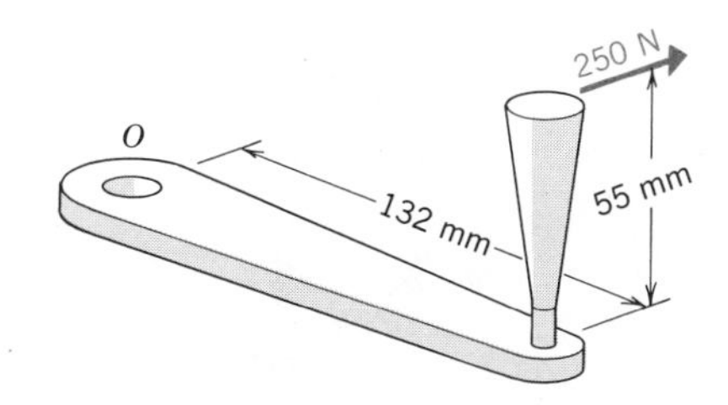

2-83 The automobile is free to roll along the horizontal surface. (The transmission is disengaged and the brakes are not set.) What would be the effect of the couple applied to the frame of the car, as depicted in (a)? What would be the effect of the same couple applied to the wheel, as depicted in (b)? Are the two force systems equivalent? Explain carefully.

M
M

3

VECTOR GEOMETRY AND ALGEBRA

In applications of the basic ideas introduced in the previous chapter and in later chapters, we may encounter geometric aspects of problems that complicate our analysis. The geometry and algebra of vectors can greatly simplify the handling of such complexities. In the first four sections of this chapter, we will examine the basic definitions and some of the resulting relationships that will be useful in mechanics; in Section 3-5 we will cast the ideas introduced in the previous chapter into the language of vector analysis.

3-1 BASIC CONCEPTS

We call a quantity that has only magnitude (and, hence, can be represented completely by a single real number) a *scalar*. Examples are mass of a body and temperature at some location. A *vector* is a quantity that has magnitude and direction relative to some reference frame. Arrows are used to depict vectors. The length of an arrow is made proportional to the magnitude of the vector, and the direction is given by its orientation. Symbols used to represent vectors in this book will appear in boldface type, for instance, **A**, **B**, **a**, **b**, and so forth. The magnitude of a vector (a scalar) will be denoted either by vertical bars or by a corresponding letter without boldface, for example, $|\mathbf{A}|$ or A.

Nothing about location of a vector in the reference frame is implied in its specification. For example, the vector specifying the 42-km/hr southward velocity of a sportscar on East Elm Street is the vector equal of the 42-km/hr southward velocity of a limousine on West Walnut Street. Thus, if some physical quantity has, in addition to magnitude and direction, some "point of application" or "line of action," we will use a *second* vector to specify the location of this point or line.

The *projection* of a vector onto a line is the vector whose initial and terminal points are the projections of the initial and terminal points, respectively, of the given vector. We denote this with a subscript indicating the line onto which the projection is made, for example, the projection of **A** onto the line x is denoted by $\mathbf{A}_x$. The length of the projection is denoted as $|\mathbf{A}_x| = A_x$. It can be shown that the length of this projection is independent of the position, relative to the line, of the arrow representing **A**. When the two directions along the line are defined as *positive* and *negative* (as with a coordinate axis), we define the scalar A_x as $\pm|\mathbf{A}_x|$, with the sign corresponding to whether $\mathbf{A}_x$ is directed in the positive or negative x-direction.

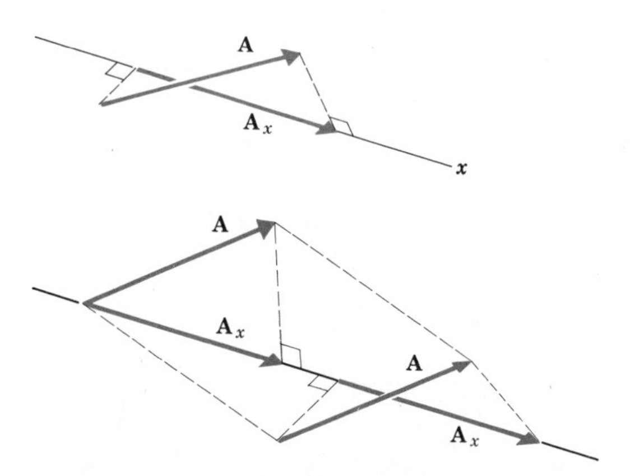

Similarly, we denote the projection of a vector **A** onto a line parallel to a second vector **B** as $\mathbf{A}_B$. The scalar quantity $A_B = \pm|\mathbf{A}_B|$ is defined as positive if $\mathbf{A}_B$ is in the direction of **B** and negative if $\mathbf{A}_B$ is in the direction opposite to that of **B**.

PROBLEMS

3-1 Determine the magnitudes of the projections A_B and B_A.

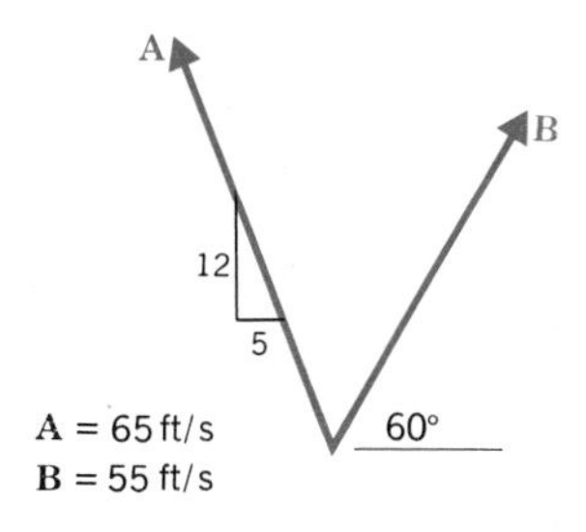

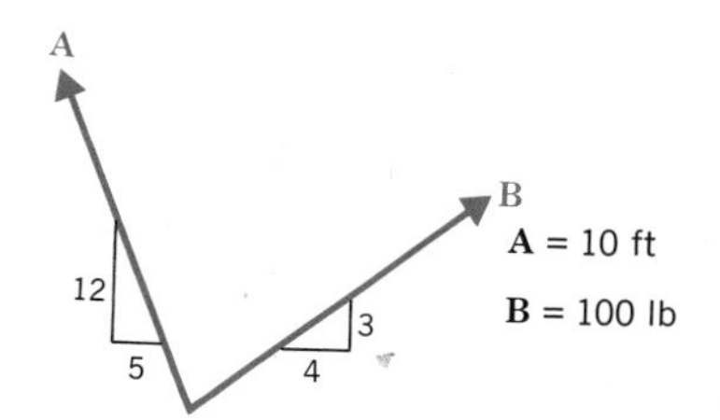

3-2 Two vectors **A** and **B** have magnitudes $A = 10$, $B = 15$. Further, they are oriented such that $B_A = 8$. Determine A_B. *Ans.* $\frac{16}{3}$.

3-3 The projection of **A** onto x_1 is $0.8A$. What is the projection of **A** onto x_2?

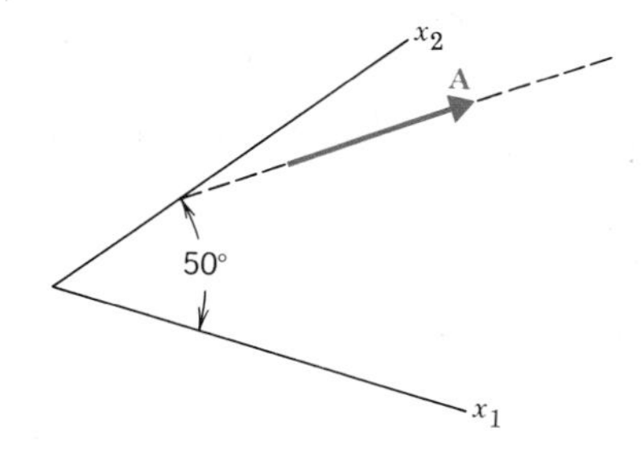

3-4 The x, y, and z axes are mutually perpendicular. Evaluate the projection of **A** onto the line PQ.

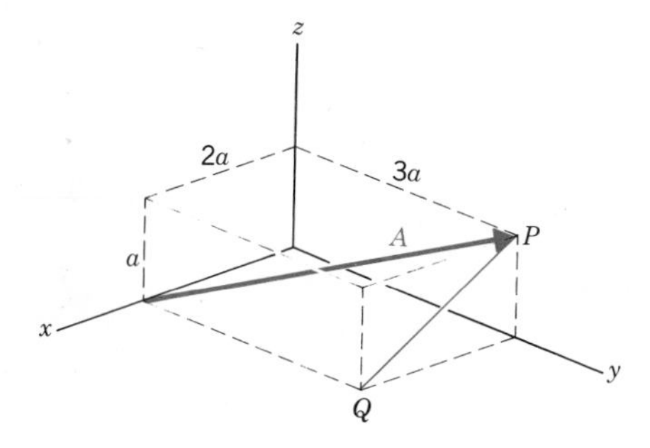

3-5 The x, y, and z axes are mutually perpendicular. Evaluate B_A and A_B. *Ans.* $-\frac{16}{5}$; $-\frac{4}{\sqrt{10}}$.

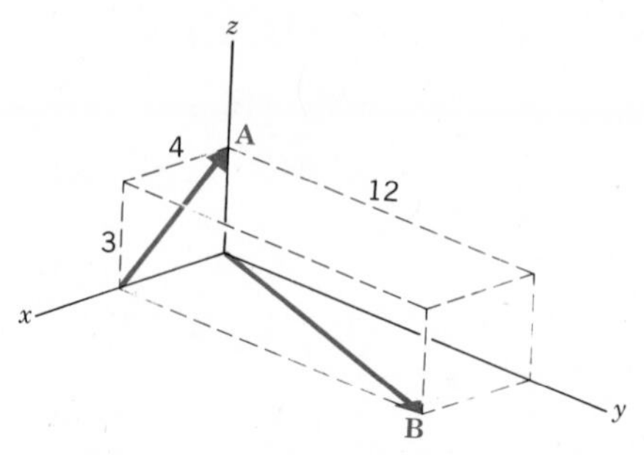

3-6 The projection of a 50-lb force vector onto a line has a magnitude of 38 lb. What is the angle between the vector and the line?

3-7 The projection of a 50-N force vector onto a 0.8-m displacement vector is a 25-N force vector in the direction opposite to that of the displacement vector. What is the angle between the 50-N force vector and the 0.8-m displacement vector? *Ans.* 120°

3-8 An airplane is flying in a southwesterly direction at a speed of 800 km/h and an altitude of 7 km. What is the projection of its velocity onto a highway that runs
(a) North and South?
(b) Northwest and Southeast?

3-9 An airplane is flying at an angle of 20° from the horizontal. Its rate of climb (the projection of its velocity onto the vertical) is 6000 ft/min.
(a) What is the speed of the aircraft?
(b) What is the magnitude of the projection of the airplane's velocity onto the horizontal?

3-10 The projection of **M** onto the x-axis has a magnitude of 60 N·m. What is the magnitude of **M**?

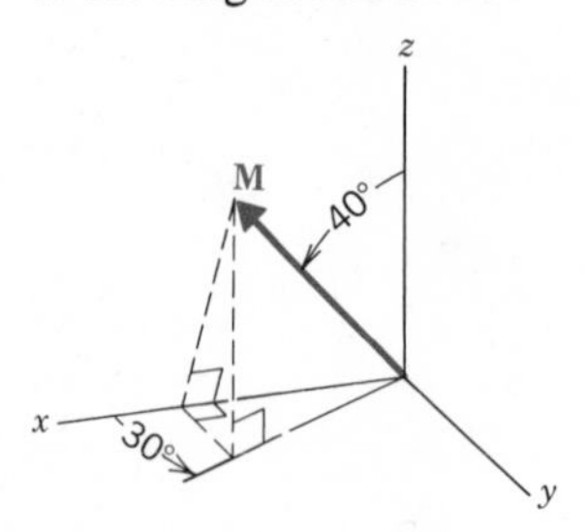

3-11 What is the angle between **M** and the x-axis of the previous problem?

3-12 Determine the magnitudes of the projections of **f** onto the x, y, and z axes, in terms of the angles ϕ and χ and the magnitude f.

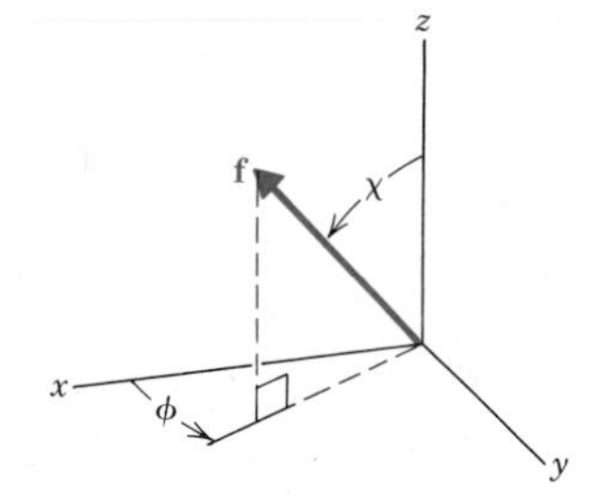

3-13 Show that the projection of a vector on a line is independent of the position relative to the line, of the arrow representing the vector.

3-14 The value of a force may be completely defined by a vector, since this gives its magnitude and direction. Is the vector value of a force sufficient to determine the response it will effect on a body? Explain carefully.

3-2 ADDITION AND SUBTRACTION

The *sum* of two vectors is defined according to the *parallelogram law* indicated in Figure 3-1*a*. This is equivalent to the tail-to-head placement indicated in Figure 3-1*b* and *c*. Successive application of the tail-to-head rule gives the sum of several vectors, as indicated in Figure 3-1*d*. A brief consideration of the figure reveals that vector addition is

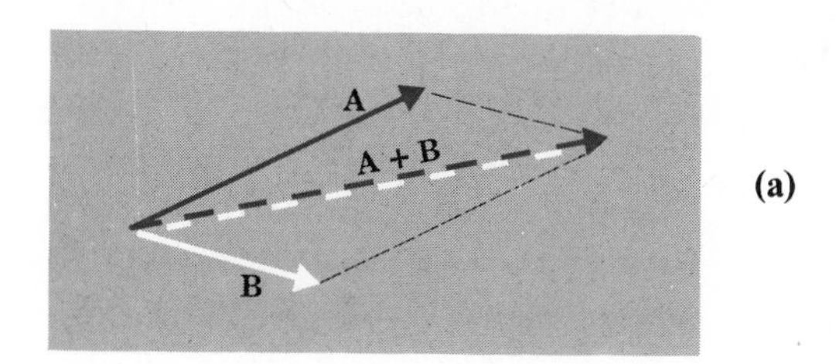

(a)

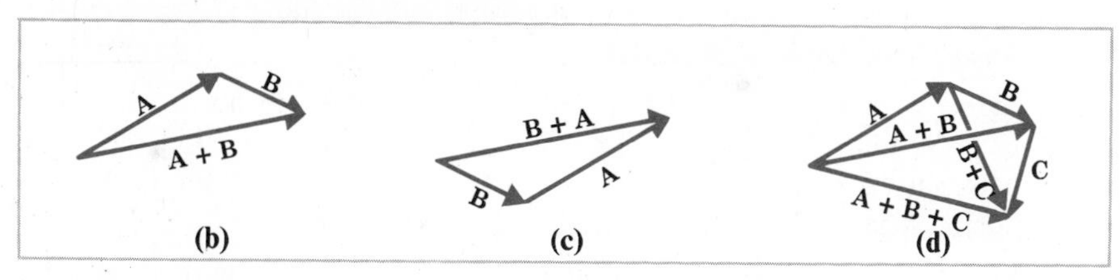

FIGURE 3-1

$$\textit{commutative}\colon \mathbf{B} + \mathbf{A} = \mathbf{A} + \mathbf{B} \tag{3-1}$$

and

$$\textit{associative}\colon \mathbf{A} + (\mathbf{B} + \mathbf{C}) = (\mathbf{A} + \mathbf{B}) + \mathbf{C} \tag{3-2}$$

From these laws, it follows that sums of any number of vectors can be formed, independent of the order of the tail-to-head placement.

The *negative* of a vector **A** is denoted by $-\mathbf{A}$ and is defined as the vector having a magnitude equal to that of **A** and direction opposite to that of **A**. Vector subtraction is then defined by

$$\mathbf{A} - \mathbf{B} = \mathbf{A} + (-\mathbf{B})$$

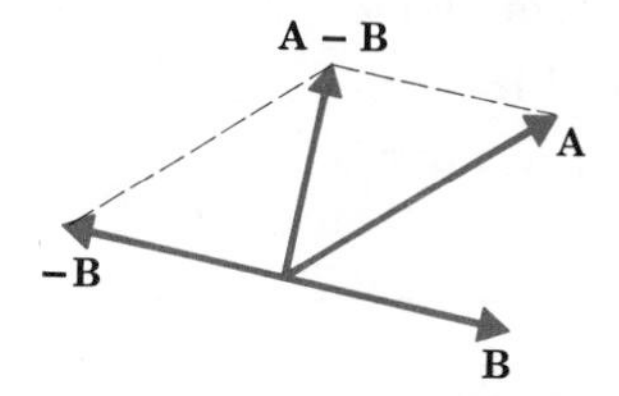

PROBLEMS

3-15 Consider the parallelepiped formed from the three vectors **A**, **B**, and **C**, as indicated. Reconstruct the sketch and show, on the sketch, $\mathbf{A} + \mathbf{B}$, $\mathbf{B} + \mathbf{C}$, $\mathbf{C} + \mathbf{A}$, and $\mathbf{A} + \mathbf{B} + \mathbf{C}$.

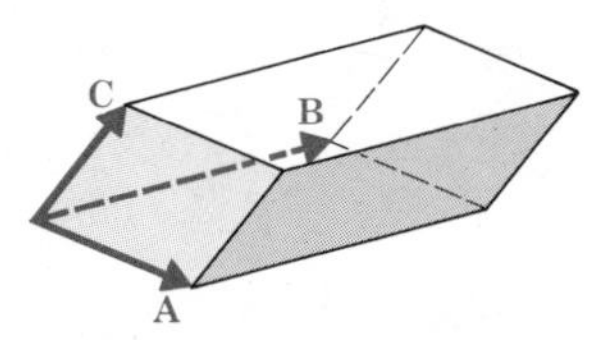

3-16 A ship travels 100 km north, then 50 km southwest. Represent these displacements graphically and determine the resultant displacement. *Ans.* 73.68 km; heading 331.3°.

3-17 After leaving port, ship *A* travels 420 nautical miles north, then 350 nautical miles northwest, where it sets anchor. After leaving the same port, ship *B* travels 420 nautical miles northeast, then 350 nautical miles north. What heading should ship *B* then take to reach ship *A*?

3-18 Referring to the previous problem, how far must ship *B* travel on the final leg in order to reach ship *A*?

For Problems 3-19 and 3-20, determine the vector sum of forces in each set, and indicate whether the sets of forces are equivalent with regard to response of a rigid body.

3-19

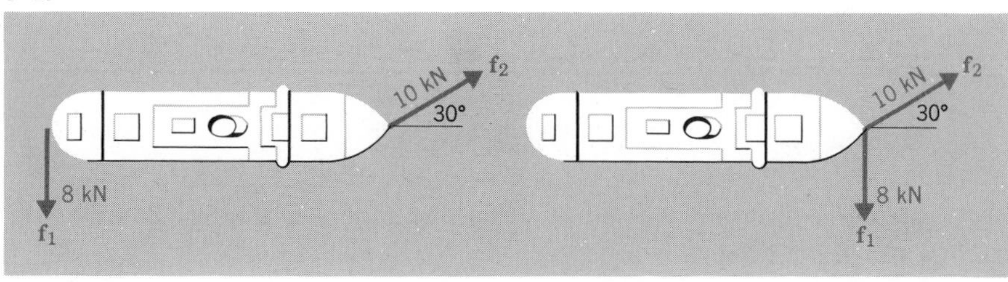

$$\mathbf{f}_1 + \mathbf{f}_2 = \begin{cases} \_\_\_\_\_\_ \text{ kN} \\ \_\_\_\_\_\_ ^\circ \text{ from} \\ \text{longitudinal ship axis} \end{cases}$$

(a)

$$\mathbf{f}_1 + \mathbf{f}_2 = \begin{cases} \_\_\_\_\_\_ \text{ kN} \\ \_\_\_\_\_\_ ^\circ \text{ from} \\ \text{longitudinal ship axis} \end{cases}$$

(b)

Equivalent? ______

3-20

(a)

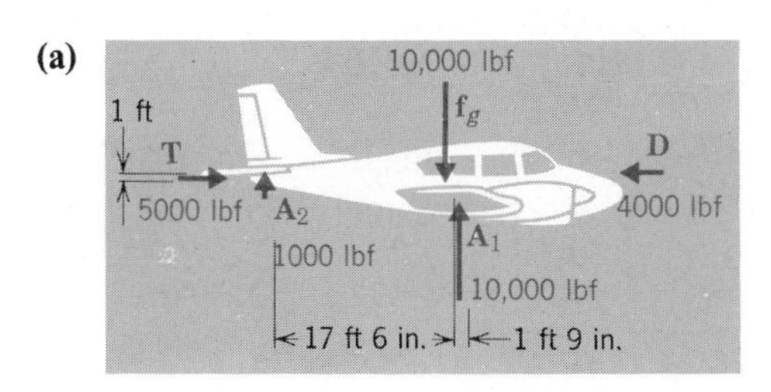

$$\mathbf{A}_1 + \mathbf{A}_2 + \mathbf{f}_g + \mathbf{T} + \mathbf{D} = \begin{cases} \_\_\_\_\_\_ \text{ lbf} \\ \_\_\_\_\_\_ ^\circ \text{ from} \\ \text{horizontal} \end{cases}$$

(b) Same as (a), except the line of action of f_g is moved 2 ft to the rear.

$$\mathbf{A}_1 + \mathbf{A}_2 + \mathbf{f}_g + \mathbf{T} + \mathbf{D} = \begin{cases} \_\_\_\_\_\_ \text{ lbf} \\ \_\_\_\_\_\_ ^\circ \text{ from} \\ \text{horizontal} \end{cases}$$

Equivalent? ______

3-21 Show graphically that $-(\mathbf{A} - \mathbf{B}) = -\mathbf{A} + \mathbf{B}$.

3-22 Show that $|\mathbf{A} + \mathbf{B}| \leqslant |\mathbf{A}| + |\mathbf{B}|$. Under what circumstance does the equality hold?

3-23 Show that $|\mathbf{A} - \mathbf{B}| \leqslant |\mathbf{A}| + |\mathbf{B}|$. Under what circumstance does the equality hold?

3-3 PRODUCTS

We have applications in mechanics for three kinds of multiplications with vectors, which are defined below.

Products with Scalars

Multiplication of a vector **A** by a scalar p results in a vector $p\mathbf{A}$ that has a magnitude pA and direction the same or opposite that of **A** according as p is positive or negative. The following rules result from this definition:

$$p(q\mathbf{A}) = (pq)\mathbf{A} \tag{3-3}$$

$$(p + q)\mathbf{A} = p\mathbf{A} + q\mathbf{A} \tag{3-4}$$

$$p(\mathbf{A} + \mathbf{B}) = p\mathbf{A} + p\mathbf{B} \tag{3-5}$$

Dot (or Scalar) Products

Dot (or *scalar*) multiplication of two vectors results in a scalar that is equal to the product of the magnitudes of the two vectors and the cosine of the angle between them. This is written as

$$\boxed{\mathbf{A}\cdot\mathbf{B} = AB\cos \measuredangle_{\mathbf{A}}^{\mathbf{B}}} \tag{3-6}$$

Note that $B\cos \measuredangle_{\mathbf{A}}^{\mathbf{B}}$ is the length of the projection of B onto a line in the direction of **A**. This is written as

$$B\cos \measuredangle_{\mathbf{A}}^{\mathbf{B}} = B_A$$

Thus the dot product may be interpreted in terms of length of **A** and the length of the projection of **B** onto **A**:

$$\mathbf{A}\cdot\mathbf{B} = AB_A$$

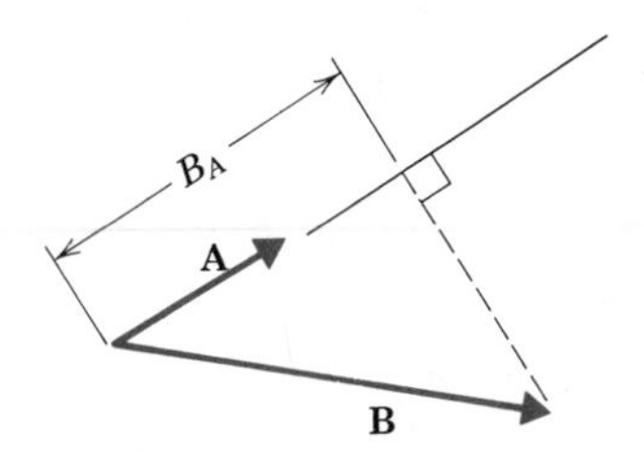

Furthermore, the important commutative property,

$$\mathbf{A}\cdot\mathbf{B} = \mathbf{B}\cdot\mathbf{A} \tag{3-7}$$

which follows from the definition (3-6), indicates that the projection interpretation may be made the other way around:

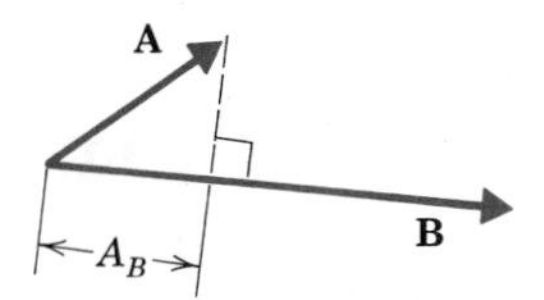

$$\mathbf{A}\cdot\mathbf{B} = A_B B$$

Observe from the definition (3.6) that the dot product may be positive, zero, or negative, depending on the angle between the two vectors. If the angle is acute, $\mathbf{A}\cdot\mathbf{B}$ will be positive, and if the angle is obtuse, $\mathbf{A}\cdot\mathbf{B}$ will be negative. If **A** and **B** are mutually perpendicular, $\mathbf{A}\cdot\mathbf{B}$ will be zero.

For the special case $\mathbf{B} = \mathbf{A}$, the dot product yields the square of the magnitude of **A**:

$$\mathbf{A}\cdot\mathbf{A} = AA\cos 0 = A^2$$

From this we obtain a useful way of calculating the magnitude of a given vector:

$$A = \sqrt{\mathbf{A}\cdot\mathbf{A}} \tag{3-8}$$

A consequence of the parallelogram law of addition is that the length of the projection of $\mathbf{A} + \mathbf{B}$ on **C** is equal to the sum of the lengths of the projections of **A** on **C** and **B** on **C**:

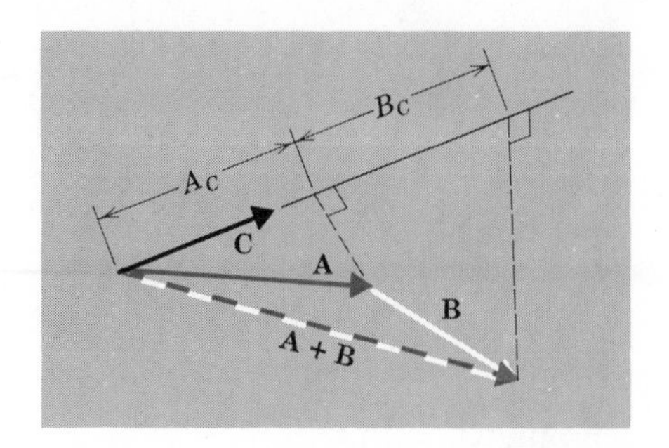

$$|(\mathbf{A} + \mathbf{B})_C| = A_C + B_C$$

This in turn implies that the dot product is *distributive*:

$$(\mathbf{A} + \mathbf{B}) \cdot \mathbf{C} = \mathbf{A} \cdot \mathbf{C} + \mathbf{B} \cdot \mathbf{C} \tag{3-9}$$

Finally, it follows readily from the definitions that the dot product and multiplication by a scalar are *associative*:

$$(p\mathbf{A}) \cdot (q\mathbf{B}) = (pq)(\mathbf{A} \cdot \mathbf{B}) \tag{3-10}$$

Cross (or Vector) Products

Cross (or vector) multiplication of two vectors **A** and **B** results in a vector, written as $\mathbf{A} \times \mathbf{B}$. This vector is defined to be perpendicular to **A** and **B**, to have magnitude equal to the product of the magnitudes A and B and the sine of the angle between **A** and **B**, and to have direction determined by the

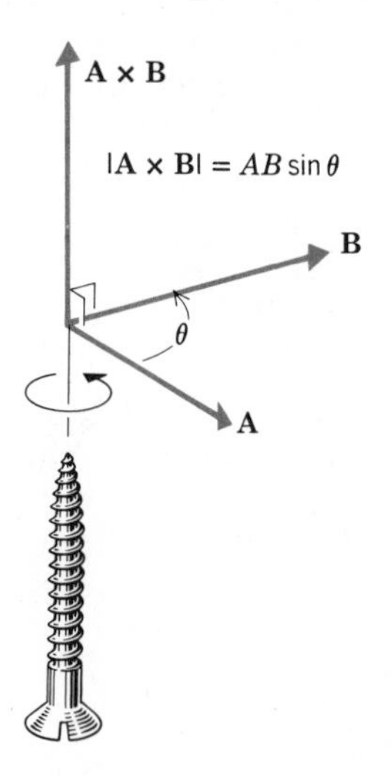

right-hand rule; that is, its direction coincides with that of the advancement of a right-hand screw, the axis of the screw being oriented perpendicular to **A** and **B**, and turned in the direction **A** toward **B**.

Note that from this definition the cross product is *not* commutative; instead,

$$\mathbf{B} \times \mathbf{A} = -\mathbf{A} \times \mathbf{B} \tag{3-11}$$

However, as is indicated by the solution to Problem 3-47, it is *distributive*:

$$\mathbf{A} \times (\mathbf{B} + \mathbf{C}) = \mathbf{A} \times \mathbf{B} + \mathbf{A} \times \mathbf{C} \tag{3-12}$$

It can also be readily shown to be *associative* with scalar multiplication:

$$(p\mathbf{A}) \times (q\mathbf{B}) = (pq)\mathbf{A} \times \mathbf{B} \tag{3-13}$$

Unit Vectors

Frequently we have use for vectors that have unit magnitude. A unit vector in the direction of a given vector **A** can be constructed by dividing the vector by its magnitude; we write this as

$$\mathbf{u}_A = \frac{\mathbf{A}}{A} \tag{3-14}$$

Components Parallel and Perpendicular to a Given Direction

As we saw in the last chapter, a vector may be resolved in many ways. In a number of applications, we will find it useful to resolve a given vector **A** into a component parallel to another given vector **B**, and a component perpendicular to **B**. This can be formally expressed as

$$\mathbf{A} = \mathbf{A}_{\parallel} + \mathbf{A}_{\perp} \tag{a}$$

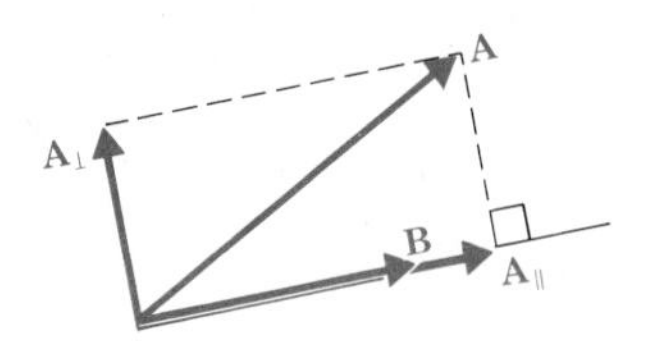

The component parallel to **B** will be the product of a unit vector in the direction of **B** and the projection of **A** onto **B**:

$$\mathbf{A}_{\parallel} = (A \cos \measuredangle_{\mathbf{B}}^{\mathbf{A}})\mathbf{u}_B$$

$$= (\mathbf{u}_B \cdot \mathbf{A})\,\mathbf{u}_B \tag{b}$$

The component perpendicular to **B** can be determined from Equations (a) and (b), as

$$\mathbf{A}_\perp = \mathbf{A} - (\mathbf{u}_B \cdot \mathbf{A})\,\mathbf{u}_B \tag{c}$$

Alternatively, the perpendicular component can be calculated by*

$$\mathbf{A}_\perp = (\mathbf{u}_B \times \mathbf{A}) \times \mathbf{u}_B \tag{c$'$}$$

By combining the above relationships, the resolution of **A** into components parallel and perpendicular to **B** can be written as

$$\mathbf{A} = (\mathbf{u}_B \cdot \mathbf{A})\,\mathbf{u}_B + (\mathbf{u}_B \times \mathbf{A}) \times \mathbf{u}_B \tag{3-15a}$$

or as

$$\mathbf{A} = \frac{(\mathbf{B} \cdot \mathbf{A})\,\mathbf{B}}{B^2} + \frac{(\mathbf{B} \times \mathbf{A}) \times \mathbf{B}}{B^2} \tag{3-15b}$$

Some Identities

In addition to the relationships pointed out in the foregoing, the following identities are sometimes useful. It is also instructive to verify them.

$$\mathbf{A} \times (\mathbf{B} \times \mathbf{C}) = (\mathbf{C} \cdot \mathbf{A})\mathbf{B} - (\mathbf{A} \cdot \mathbf{B})\mathbf{C} \tag{3-16}$$

$$\mathbf{A} \cdot (\mathbf{B} \times \mathbf{C}) = \mathbf{B} \cdot (\mathbf{C} \times \mathbf{A}) = \mathbf{C} \cdot (\mathbf{A} \times \mathbf{B}) \tag{3-17}$$

$$(\mathbf{A} \times \mathbf{B}) \cdot (\mathbf{C} \times \mathbf{D}) = (\mathbf{A} \cdot \mathbf{C})(\mathbf{B} \cdot \mathbf{D}) - (\mathbf{B} \cdot \mathbf{C})(\mathbf{D} \cdot \mathbf{A}) \tag{3-18}$$

PROBLEMS

3-24 Indicate geometrically how any vector **A** that lies in the plane formed by two given, noncollinear vectors $\mathbf{a}_1$ and $\mathbf{a}_2$ can be formed in terms of these two vectors as

$$\mathbf{A} = A_1 \mathbf{a}_1 + A_2 \mathbf{a}_2$$

For given **A**, $\mathbf{a}_1$, and $\mathbf{a}_2$, are the coefficients A_1 and A_2 unique?

3-25 Indicate geometrically how any vector **A** can be formed in terms of three given, noncoplanar vectors $\mathbf{a}_1$, $\mathbf{a}_2$ and $\mathbf{a}_3$, as

*See Problems 3-39 and 3-40.

$$\mathbf{A} = A_1 \mathbf{a}_1 + A_2 \mathbf{a}_2 + A_3 \mathbf{a}_3$$

For given $\mathbf{A}$, $\mathbf{a}_1$, $\mathbf{a}_2$, and $\mathbf{a}_3$, are the coefficients A_1, A_2, and A_3 unique?

3-26 Under what circumstances will $\mathbf{A} \times \mathbf{B}$ be zero?

3-27 If $\mathbf{A} \cdot \mathbf{C} = \mathbf{B} \cdot \mathbf{C}$, does it follow that $\mathbf{A} = \mathbf{B}$? Explain.

3-28 If $\mathbf{A} \times \mathbf{C} = \mathbf{B} \times \mathbf{C}$, does it follow that $\mathbf{A} = \mathbf{B}$? If not, how are $\mathbf{A}$ and $\mathbf{B}$ related?

3-29 Given $\mathbf{B}$ and $\mathbf{C}$, does the equation $\mathbf{C} = \mathbf{A} \times \mathbf{B}$ uniquely define $\mathbf{A}$? Explain.

3-30 Is $\mathbf{A} \times (\mathbf{B} \times \mathbf{C}) = (\mathbf{A} \times \mathbf{B}) \times \mathbf{C}$?

3-31 Show that $\mathbf{B} \times (\mathbf{A} \times \mathbf{B}) = (\mathbf{B} \times \mathbf{A}) \times \mathbf{B}$.

3-32 Show that the area A of the parallelogram with adjacent sides $\mathbf{a}$ and $\mathbf{b}$ is given by

$$A = |\mathbf{a} \times \mathbf{b}|$$

3-33 Show that $|\mathbf{A} \cdot (\mathbf{B} \times \mathbf{C})|$ is equal to the volume of the parallelepiped having $\mathbf{A}$, $\mathbf{B}$, and $\mathbf{C}$ as adjacent edges.

3-34 Show that the necessary and sufficient condition that $\mathbf{A}$, $\mathbf{B}$, and $\mathbf{C}$ all lie in parallel planes is that

$$\mathbf{A} \cdot (\mathbf{B} \times \mathbf{C}) = 0$$

3-35 The edges OP, OQ, and OR of the rectangular pyramid are mutually perpendicular. Show that the ratio of the area A_i normal to the i-axis, to the area A of the triangle PQR, is equal to the direction cosine of the normal to PQR with the i-axis, that is,

$$\frac{A_i}{A} = \cos \measuredangle_i^{\text{normal}} \qquad (i = x, y, z)$$

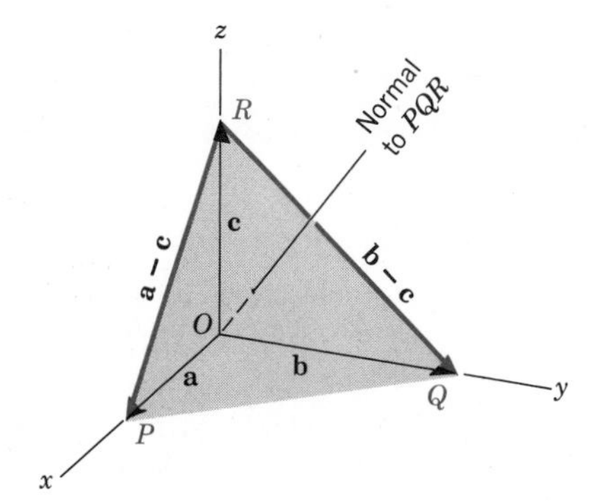

3-36 Use geometric definitions to derive Equation 3-16.

3-37 Derive Equation 3-17. *Suggestion.* Refer to the result of Problem 3-33.

3-38 Derive Equation 3-18. *Suggestion.* Use Equations 3-17 and 3-16.

3-39 Show how to obtain Equation c′ (p. 64) from the geometric definition of the cross product.

3-40 Use the identity 3-16 to derive Equation 3-15.

3-41 Show that $|\mathbf{A} \times \mathbf{B}| = \sqrt{A^2B^2 - (\mathbf{A}\cdot\mathbf{B})^2}$

(a) From the basic definitions of the dot and cross products.

(b) By means of the identity 3-18.

3-42 Show that the minimum distance between the two nonparallel lines OP and $O'P'$ is given by

$$d = \frac{|(\mathbf{a} \times \mathbf{b})\cdot\mathbf{r}|}{|\mathbf{a} \times \mathbf{b}|}$$

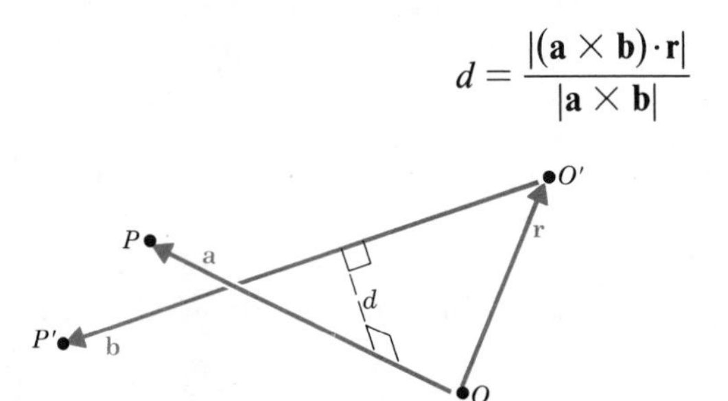

3-43 A rigid body rotates about an axis through O at a rate of ω radians per unit time. Let $\mathbf{r}$ be a position vector from point O to a point P on the body. Show that the velocity of point P is given by

$$\mathbf{v}_P = \boldsymbol{\omega} \times \mathbf{r}$$

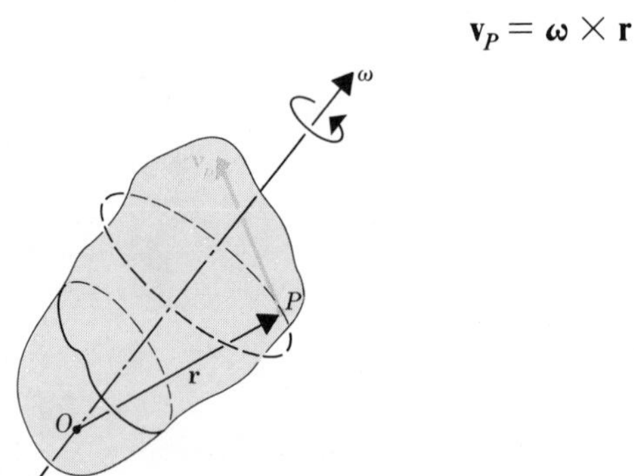

where the vector $\boldsymbol{\omega}$ is in the direction of the axis of rotation, has magnitude ω, and sense determined by the direction of rotation according to the right-hand rule.

3-44 Let **a** and **b** be position vectors from a point O to points A and B, respectively. Show that if P is a point that divides the line AB into segments that are in the ratio $\alpha : \beta$, the position vector from O to P is given by

$$\mathbf{r} = \frac{\alpha\mathbf{a} + \beta\mathbf{b}}{\alpha + \beta}$$

3-45 Show that the equation of a plane passing through the three points A, B, and C can be written in terms of position vectors **a**, **b**, and **c** from an origin O to A, B, and C, respectively, as

$$\mathbf{r} = \frac{\xi\mathbf{a} + \eta\mathbf{b} + \zeta\mathbf{c}}{\xi + \eta + \zeta}$$

in which ξ, η, and ζ are any scalars, which define **r** as the position vector from O to a point in the plane.

3-46 Show that if **c** is a constant vector and **r** is a position vector from a fixed point to a varying point, the equation

$$\mathbf{c}\cdot\mathbf{r} = 0$$

defines a plane. What is the geometric significance of **c** in this case?

3-47 (a) Show that if **A** is perpendicular to both **B** and **C**, then $\mathbf{A} \times (\mathbf{B} + \mathbf{C}) = \mathbf{A} \times \mathbf{B} + \mathbf{A} \times \mathbf{C}$.

(b) Show that if **B** is resolved into a component vector $\mathbf{B}_{\parallel}$, which is parallel to **A**, and a component vector $\mathbf{B}_{\perp}$, which is perpendicular to **A**, that $\mathbf{A} \times \mathbf{B} = \mathbf{A} \times \mathbf{B}_{\perp}$.

(c) Show that the distributive law (3-12) is valid in the general case, where there are no restrictions on **A**, **B**, and **C**. [Note that if **C** is resolved in the same way as **B** then, from the associative law for addition,

$$\mathbf{B} + \mathbf{C} = (\mathbf{B}_{\parallel} + \mathbf{B}_{\perp}) + (\mathbf{C}_{\parallel} + \mathbf{C}_{\perp}) = (\mathbf{B}_{\parallel} + \mathbf{C}_{\parallel}) + (\mathbf{B}_{\perp} + \mathbf{C}_{\perp})$$
$$= (\mathbf{B} + \mathbf{C})_{\parallel} + (\mathbf{B} + \mathbf{C})_{\perp}]$$

3-4 RECTANGULAR CARTESIAN COMPONENTS

To *resolve* a vector is to replace it with two or more component vectors whose sum is the original vector. Of the infinitely many possible *resolutions* for a vector, perhaps the most useful is a set of three mutually perpendicular component vectors $\mathbf{A}_x$, $\mathbf{A}_y$, and $\mathbf{A}_z$. The equivalence is written as

$$\mathbf{A} = \mathbf{A}_x + \mathbf{A}_y + \mathbf{A}_z$$

in which the subscripts refer to a selected set of mutually perpendicular axes, as shown in Figure 3-2. The vectors $\mathbf{A}_x$, $\mathbf{A}_y$, and $\mathbf{A}_z$ form a set of orthogonal *vector components*. It is also helpful to define a set of three vectors, with unit magnitudes and in the directions of the component vectors, as

$$\mathbf{u}_x = \frac{\mathbf{A}_x}{A_x} \qquad \mathbf{u}_y = \frac{\mathbf{A}_y}{A_y} \qquad \mathbf{u}_z = \frac{\mathbf{A}_z}{A_z}$$

and to write the resolution of the vector as

$$\mathbf{A} = A_x\mathbf{u}_x + A_y\mathbf{u}_y + A_z\mathbf{u}_z \tag{3-19}$$

The $\mathbf{u}_i$ $(i = x, y, z)$ are called the *unit base vectors* for this resolution, and the A_i are called *rectangular Cartesian components* of $\mathbf{A}$.

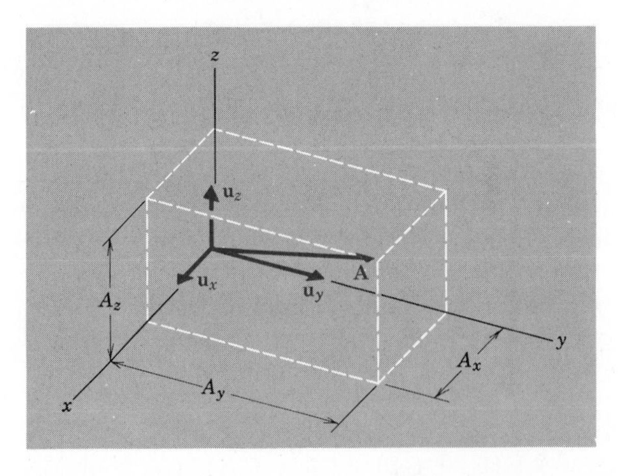

FIGURE 3-2

Equations

Analysis with vectors frequently leads to a result of the form

$$\mathbf{f} = \mathbf{g}$$

which may also be expressed as

$$f_x\mathbf{u}_x + f_y\mathbf{u}_y + f_z\mathbf{u}_z = g_x\mathbf{u}_x + g_y\mathbf{u}_y + g_z\mathbf{u}_z$$

Most applications require the use of the *component equivalent* of such a vector equation. This equivalent expresses the fact that the projections of two equal vectors onto the same axis are equal; that is, the above implies the three equations:

$$f_x = g_x$$

$$f_y = g_y$$

$$f_z = g_z$$

Addition and Subtraction

Sums and differences of two or more vectors can be carried out in terms of the corresponding components in a set of Cartesian coordinate directions. From the laws (3-2) and (3-4),

$$\begin{aligned}
\mathbf{A} \pm \mathbf{B} \pm \mathbf{C} \cdots &= (A_x \mathbf{u}_x + A_y \mathbf{u}_y + A_z \mathbf{u}_z) \\
&\quad \pm (B_x \mathbf{u}_x + B_y \mathbf{u}_y + B_z \mathbf{u}_z) \\
&\quad \pm (C_x \mathbf{u}_x + C_y \mathbf{u}_y + C_z \mathbf{u}_z) \\
&\quad \pm \cdots \\
&= (A_x \pm B_x \pm C_x \pm \cdots) \mathbf{u}_x \\
&\quad + (A_y \pm B_y \pm C_y \pm \cdots) \mathbf{u}_y \\
&\quad + (A_z \pm B_z \pm C_z \pm \cdots) \mathbf{u}_z
\end{aligned} \tag{3-20}$$

Thus the ith component of the sum or difference is equal to the sum or difference of the ith components.

Dot Product

By using the expression (3-19) and algebraic properties established in the previous section, useful formulas for the dot and cross products can be derived. These formulas also depend on the following relationships, which are evident from Figure 3-2:

$$\mathbf{u}_x \cdot \mathbf{u}_x = \mathbf{u}_y \cdot \mathbf{u}_y = \mathbf{u}_z \cdot \mathbf{u}_z = 1 \tag{3-21a}$$

$$\mathbf{u}_x \cdot \mathbf{u}_y = \mathbf{u}_y \cdot \mathbf{u}_z = \mathbf{u}_z \cdot \mathbf{u}_x = 0 \tag{3-21b}$$

$$\mathbf{u}_x \times \mathbf{u}_x = \mathbf{u}_y \times \mathbf{u}_y = \mathbf{u}_z \times \mathbf{u}_z = \mathbf{0} \tag{3-22a}$$

$$\mathbf{u}_x \times \mathbf{u}_y = \mathbf{u}_z \qquad \mathbf{u}_y \times \mathbf{u}_z = \mathbf{u}_x \qquad \mathbf{u}_z \times \mathbf{u}_x = \mathbf{u}_y \tag{3-22b}$$

By using Equations 3-19, 3-9, 3-10, and 3-21, the dot product can be written as

$$\begin{aligned}
\mathbf{A}\cdot\mathbf{B} &= (A_x\mathbf{u}_x + A_y\mathbf{u}_y + A_z\mathbf{u}_z)\cdot(B_x\mathbf{u}_x + B_y\mathbf{u}_y + B_z\mathbf{u}_z) \\
&= A_xB_x\mathbf{u}_x\cdot\mathbf{u}_x + A_xB_y\mathbf{u}_x\cdot\mathbf{u}_y + A_xB_z\mathbf{u}_x\cdot\mathbf{u}_z \\
&\quad + A_yB_x\mathbf{u}_y\cdot\mathbf{u}_x + A_yB_y\mathbf{u}_y\cdot\mathbf{u}_y + A_yB_z\mathbf{u}_y\cdot\mathbf{u}_z \\
&\quad + A_zB_x\mathbf{u}_z\cdot\mathbf{u}_x + A_zB_y\mathbf{u}_z\cdot\mathbf{u}_y + A_zB_z\mathbf{u}_z\cdot\mathbf{u}_z
\end{aligned}$$

$$\boxed{\mathbf{A}\cdot\mathbf{B} = A_xB_x + A_yB_y + A_zB_z} \tag{3-23}$$

This formula has enough applications to warrant its memorization. As we saw on page 61, the special case $\mathbf{B} = \mathbf{A}$ leads to the formula (3-8) for the magnitude of a vector. In terms of rectangular Cartesian components, this becomes

$$\boxed{A = \sqrt{A_x^{\,2} + A_y^{\,2} + A_z^{\,2}}} \tag{3-24}$$

which agrees with the result of applying the formula of Pythagoras to the diagram of Figure 3-2.

EXAMPLE

Determine the lengths of the guy lines $0'P$ and $O'Q$, shown in Figure 3-3, and the angle between them.

Let position vectors $\mathbf{A}$ and $\mathbf{B}$ run from point O' to points P and Q, respectively, and resolve these into components along the x, y, z axes shown.

$$\mathbf{A} = (5\text{ m})\,\mathbf{u}_x + (4\text{ m})\,\mathbf{u}_y + (8\text{ m})\,\mathbf{u}_z$$

$$\mathbf{B} = (-3\text{ m})\,\mathbf{u}_x + (4\text{ m})\,\mathbf{u}_y + (8\text{ m})\,\mathbf{u}_z$$

The required lengths can then be determined using Equation 3-24:

$$A = \sqrt{(5\text{ m})^2 + (4\text{ m})^2 + (8\text{ m})^2} = 10.25\text{ m}$$

$$B = \sqrt{(-3\text{ m})^2 + (4\text{ m})^2 + (8\text{ m})^2} = 9.43\text{ m}$$

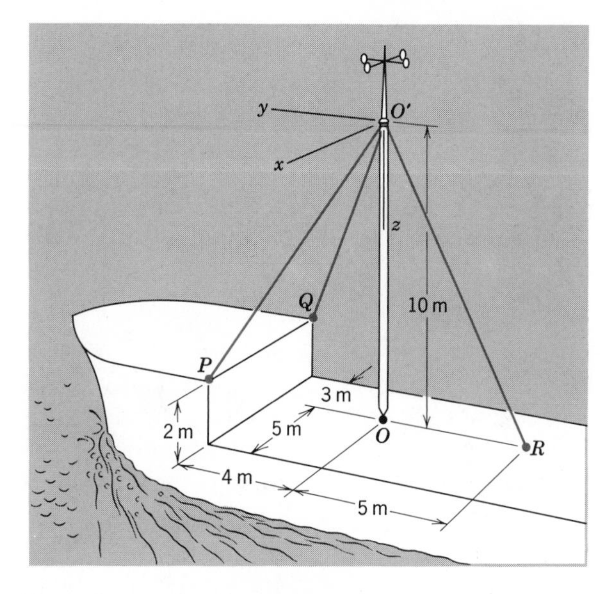

FIGURE 3-3

Now, the required angle is a factor in the definition of $\mathbf{A}\cdot\mathbf{B}$, and in fact will be the only unknown in this equation after $\mathbf{A}\cdot\mathbf{B}$ is evaluated. From Equation 3-23 we have

$$\mathbf{A}\cdot\mathbf{B} = (5\text{ m})(-3\text{ m}) + (4\text{ m})(4\text{ m}) + (8\text{ m})(8\text{ m})$$

$$= 65.00\text{ m}^2$$

Then the definition (3-6) gives

$$\cos\theta = \frac{\mathbf{A}\cdot\mathbf{B}}{AB}$$

$$= \frac{65.0\text{ m}^2}{(10.25\text{ m})(9.43\text{ m})}$$

$$= 0.672$$

from which

$$\theta = 47.75°$$

Direction Cosines

A useful way to specify *direction* relative to a set of rectangular Cartesian axes is to give the components of a unit vector in the desired direction. For example, we specify the direction of the vector **A** in Figure 3-2 by giving the components of the unit vector in the direction of **A**:

$$\mathbf{u}_A = \frac{\mathbf{A}}{A}$$

$$= \frac{1}{A}(A_x \mathbf{u}_x + A_y \mathbf{u}_y + A_z \mathbf{u}_z)$$

$$= \frac{A_x}{A}\mathbf{u}_x + \frac{A_y}{A}\mathbf{u}_y + \frac{A_z}{A}\mathbf{u}_z$$

Observe that the magnitude of the ith component A_i/A of this unit vector is the length of the projection of the unit vector onto the i axis; therefore, it is equal to the cosine of the angle between **A** and the i axis. The quantities A_x/A, A_y/A, and A_z/A are called the *direction cosines*, which specify the direction of **A**.

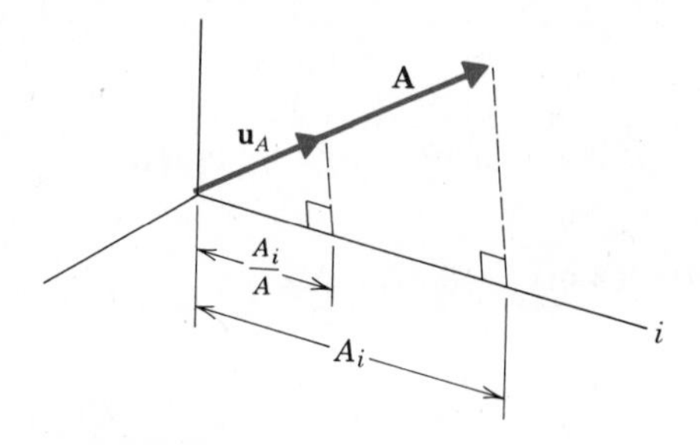

EXAMPLE

Determine the direction cosines of the guy lines $O'P$ and $O'Q$ relative to the rectangular Cartesian axes shown in Figure 3-3.

The length of line $O'P$ is $A = 10.247$ m. Then

$$\mathbf{u}_A = \frac{1}{10.247\ \text{m}}\left[(5\ \text{m})\,\mathbf{u}_x + (4\ \text{m})\,\mathbf{u}_y + (8\ \text{m})\,\mathbf{u}_z\right]$$

$$= 0.4880\,\mathbf{u}_x + 0.3904\,\mathbf{u}_y + 0.7807\,\mathbf{u}_z$$

The length of line $O'Q$ is $B = 9.434$ m. Then,

$$\mathbf{u}_B = \frac{1}{9.434\ \text{m}}\left[(-3\ \text{m})\,\mathbf{u}_x + (4\ \text{m})\,\mathbf{u}_y + (8\ \text{m})\,\mathbf{u}_z\right]$$

$$= -0.3180\,\mathbf{u}_x + 0.4240\,\mathbf{u}_y + 0.8480\,\mathbf{u}_z$$

Thus the direction cosines are

$$O'P: \qquad (0.4880, 0.3904, 0.7807)$$

$$O'Q: \qquad (-0.3180, 0.4240, 0.8480)$$

Cross Product

Steps similar to those that led to Equation (3-23) lead to the following formula for the cross product.

$$\boxed{\begin{aligned}\mathbf{A}\times\mathbf{B} = &\left(A_yB_z - A_zB_y\right)\mathbf{u}_x\\ &+\left(A_zB_x - A_xB_z\right)\mathbf{u}_y\\ &+\left(A_xB_y - A_yB_x\right)\mathbf{u}_z\end{aligned}} \tag{3-25}$$

It is important to realize that this result depends on the coordinate directions being *right-handed*, that is, on Equations 3-22 being satisfied. As an aid to memorizing Equation 3-25, observe that groups of subscripts in the order $\overset{x}{_{z}\circlearrowright_{y}}$ are associated with positive terms and that those with subscripts in the order $\overset{x}{_{z}\circlearrowleft_{y}}$ are associated with negative terms.

Scalar Triple Product

The product indicated in Equation 3-17 may be computed in terms of rectangular Cartesian components by combining Equations 3-23 and 3-25. A convenient form of the result is the determinant

$$\mathbf{A}\cdot(\mathbf{B}\times\mathbf{C}) = \begin{vmatrix} A_x & A_y & A_z \\ B_x & B_y & B_z \\ C_x & C_y & C_z \end{vmatrix} \tag{3-26}$$

Applications of the geometric concepts and definitions of the first three sections and the computational formulas of this section are presented in the next section.

PROBLEMS

For Problems 3-48 through 3-55, evaluate the unit vector in the direction of **A**, **A**·**B**, the angle between **A** and **B**, **A** × **B**, **A**·(**B** × **C**), **A** × (**B** × **C**), the components of **A** parallel to **B** and perpendicular to **B**, $\mathbf{A}\cdot(\mathbf{B}+\mathbf{C})$, and $\mathbf{A}\times(\mathbf{C}-\mathbf{B})$.

3-48 $\mathbf{A} = 3\mathbf{u}_x + \mathbf{u}_y - 2\mathbf{u}_z$
$\mathbf{B} = 2\mathbf{u}_x + 2\mathbf{u}_y + \mathbf{u}_z$
$\mathbf{C} = \mathbf{u}_x + \mathbf{u}_y - \mathbf{u}_z$

3-49 $\mathbf{A} = \frac{3}{13}\mathbf{u}_x + \frac{4}{13}\mathbf{u}_y - \frac{12}{13}\mathbf{u}_z$
$\mathbf{B} = \mathbf{u}_x - 2\mathbf{u}_y - \mathbf{u}_z$
$\mathbf{C} = -2\mathbf{u}_x + 4\mathbf{u}_y + 2\mathbf{u}_z$

3-50 $\mathbf{A} = -\mathbf{u}_x + \mathbf{u}_y + 2\mathbf{u}_z$
$\mathbf{B} = \mathbf{u}_x + 3\mathbf{u}_y - \mathbf{u}_z$
$\mathbf{C} = -\mathbf{u}_x - 3\mathbf{u}_y + \mathbf{u}_z$

3-51 $\mathbf{A} = \frac{7}{25}\mathbf{u}_x - \frac{24}{25}\mathbf{u}_y$
$\mathbf{B} = \mathbf{u}_x - \mathbf{u}_y + \mathbf{u}_z$
$\mathbf{C} = \mathbf{u}_x + \mathbf{u}_y + \mathbf{u}_z$

3-52 $\mathbf{A} = \mathbf{u}_x + \mathbf{u}_y + \mathbf{u}_z$
$\mathbf{B} = \mathbf{u}_x + \mathbf{u}_y - \mathbf{u}_z$
$\mathbf{C} = \mathbf{u}_x - \mathbf{u}_y - \mathbf{u}_z$

3-53 $\mathbf{A} = \frac{1}{2}\mathbf{u}_x - \frac{1}{2}\mathbf{u}_y + \frac{\sqrt{2}}{2}\mathbf{u}_z$
$\mathbf{B} = \frac{\sqrt{2}}{2}\mathbf{u}_x + \frac{\sqrt{2}}{2}\mathbf{u}_y$
$\mathbf{C} = -\frac{1}{2}\mathbf{u}_x + \frac{1}{2}\mathbf{u}_y + \frac{\sqrt{2}}{2}\mathbf{u}_z$

3-54

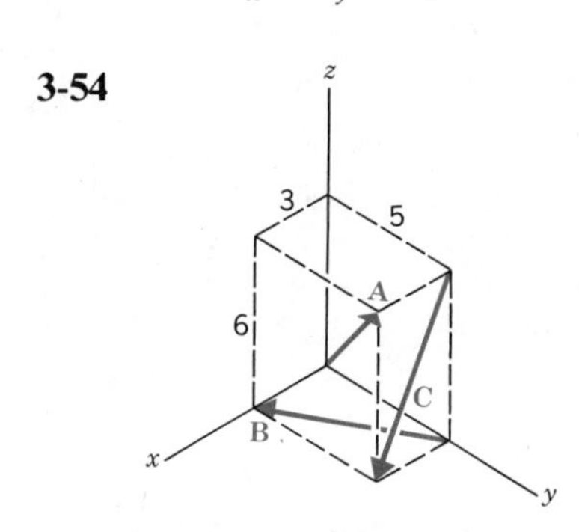

3-55

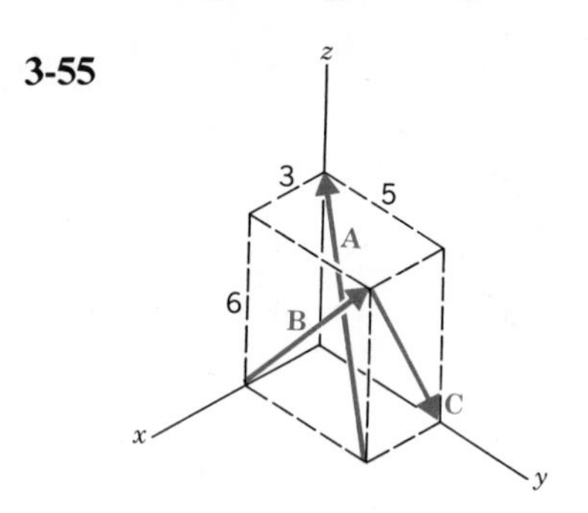

3-56 Evaluate A_x and A_y

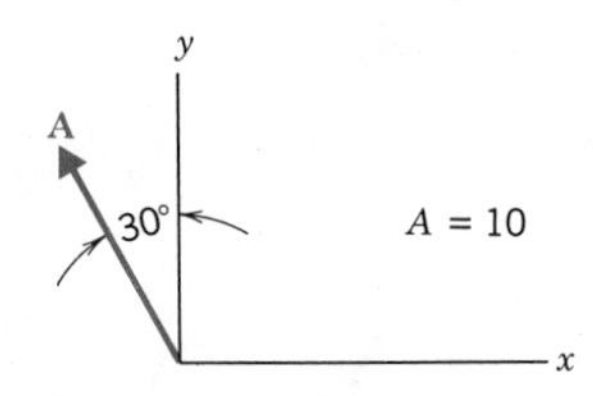

3-57 Evaluate A_x, A_y, and A_z. *Ans.* 5.000; 6.428; 5.804

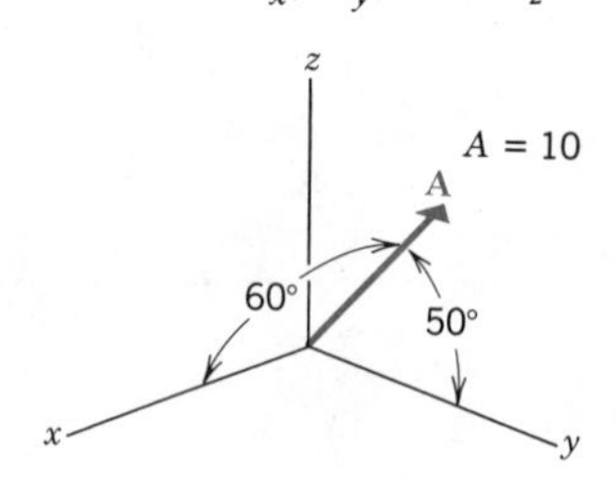

3-58 What is the angle between the vector **A** and the z-axis of the previous problem?

3-59 Write the vector **f** of Problem 3-12 (p. 57) in terms of $\mathbf{u}_x$, $\mathbf{u}_y$, $\mathbf{u}_z$, f, ϕ, and χ. Verify that $f = \sqrt{f_x^2 + f_y^2 + f_z^2}$.

3-60 Referring to Figure 3-3, determine the minimum distance between the lines $O'Q$ and OP. *Suggestion.* Refer to the result of Problem 3-42 (p. 66).

3-61 Referring to Figure 3-3, determine the minimum distance between the lines OQ and $O'P$. *Suggestion.* Refer to the result of Problem 3-42 (p. 66).

3-62 Express $\mathbf{u}_x$ and $\mathbf{u}_y$ in terms of $\bar{\mathbf{u}}_x$, $\bar{\mathbf{u}}_y$, and ϕ. Substitute these into the equation $\mathbf{A} = A_x\mathbf{u}_x + A_y\mathbf{u}_y$. By comparing the result with $\mathbf{A} = \bar{A}_x\bar{\mathbf{u}}_x + \bar{A}_y\bar{\mathbf{u}}_y$, write expressions for the components $\bar{A}_x$ and $\bar{A}_y$ in terms of A_x, A_y, and ϕ.

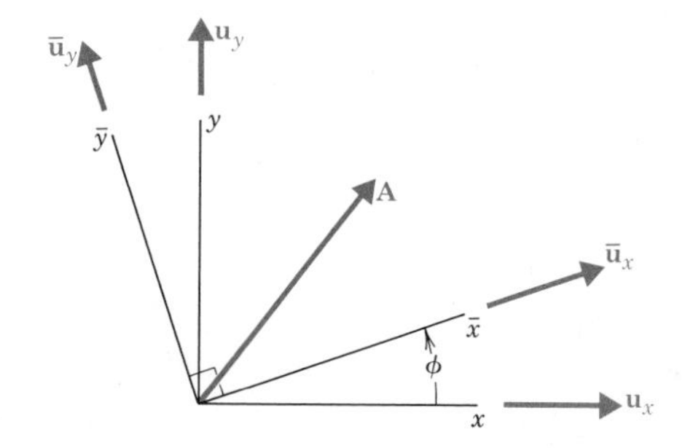

3-63 Given $\mathbf{A} = 3\mathbf{a} + 2\mathbf{b}$, determine the coefficients A_x and A_y in the expression

$$\mathbf{A} = A_x\mathbf{u}_x + A_y\mathbf{u}_y$$

Ans. 2.985; 2.735.

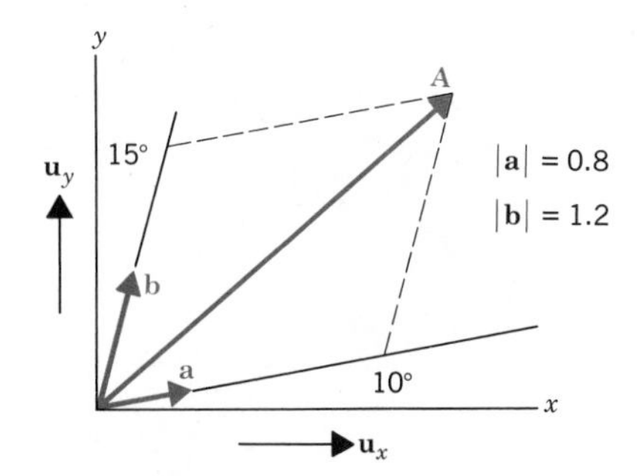

3-64 With **a**, **b**, $\mathbf{u}_x$ and $\mathbf{u}_y$ as given in the previous problem, and

$$\mathbf{B} = 3\,\mathbf{u}_x + 2\,\mathbf{u}_y$$

determine the coefficients α and β in the expression

$$\mathbf{B} = \alpha\mathbf{a} + \beta\mathbf{b}$$

3-65 Write an expression for a vector normal to the plane PQR, in terms of $\mathbf{u}_x$, $\mathbf{u}_y$, and $\mathbf{u}_z$. *Ans.* $C(6\,\mathbf{u}_x + 4\,\mathbf{u}_y + 3\,\mathbf{u}_z)$ where C = any number.

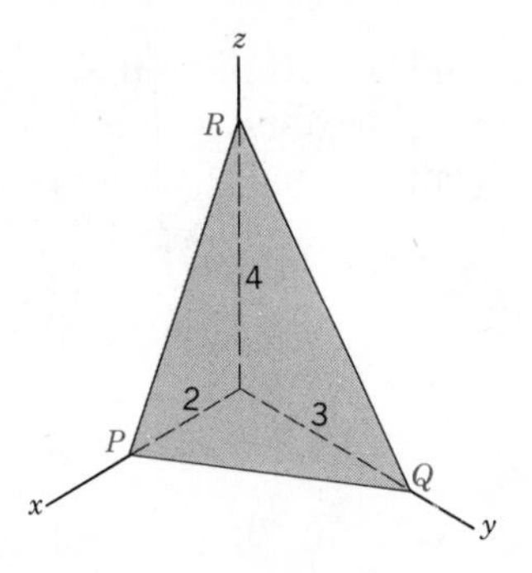

3-66 Write an equation for the plane PQR of the previous problem in terms of x, y, and z.

3-67 Determine the area of the triangle PQR of Problem 3-65.

3-68 The line of action of the force $f = (60\text{ N})\,\mathbf{u}_x - (20\text{ N})\,\mathbf{u}_y - (15\text{ N})\,\mathbf{u}_z$ passes through the point $O'(-12\text{ m}, 6\text{ m}, 4\text{ m})$. What are the coordinates (y_0, z_0) of the point where the line of action crosses the y-z plane?

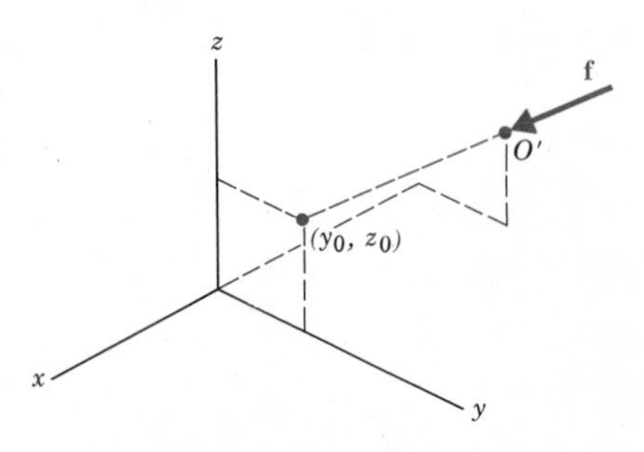

3-69 Verify that Equation 3-23 is consistent with Equation 3-7.

3-70 Derive Equation 3-25.

3-71 Verify that Equation 3-25 is consistent with Equation 3-11.

3-72 Verify the consistency of Equations 3-23 and 3-25 with
(a) Equation 3-16.
(b) Equation 3-17.
(c) Equation 3-18.

3-73 Sketch a left-handed triad of orthogonal unit base vectors and derive the corresponding formula for the cross product in terms of the rectangular Cartesian components. Compare the result with Equation 3-25.

3-74 Determine the angle $QO'R$ in Figure 3-3.

3-75 Determine the minimum distance from point Q to the line $O'R$ in Figure 3-3. *Ans.* 7.759 m.

3-76 Determine the minimum distance from point P to the plane $O'QR$ in Figure 3-3. *Ans.* 7.738 m.

3-77 Calculate the sum of the squares of the direction cosines for the line $O'P$ in Figure 3-3 (p. 71). Repeat the calculation for the line $O'Q$. Will the result be true in general?

3-5 COMPOSITION AND RESOLUTION OF FORCE SYSTEMS

When the geometry of a force system becomes more complex than in the examples we encountered in the previous chapter, vector analysis provides the most straightforward means for handling the computations. It also facilitates derivations of a number of important relationships among force and moment resultants. Accordingly, we now reexamine the ideas of the previous chapter using vectors to carry out the analysis.

Moments of Forces

The most important application in statics for the cross product is the evaluation of *moments*. A moment of a force is a vector with magnitude and direction as defined in Section 2-3. Examination of the definitions of moment of force and the cross product reveals that *the moment about a point O of the force* **f** *is given by*

$$\boxed{\mathbf{M}_O = \mathbf{r} \times \mathbf{f}} \tag{3-27}$$

where **r** *is a position vector from O to any point on the line of action of* **f**.

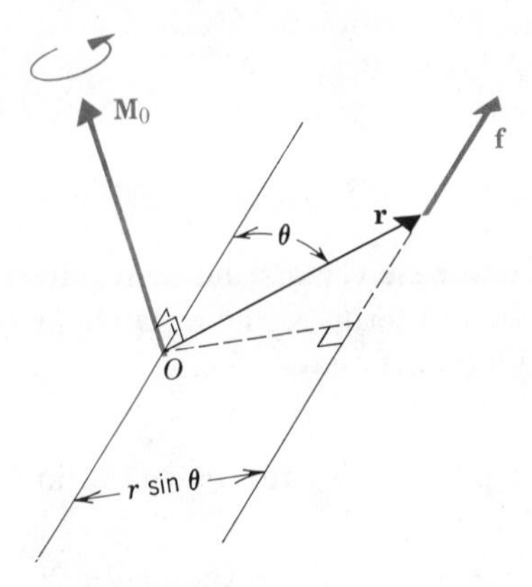

EXAMPLE

Suppose the guy line $O'P$ in Figure 3-3 has a tension of 800 N. What is the moment about O of the force from this cable acting on the mast?

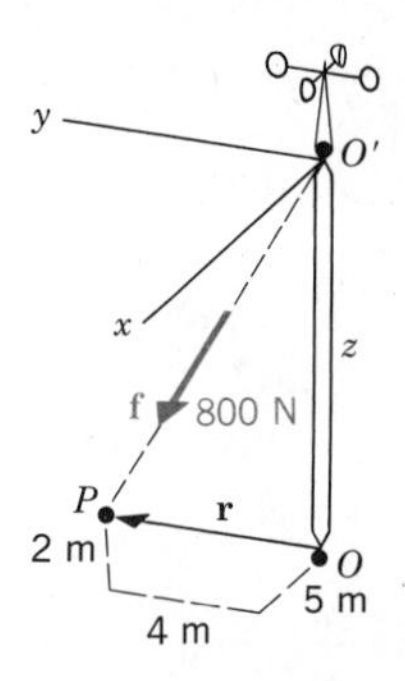

Since point P is on the line of action of this force, a position vector **r** from O to P can be used for evaluation of the moment about O. Referring to the axes shown in Figure 3-3, this vector has the resolution

$$\mathbf{r} = (5\text{ m})\,\mathbf{u}_x + (4\text{ m})\,\mathbf{u}_y + (-2\text{ m})\,\mathbf{u}_z$$

The resolution of the force can be determined by multiplying its magnitude by the unit vector in the direction of $O'P$, components of which were evaluated in the previous example:

$$\mathbf{f} = 800\text{ N}\,(0.4880\,\mathbf{u}_x + 0.3904\,\mathbf{u}_y + 0.7807\,\mathbf{u}_x)$$

$$= (3904\ \text{N})\,\mathbf{u}_x + (312.3\ \text{N})\,\mathbf{u}_y + (624.6\ \text{N})\,\mathbf{u}_z.$$

The moment can now be evaluated with reference to Equation 3-25 (p. 73):

$$\begin{aligned}\mathbf{M}_O = &\left[(4\ \text{m})(624.6\ \text{N}) - (-2\ \text{m})(312.3\ \text{N})\right]\mathbf{u}_x \\ &+\left[(-2\ \text{m})(390.4\ \text{N}) - (5\ \text{m})(624.6\ \text{N})\right]\mathbf{u}_y \\ &+\left[(5\ \text{m})(312.3\ \text{N}) - (4\ \text{m})(390.4\ \text{N})\right]\mathbf{u}_z \\ = &\ (3123\ \text{N}\cdot\text{m})\,\mathbf{u}_x - (3904\ \text{N}\cdot\text{m})\,\mathbf{u}_y\end{aligned}$$

Observe that point O' is also on the line of action of $\mathbf{f}$, so that a position vector $\mathbf{r} = (-10\ \text{m})\,\mathbf{u}_z$ could have been used instead of the one above, and with less arithmetic.

The *moment about an axis Oi* is defined as the projection onto the axis of the moment about some point on the axis. To express this analytically, we define the positive sense along the axis with the unit vector $\mathbf{u}_i$ and write

$$\boxed{\mathbf{M}_{Oi} = (\mathbf{M}_O\cdot\mathbf{u}_i)\,\mathbf{u}_i} \tag{3-28a}$$

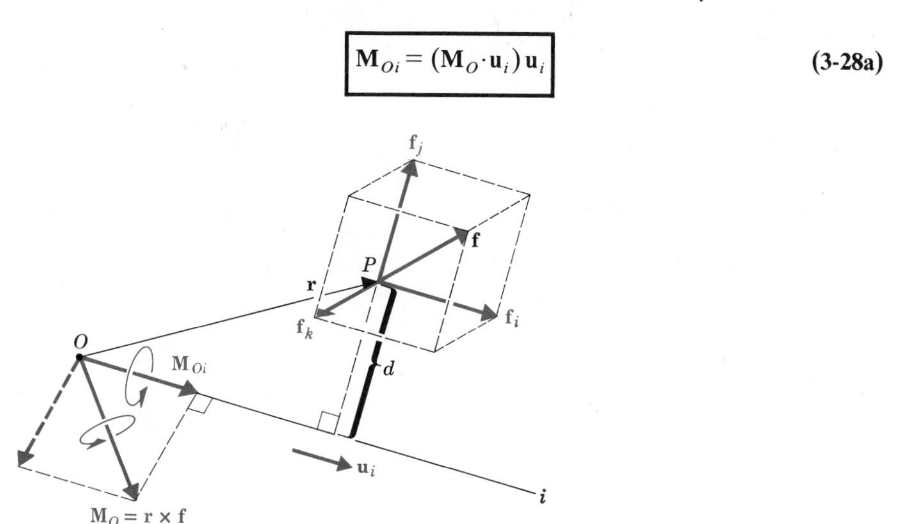

Although the moment about the axis can be computed according to this definition, an alternative form of this equation is often easier to use, and provides a different and very useful interpretation. To obtain this alternative form, we first insert the definition of $\mathbf{M}_O$ in terms of $\mathbf{r}$ and $\mathbf{f}$, and re-order the resulting triple product according to the identity 3-17 (p. 64):

$$\mathbf{M}_{Oi} = [(\mathbf{r} \times \mathbf{f}) \cdot \mathbf{u}_i]\, \mathbf{u}_i$$
$$= [(\mathbf{u}_i \times \mathbf{r}) \cdot \mathbf{f}]\, \mathbf{u}_i \quad \textbf{(3-28b)}$$

Next, we resolve **f** into a component $\mathbf{f}_i$ parallel to the axis Oi, a component $\mathbf{f}_j$ perpendicular to Oi and in the plane of **r** and Oi, and a component $\mathbf{f}_k$ perpendicular to this plane, as indicated in the sketch. Noting that both $\mathbf{f}_i$ and $\mathbf{f}_j$ are perpendicular to $\mathbf{u}_i \times \mathbf{r}$, we see from (3-28b) that these two components do not contribute to $\mathbf{M}_{Oi}$. Also, because $\mathbf{u}_i \times \mathbf{r}$ has the magnitude $d = r \sin \measuredangle_i^{\mathbf{r}}$, we can express the magnitude of the moment component as

$$\boxed{|\mathbf{M}_{Oi}| = f_k d} \quad \textbf{(3-28c)}$$

where d is the perpendicular distance from point P to the axis Oi. The sense of $\mathbf{M}_{Oi}$ (i.e., whether it is directed in the positive or negative i-direction) is readily determined from the direction of $\mathbf{f}_k$ and the right-hand rule. Alternatively, the sense may be determined by the sign of the factor $\mathbf{M}_O \cdot \mathbf{u}_i$ in Equation 3-28a.

Observe that the same value of d would be obtained regardless of where point O is placed on the i-axis. Also recall that the position vector **r** in the definition $\mathbf{M}_O = \mathbf{r} \times \mathbf{f}$ can be from O to *any* point on the line of action of **f**. Therefore, Equation 3-28b will yield the value of $\mathbf{M}_{Oi}$ with **r** as a position vector from *any* point on the axis Oi to *any* point on the line of action of **f**.

The moment about an axis is a measure of the tendency of the force(s) to cause rotation about the axis. For example, if a rotor is mounted in bearings and subjected to a set of forces, the moment of these forces about the axis of the bearings is found to be directly related to the rate of increase of rotational speed. Neither forces parallel to the axis nor forces with lines of action passing through the axis will affect the rotation of the rotor.

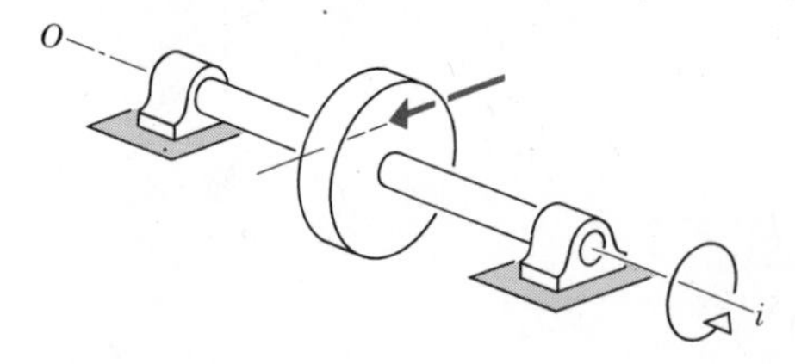

EXAMPLE

With the x, y, and z axes placed as shown, determine the moment about the axis Oy of the 800-N force acting on the mast of the previous example.

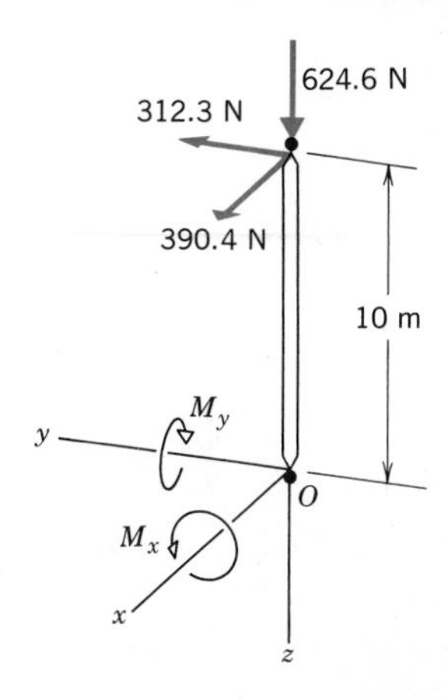

The x, y, and z components of the force were calculated previously and are shown acting at the top of the mast. (Any other point on the line of action of the force could also be used, but this particular point is the most convenient for this calculation.) The 312.3-N component does not contribute to $\mathbf{M}_y$ because it is parallel to the y-axis and the 624.6-N component does not contribute to $\mathbf{M}_y$ because its line of action passes through the y-axis. The 390.4-N component is perpendicular to the y-axis and its line of action is 10 meters from the y-axis. Therefore the magnitude of the moment about the y-axis is $f_x d = (390.4 \text{ N})(10 \text{ m})$. Referring to the diagram and applying the right-hand rule, we see that this moment component is directed opposite to the positive y-direction, so that M_y is negative. Accordingly, we write

$$M_y = -3904 \text{ N} \cdot \text{m}$$

Observe that this is identical to the coefficient of $\mathbf{u}_y$ in the result on page 79.

EXAMPLE

In terms of the tension T_P in the guy line $O'P$ of Figure 3-3, evaluate the moment about the axis RQ of the force exerted by the line $O'P$ on the mast.

Unlike that of the previous example, the value of the distance d for use in Equation 3-28c is not readily apparent from the figure, although it could be calculated using the result of Problem 3-42. Rather than calculate d, let us use Equation 3-28b, with the x-y-z components indicated in the diagram.

The unit vector in the direction of the axis RQ can be expressed as

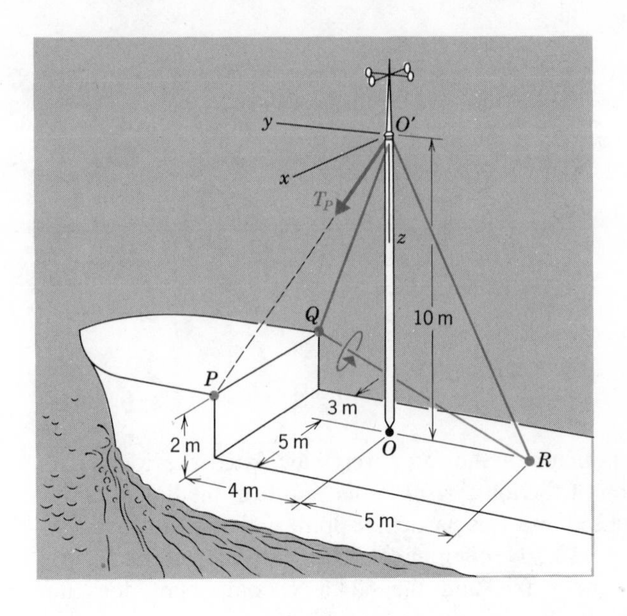

$$\mathbf{u}_{RQ} = \frac{-3\,\mathbf{u}_x + 9\,\mathbf{u}_y - 2\,\mathbf{u}_z}{\sqrt{(3)^2 + (9)^2 + (2)^2}}$$

$$= -0.3094\,\mathbf{u}_x + 0.9283\,\mathbf{u}_y - 0.2063\,\mathbf{u}_z$$

and the force as

$$\mathbf{T}_P = T_P(0.4880\,\mathbf{u}_x + 0.3904\,\mathbf{u}_y + 0.7807\,\mathbf{u}_z)$$

Now, the position vector $\mathbf{r}$ can be from any point on the axis to any point on the line of action of the force; let us use that from R to O':

$$\mathbf{r}_{RO'} = (5\text{ m})\,\mathbf{u}_y - (10\text{ m})\,\mathbf{u}_z$$

Referring to Equation 3-26, (p. 73) we calculate the desired moment as

$$\mathbf{M}_{RQ} = \left[(\mathbf{r}_{RO'} \times \mathbf{T}_P)\cdot\mathbf{u}_{RQ}\right]\mathbf{u}_{RQ}$$

$$= \begin{vmatrix} -0.3094 & 0.9283 & -0.2063 \\ 0 & 5\text{ m} & -10\text{ m} \\ 0.4880T_P & 0.3904T_P & 0.7807T_P \end{vmatrix}\mathbf{u}_{RQ}$$

$$= -(6.442\text{ m})T_P\,\mathbf{u}_{RQ}$$

The negative sign indicates that the moment opposes the direction of $\mathbf{u}_{RQ}$; that is, the force tends to rotate the mast in the direction indicated by the curved arrow in the sketch above. Observe that the calculation may be shortened slightly by using a position vector

$$\mathbf{r}_{QP} = (8 \text{ m})\,\mathbf{u}_x$$

instead of that used above.

The *resultant moment about point O of a set of forces* $\mathbf{f}_1, \mathbf{f}_2, \mathbf{f}_3, \ldots$ is defined as the sum of the moments of the individual forces in the set:

$$\mathbf{M}_O = \mathbf{r}_1 \times \mathbf{f}_1 + \mathbf{r}_2 \times \mathbf{f}_2 + \mathbf{r}_3 \times \mathbf{f}_3 + \cdots \tag{3-29}$$

where $\mathbf{r}_i$ is a position vector from point O to any point on the line of action of $\mathbf{f}_i$ $(i = 1, 2, 3, \ldots)$.

The *resultant moment about an axis Oi of a set of forces* is defined as

$$\begin{aligned} \mathbf{M}_{Oi} &= \left[(\mathbf{r}_1 \times \mathbf{f}_1 + \mathbf{r}_2 \times \mathbf{f}_2 + \mathbf{r}_3 \times \mathbf{f}_3 + \cdots)\cdot\mathbf{u}_i\right]\mathbf{u}_i \\ &= \left[(\mathbf{u}_i \times \mathbf{r}_1)\cdot\mathbf{f}_1 + (\mathbf{u}_i \times \mathbf{r}_2)\cdot\mathbf{f}_2 + \cdots\right]\mathbf{u}_i \end{aligned} \tag{3-30}$$

By comparing the individual terms of (3-30) with (3-28b), we see that the resultant moment about an axis can be computed as the sum of the moments of the individual forces about the axis.

EXAMPLE

The forces T_A, T_B, R_x, and R_z are those imparted to the sign pole in Figure 3-4 by the supports. Evaluate the resultant moment of all the forces about the axis AB.

Since the lines of action of T_A, T_B, and R_z all pass through this axis, these forces will not contribute to the moment about AB. In Figure 3-4*b* we can see the components of the gravitational force in directions parallel and perpendicular to AB, and the moment arm d for the force R_x. The resultant moment about the axis AB is then

$$\begin{aligned} M_{AB} &= f_g \cos\theta(16 \text{ ft}) - R_x d \\ &= \left[(16 \text{ ft})f_g - (10 \text{ ft})R_x\right]\cos\theta \\ &= (15.52 \text{ ft})f_g - (9.70 \text{ ft})R_x \end{aligned}$$

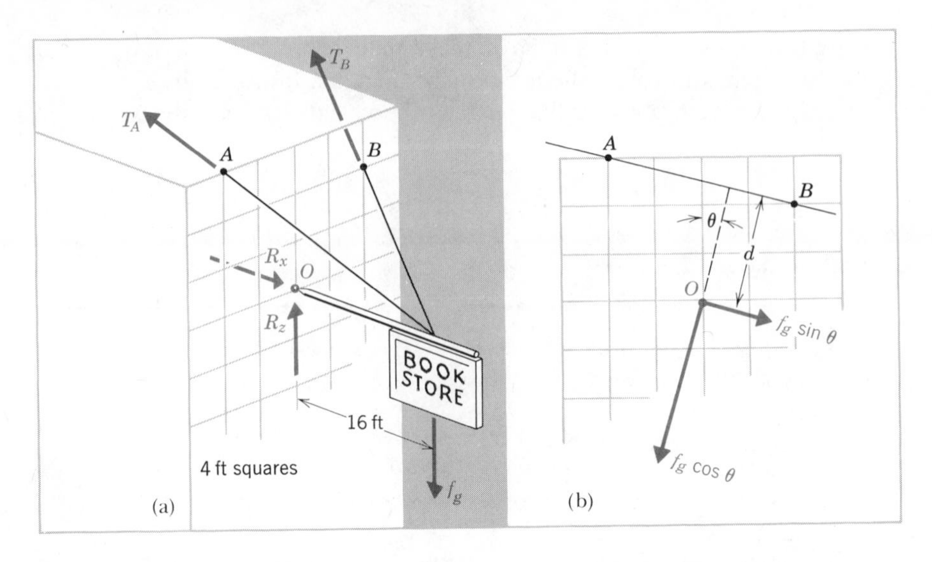

FIGURE 3-4

As a check on this result, we may define position vectors from point A to each of the force lines of action, compute the sum of the moments about point A, and dot multiply the sum of these with a unit vector in the direction of AB, in accordance with Equation 3-28a.

Varignon's Theorem

Suppose the lines of action of several forces intersect at some point P. Then the same position vector $\mathbf{r}$, from point O to point P, can be used to evaluate the moment about O of each of the forces, and the resultant moment about O of all the forces can be written as

$$\mathbf{M}_O = \mathbf{r} \times \mathbf{f}_1 + \mathbf{r} \times \mathbf{f}_2 + \mathbf{r} \times \mathbf{f}_3 + \cdots$$

But in view of Equation 3-12, this may also be written as

$$\mathbf{M}_O = \mathbf{r} \times (\mathbf{f}_1 + \mathbf{f}_2 + \mathbf{f}_3 + \cdots)$$

Therefore, *if the lines of action of a set of forces intersect, the sum of the moments of the forces is equal to the moment of the resultant force acting through the point of intersection of the lines of action*. This theorem was given by Pierre Varignon in his book, *Nouvelle Mechanique ou Statique* (1725). It is a generalization of the statement made and illustrated on page 36, where we dealt with two mutually perpendicular components in the plane of the force and the moment reference point. The example on page 80 indicates how the components of the resultant moment can be determined in terms of a set of components of the resultant force.

Moments about Different Points

Consider the resultant moment of a set of forces about two different points, O and O'. Let us denote by $\boldsymbol{\rho}$ the position vector from O to O' and by $\mathbf{r}_i$ and $\mathbf{r}_i'$, respectively, the position vectors from O and O' to a point on the line of action of the ith force of the set. The resultant moment about O is then

$$\begin{aligned}\mathbf{M}_O &= \mathbf{r}_1 \times \mathbf{f}_1 + \mathbf{r}_2 \times \mathbf{f}_2 + \cdots + \mathbf{r}_m \times \mathbf{f}_m \\ &= (\mathbf{r}_1' + \boldsymbol{\rho}) \times \mathbf{f}_1 + (\mathbf{r}_2' + \boldsymbol{\rho}) \times \mathbf{f}_2 + \cdots + (\mathbf{r}_m' + \boldsymbol{\rho}) \times \mathbf{f}_m \\ &= \mathbf{r}_1' \times \mathbf{f}_1 + \mathbf{r}_2' \times \mathbf{f}_2 + \cdots + \mathbf{r}_m' \times \mathbf{f}_m + \boldsymbol{\rho} \times (\mathbf{f}_1 + \mathbf{f}_2 + \cdots + \mathbf{f}_m) \\ &= \mathbf{M}_{O'} + \boldsymbol{\rho} \times \mathbf{f} \qquad \textbf{(3-31)}\end{aligned}$$

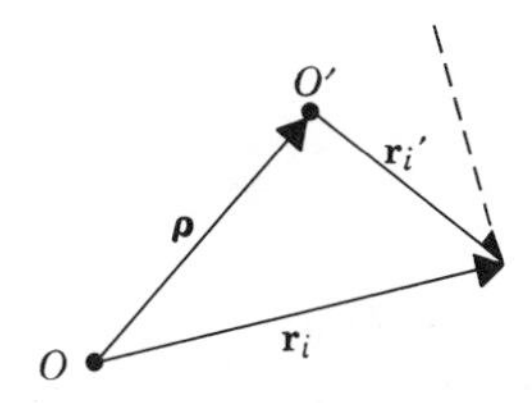

Thus, the moment about O is equal to that about O', plus the moment that a force equal to the sum of the given forces would have about O if the line of action of this force passed through O'.

EXAMPLE

A drive belt exerts the two forces indicated on the pulley. What is the resultant moment of these two forces about the point O?

The resultant force is

$$\begin{aligned}\mathbf{f} &= [0.3 \text{ kN} + (1.2 \text{ kN}) \cos 27°]\, \mathbf{u}_x + (1.2 \text{ kN}) \sin 27°\, \mathbf{u}_y \\ &= (1.369 \text{ kN})\, \mathbf{u}_x + (0.545 \text{ kN})\, \mathbf{u}_y\end{aligned}$$

and the resultant moment about O' is

$$\mathbf{M}_{O'} = (0.3 \text{ kN})(0.1 \text{ m})\, \mathbf{u}_z - (1.2 \text{ kN})(0.1 \text{ m})\, \mathbf{u}_z$$

$$= -(0.090\ \text{kN}\cdot\text{m})\,\mathbf{u}_z$$

Now, according to Equation 3-31, the moment about O is

$$\begin{aligned}\mathbf{M}_O &= -(0.090\ \text{kN}\cdot\text{m})\,\mathbf{u}_z \\ &\quad + (0.18\ \text{m})\,\mathbf{u}_z \times \left[(1.369\ \text{kN})\,\mathbf{u}_x + (0.545\ \text{kN})\,\mathbf{u}_y\right] \\ &= -(0.098\ \text{kN}\cdot\text{m})\,\mathbf{u}_x + (0.246\ \text{kN}\cdot\text{m})\,\mathbf{u}_y - (0.090\ \text{kN}\cdot\text{m})\,\mathbf{u}_z\end{aligned}$$

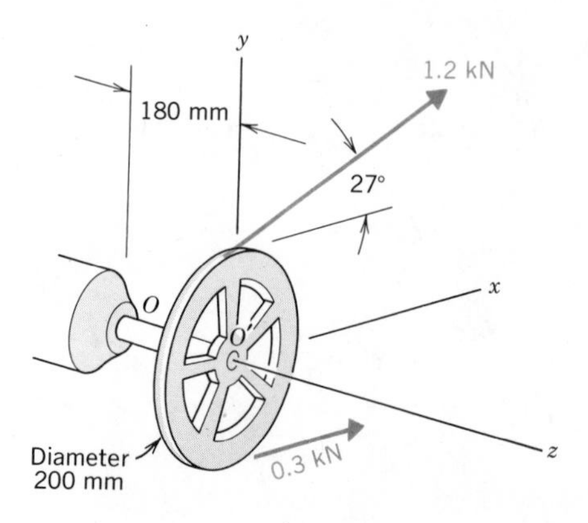

It is instructive to compare this calculation with one that uses position vectors from point O to points on the lines of action of the two forces.

Equivalent Force Systems

For the analysis of *rigid* bodies, two sets of forces are said to be equivalent if the sum of the forces in one set is equal to that of the other, and if, about any point, the resultant moment of the forces in one set is equal to that of the other.

To check equivalence, we must compare the sums of the force vectors and the moment resultant with respect to just one, arbitrarily selected point. For example, we checked the equivalence of the force system in Figure 2-8 in terms of the moments about the corner of the bracket and then observed that the moment about the free end was *also* the same for each set of forces. A further check of the moments about the point of attachment at the wall will show agreement again.

To demonstrate this in general, let us denote two different sets of forces as follows:

	Set 1	Set 2
Forces	$\mathbf{f}_1, \mathbf{f}_2, \ldots \mathbf{f}_m$	$\mathbf{g}_1, \mathbf{g}_2, \ldots \mathbf{g}_n$
Position vectors emanating from point O	$\mathbf{r}_1, \mathbf{r}_2, \ldots \mathbf{r}_m$	$\mathbf{s}_1, \mathbf{s}_2, \ldots \mathbf{s}_n$
Resultant force	$\mathbf{f} = \mathbf{f}_1 + \mathbf{f}_2 + \cdots + \mathbf{f}_m$	$\mathbf{g} = \mathbf{g}_1 + \mathbf{g}_2 + \cdots + \mathbf{g}_n$
Resultant moment about O	$\mathbf{M}_O = \mathbf{r}_1 \times \mathbf{f}_1 + \mathbf{r}_2 \times \mathbf{f}_2 + \cdots + \mathbf{r}_m \times \mathbf{f}_m$	$\mathbf{N}_O = \mathbf{s}_1 \times \mathbf{g}_1 + \mathbf{s}_2 \times \mathbf{g}_2 + \cdots + \mathbf{s}_n \times \mathbf{g}_n$

Now, if the stated requirements are met,

$$\mathbf{f} = \mathbf{g}$$

$$\mathbf{M}_O = \mathbf{N}_O$$

Now consider the moments about a different point, O', located by a position vector $\boldsymbol{\rho}$ from O to O'. The resultant moment about O' of the first set of forces is, from Equation 3-31,

$$\mathbf{M}_{O'} = \mathbf{M}_O - \boldsymbol{\rho} \times \mathbf{f}$$

Similarly, the resultant moment about O' of the second set of forces is

$$\mathbf{N}_{O'} = \mathbf{N}_O - \boldsymbol{\rho} \times \mathbf{g}$$

But since $\mathbf{N}_O = \mathbf{M}_O$ and $\mathbf{g} = \mathbf{f}$,

$$\mathbf{N}_{O'} = \mathbf{M}_{O'}$$

That is, *equality of force resultant and moment resultant about one point implies equality of moment resultant about every other point.*

Do not forget that equivalence, in the sense that it is used here, implies an equivalent response in a *rigid* body; the response in a *deformable* body depends on each individual point of application of the forces in the set.

Composition of Force Systems

The composition of a force system is the process of combining several components to form a simpler equivalent set. For example, let us consider the planar set of forces shown in Figure 3-5*a*.

FIGURE 3-5

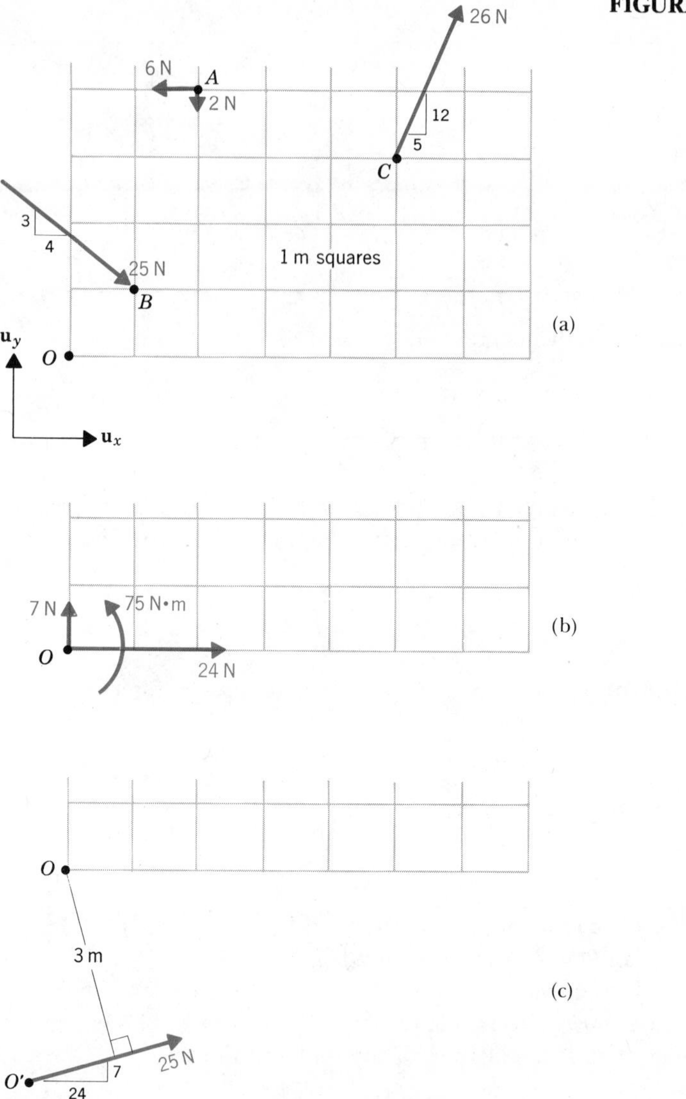

A convenient way of determining the resultant force is to resolve each force into horizontal and vertical components and add, as follows:

$$\begin{aligned}
\mathbf{f}_A &= (-6\text{ N})\,\mathbf{u}_x - (2\text{ N})\,\mathbf{u}_y \\
\mathbf{f}_B &= (20\text{ N})\,\mathbf{u}_x - (15\text{ N})\,\mathbf{u}_y \\
\mathbf{f}_C &= (10\text{ N})\,\mathbf{u}_x + (24\text{ N})\,\mathbf{u}_y \\
\hline
\mathbf{f} &= (24\text{ N})\,\mathbf{u}_x + (7\text{ N})\,\mathbf{u}_y
\end{aligned}$$

Thus the replacement of the given forces with this resultant will satisfy the first requirement for forming an equivalent force system. To effect moment equivalence, let us first evaluate the moment resultant of the given system of forces about point O :

$$\begin{aligned} \mathbf{M}_{O(A)} &= 20\ \text{N}\cdot\text{m}\,\mathbf{u}_z \\ \mathbf{M}_{O(B)} &= -35\ \text{N}\cdot\text{m}\,\mathbf{u}_z \\ \mathbf{M}_{O(C)} &= 90\ \text{N}\cdot\text{m}\,\mathbf{u}_z \\ \hline \mathbf{M}_O &= 75\ \text{N}\cdot\text{m}\,\mathbf{u}_z \end{aligned}$$

Now, an equivalent force system would be a single force $\mathbf{f} = (24\ \text{N})\mathbf{u}_x + (7\ \text{N})\mathbf{u}_y$ with its line of action passing through O (contributing no moment about O), together with a couple of moment $\mathbf{M}_O = (75\ \text{N}\cdot\text{m})\mathbf{u}_z$, as is depicted in Figure 3-5*b*. Further simplification can be achieved by replacing this system with a single force $\mathbf{f}$ with its line of action placed at a distance $\rho = (75\ \text{N}\cdot\text{m})/25\ \text{N} = 3\ \text{m}$ from point O, as shown in Figure 3-5*c*. Observe that, in constructing an equivalent force system as a single force through a selected point together with a couple (as exemplified by (*b*) and (*c*) of Figure 3-5), the value of the force vector does not depend on the point chosen for its line of action, but the moment of the added couple *does* depend on this choice.

By the process illustrated above, any planar system of forces can be reduced to an equivalent single force with a specified line of action. However, the same is not true for a general three-dimensional system of forces; in general, the nearest we can come to achieving this is to eliminate the component of moment perpendicular to the force.

There can always be found a line of action for the force such that the moment of the accompanying couple is parallel to the force. This particular force system is called a *wrench*. Given a set of forces $\mathbf{f}_1, \mathbf{f}_2, \mathbf{f}_3, \ldots$ with specified lines of action, the equivalent wrench can be determined as follows. First, the resultant force is evaluated by addition:

$$\mathbf{f} = \mathbf{f}_1 + \mathbf{f}_2 + \mathbf{f}_3 + \cdots$$

Next, a convenient point O is selected and the resultant moment about this point is computed:

$$\mathbf{M}_O = \mathbf{r}_1 \times \mathbf{f}_1 + \mathbf{r}_2 \times \mathbf{f}_2 + \mathbf{r}_3 \times \mathbf{f}_3 \cdots$$

Now if this moment vector is resolved into components parallel and perpendicular to $\mathbf{f}$ (see Equation 3-15, p. 64), we have

$$\mathbf{M}_O = \left(\frac{\mathbf{f}\cdot\mathbf{M}_O}{\mathbf{f}\cdot\mathbf{f}}\right)\mathbf{f} + \left(\frac{\mathbf{f}\times\mathbf{M}_O}{\mathbf{f}\cdot\mathbf{f}}\right)\times\mathbf{f}$$

Comparing this with Equation 3-31, which relates the moments of a set of forces about two different points,

$$\mathbf{M}_O = \mathbf{M}_{O'} + \boldsymbol{\rho} \times \mathbf{f} \qquad [3\text{-}31]^*$$

we note that the vector

$$\boldsymbol{\rho} = \frac{\mathbf{f} \times \mathbf{M}_O}{\mathbf{f}\cdot\mathbf{f}} \qquad (3\text{-}32)$$

locates a point O' relative to O such that the parallel component $[(\mathbf{f}\cdot\mathbf{M}_O)/f^2]\mathbf{f}$ is the resultant moment about O'. The wrench therefore consists of a force $\mathbf{f}$ with line of action passing through point O', together with a couple having moment given by

$$\mathbf{M}_{O'} = \frac{(\mathbf{f}\cdot\mathbf{M}_O)\mathbf{f}}{\mathbf{f}\cdot\mathbf{f}} \qquad (3\text{-}33)$$

The relationships expressed here can be understood by considering Figure 3-6, which shows the effect of translating the line of action of **f**. Figure 3-6*a* shows the force **f** passing through the point O, together with the moment $\mathbf{M}_O$ of the couple necessary to maintain equivalence with the given set of forces. In Figure

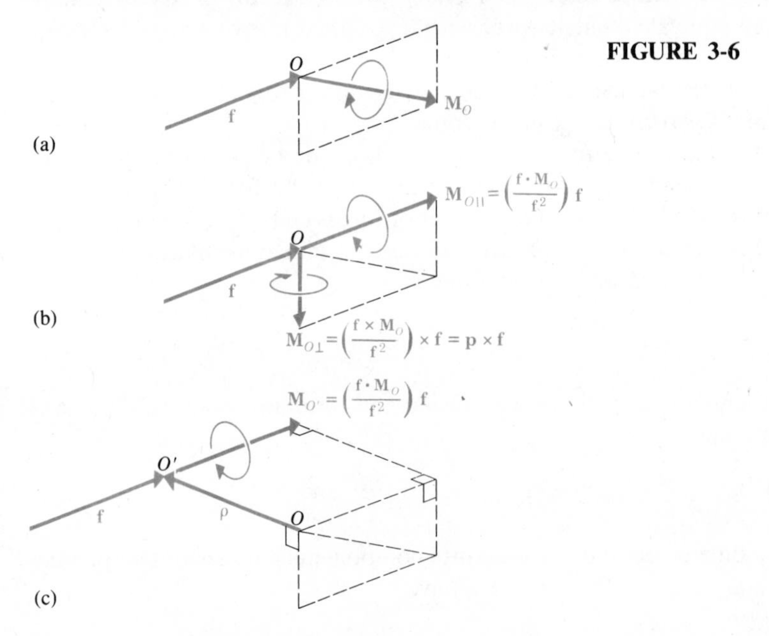

FIGURE 3-6

*Brackets signify an equation that is repeated from an earlier development.

3-6*b*, the moment of the couple has been replaced with components parallel and perpendicular to **f**. Now, if the line of action of **f** is translated so that it passes through O', as indicated in Figure 3-6*c*, the moment of the accompanying couple must be diminished by the quantity $\boldsymbol{\rho} \times \mathbf{f}$ in order to preserve equivalence. But if we select point O' such that $\boldsymbol{\rho} = (\mathbf{f} \times \mathbf{M}_O)/f^2$, the part of $\mathbf{M}_O$ that is removed is exactly $\mathbf{M}_{O\perp}$, so that the remaining moment of the couple is in the direction of **f**. The force through O' together with the couple of moment $[(\mathbf{f} \cdot \mathbf{M}_O)/f^2]\mathbf{f}$ constitutes the wrench equivalent to the original set of forces. No further reduction of the couple is possible. In the special case in which the moment vector is perpendicular to **f**, the couple can be removed entirely, as in the planar case.

EXAMPLE

Determine the wrench equivalent to the three forces shown in Figure 3-7*a*.

Resolutions of the given forces along the axes shown are

$$\mathbf{f}_A = (-113.137 \text{ N})\mathbf{u}_x + (-84.853 \text{ N})\mathbf{u}_y + (141.421 \text{ N})\mathbf{u}_z$$

$$\mathbf{f}_B = (38.587 \text{ N})\mathbf{u}_y + (64.312 \text{ N})\mathbf{u}_z$$

$$\mathbf{f}_C = (40.000 \text{ N})\mathbf{u}_x + (-30.000 \text{ N})\mathbf{u}_y$$

and position vectors locating points on the lines of action are

$$\mathbf{r}_A = (5 \text{ m})\mathbf{u}_z$$

$$\mathbf{r}_B = (4 \text{ m})\mathbf{u}_x$$

$$\mathbf{r}_C = (3 \text{ m})\mathbf{u}_y + (5 \text{ m})\mathbf{u}_z$$

The resultant force is found by summation:

$$\mathbf{f} = (-73.137 \text{ N})\mathbf{u}_x + (-76.266 \text{ N})\mathbf{u}_y + (205.733 \text{ N})\mathbf{u}_z$$

Next, the resultant moment about O is

$$\begin{aligned} \mathbf{M}_O &= \mathbf{r}_A \times \mathbf{f}_A + \mathbf{r}_B \times \mathbf{f}_B + \mathbf{r}_C \times \mathbf{f}_C \\ &= (574.26 \text{ N}\cdot\text{m})\mathbf{u}_x - (622.93 \text{ N}\cdot\text{m})\mathbf{u}_y + (34.35 \text{ N}\cdot\text{m})\mathbf{u}_z \end{aligned}$$

An equivalent system would consist of the above resultant force **f** acting through O, plus a couple having moment equal to $\mathbf{M}_O$ above, as is shown in

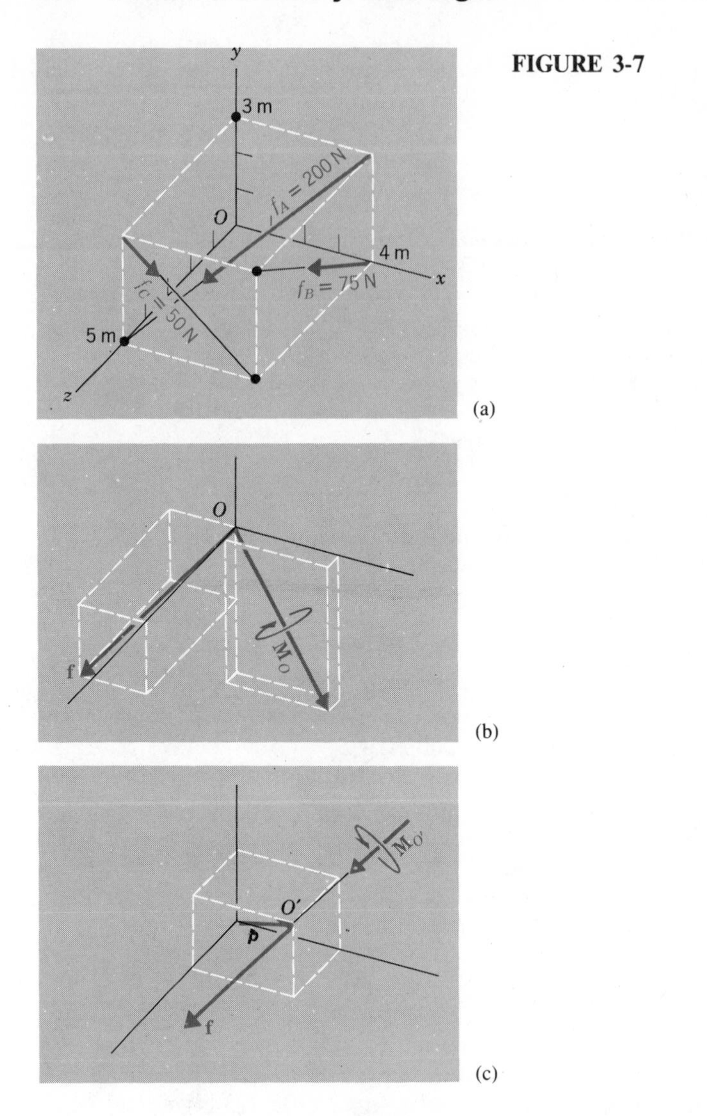

FIGURE 3-7

Figure 3-7*b*. The wrench consists of the force $\mathbf{f}$ acting through the point O', which is located relative to O by the position vector

$$\boldsymbol{\rho} = \frac{\mathbf{f} \times \mathbf{M}_O}{\mathbf{f} \cdot \mathbf{f}}$$

$$= \frac{(125\,540\ \text{N}^2\cdot\text{m})\,\mathbf{u}_x + (120\,660\ \text{N}^2\cdot\text{m})\,\mathbf{u}_y + (89\,360\ \text{N}^2\cdot\text{m})\,\mathbf{u}_z}{53\,490\ \text{N}^2}$$

$$= (2.347\ \text{m})\,\mathbf{u}_x + (2.256\ \text{m})\,\mathbf{u}_y + (1.670\ \text{m})\,\mathbf{u}_z$$

together with a couple having moment

$$\mathbf{M}_{O'} = \frac{\mathbf{f}\cdot\mathbf{M}_O}{\mathbf{f}\cdot\mathbf{f}}\mathbf{f}$$

$$= \frac{12\,575\ \text{N}^2\cdot\text{m}}{53\,490\ \text{N}^2}\left[(-73.137\ \text{N})\,\mathbf{u}_x - (76.266\ \text{N})\,\mathbf{u}_y + (205.733\ \text{N})\,\mathbf{u}_z\right]$$

$$= -(17.19\ \text{N}\cdot\text{m})\,\mathbf{u}_x - (17.93\ \text{N}\cdot\text{m})\,\mathbf{u}_y + (48.36\ \text{N}\cdot\text{m})\,\mathbf{u}_z$$

This is illustrated in Figure 3-7*c*.

PROBLEMS

3-78 The tension in the vertical line AC is 450 lbf and that in line BC is 1350 lbf. Determine the magnitude of the resultant force exerted by the two lines at C. *Ans.* 1533 lbf.

3-79 Determine the magnitude of the resultant force exerted on the tower by the two cables. The tension in each cable is 70 kN.

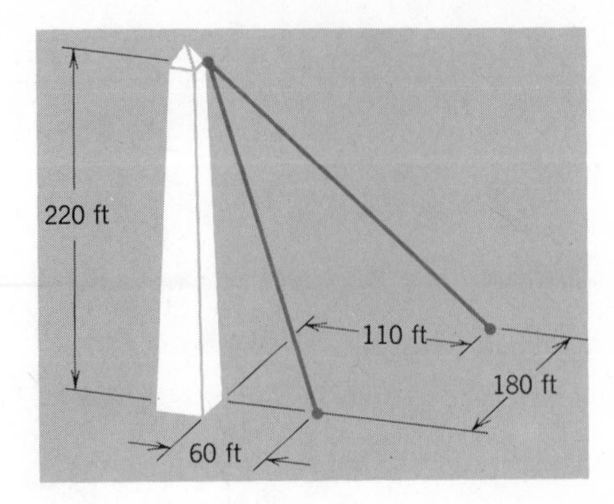

3-80 Evaluate the projection of **f** onto the line AB, and the moment of **f** about the axis AB.

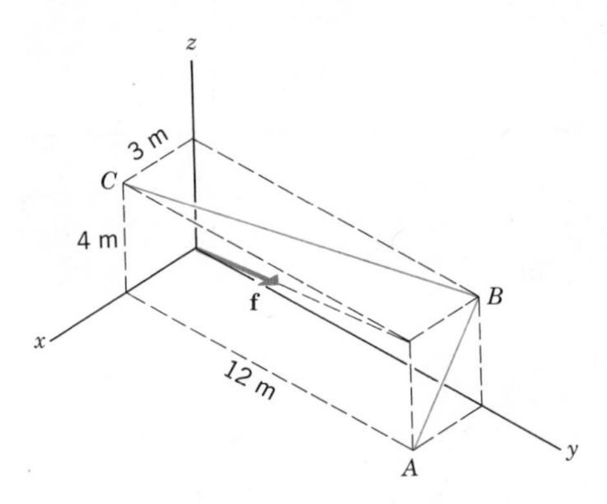

3-81 Work the previous problem with AB replaced with BC.

3-82 Evaluate the moments of the forces T_P and T_Q exerted on the mast in Figure 3-3 by the cables $O'P$ and $O'Q$, using a position vector from O to O'. Compare your result with values determined by using position vectors from O to P and O to Q.

3-83 Evaluate the moments about O of the forces T_Q and T_R exerted on the mast in Figure 3-3 by the cables $O'Q$ and $O'R$. If the resultant moment about O of the forces T_P ($= 800$ N), T_Q, and T_R is equal to **0**, what are the values of T_Q and T_R? What is the resultant of the three forces from the cables?

3-84 Referring to Problem 2-81 (p. 51), evaluate the moment about O of the given forces and of the equivalent forces.

3-85 From the sketch on page 81, calculate M_x and compare it with the coefficient of $\mathbf{u}_x$ in the result on page 79.

3-86 Evaluate the moment about the axis OP of the force T_Q exerted by the cable $O'Q$ on the mast in Figure 3-3. From this determine the perpendicular distance between the lines OP and $O'Q$. *Ans.* 5.20 m.

3-87 Evaluate the moment about the axis OQ of the force T_P exerted by the cable $O'P$ on the mast in Figure 3-3. From this determine the perpendicular distance between the lines OQ and $O'P$.

3-88 Referring to the example on page 81, what will be the moment about the axis RQ of a force of magnitude F, acting upward through point O? If the sum of the moments about RQ of F and T_P is zero, what is the ratio of F to T_P?

3-89 Carry out the check suggested at the end of the example on page 84.

3-90 Determine the resultant moment about the axis AB of the 6-kN force and the force R. What must be the magnitude R if this moment is to be zero?

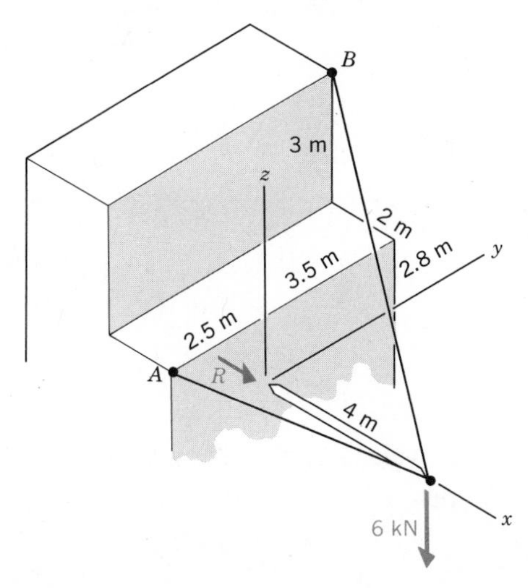

3-91 Show that the moment of a couple is independent of the point about which the moment is reckoned.

3-92 Determine the single force and its line of action, equivalent to the system in Figure 3-5a, by computing the resultant moment about point B rather than point O.

3-93 Evaluate the moments of the 220-N force about the z and $\bar{z}$ axes. The force acts perpendicular to the plane of these two axes.

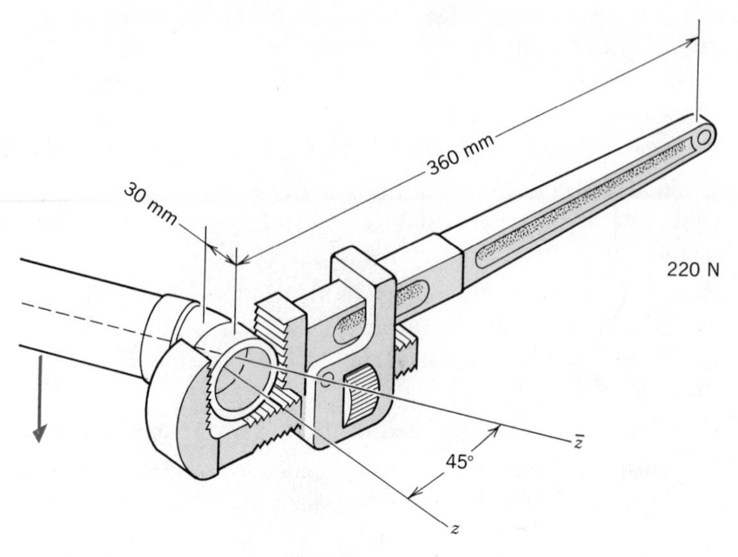

3-94 Specify a system of forces equivalent to that shown, consisting of a force at O together with a couple.

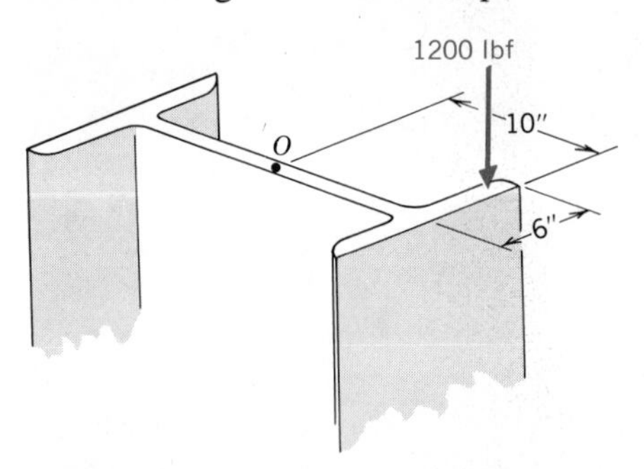

3-95 Specify a system of forces equivalent to that shown, consisting of a force at C together with a couple.

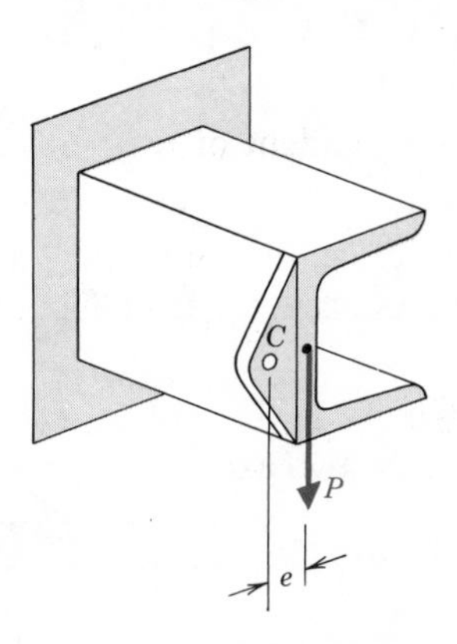

3-96 Specify a system of forces equivalent to those shown, consisting of a force at O together with a couple.

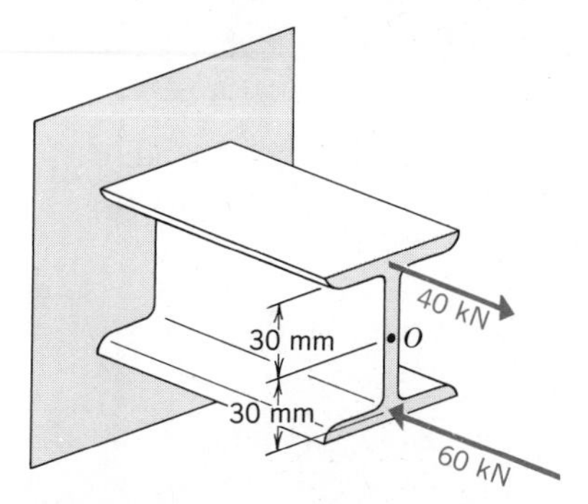

3-97 Specify a system of forces equivalent to those shown, consisting of a force at O together with a couple.

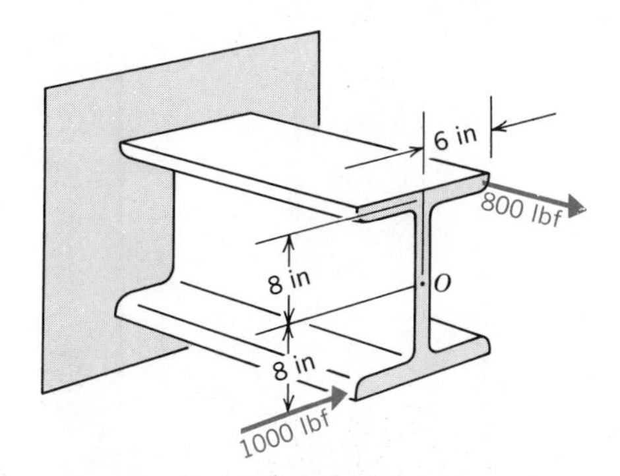

3-98 Indicate the equivalent of the two forces acting on the steering wheel as a force through O together with a couple, and give the moment of the couple.

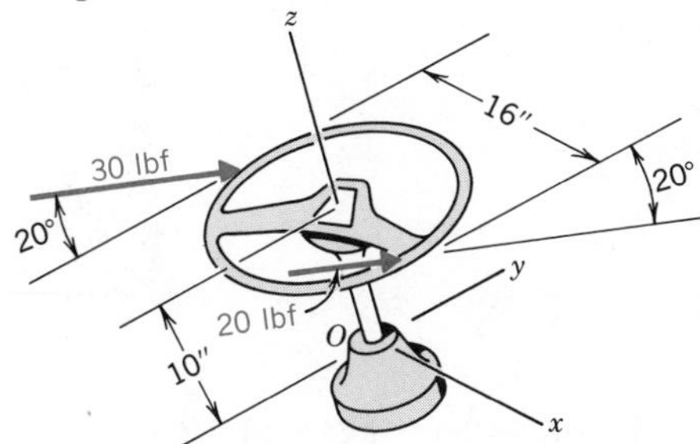

3-99 Referring to the example on page 85, compute the moment about O of the two forces on the pulley by defining position vectors from O to points on the lines of action of the two forces, and using $\mathbf{M}_O = \mathbf{r}_1 \times \mathbf{f}_1 + \mathbf{r}_2 \times \mathbf{f}_2$. How does this procedure compare in complexity with that used in the example?

3-100 In the special case in which $\mathbf{f} = \mathbf{0}$, what does Equation 3-31 tell us?

3-101 Determine the components of a force at O and a couple that, when added to the force components shown, will make the resultant force and moment about O both zero. What will then be the resultant moment about any other point?

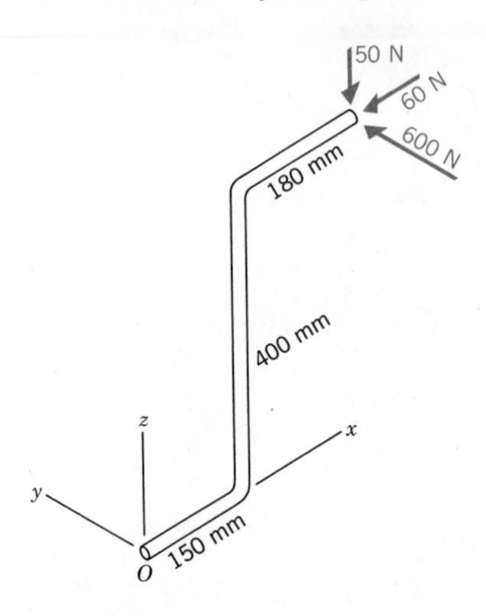

3-102 Replace the 250-N force with an equivalent force
(a) At Q together with a couple.
(b) At O together with a couple.

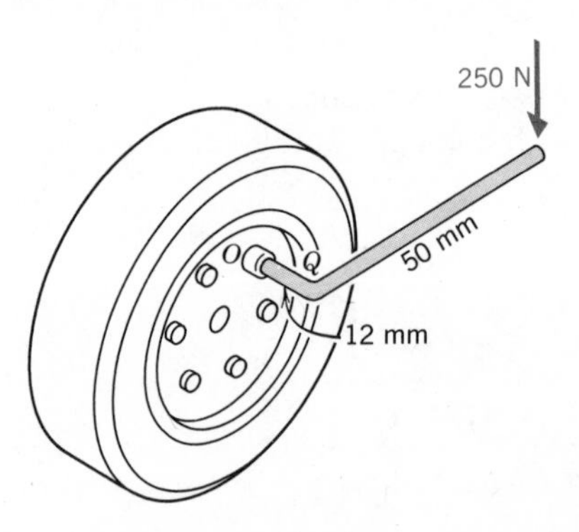

3-103 Show a force acting on the vertical bar at O and a couple that, when added to the given forces, will make the resultant force and moment about every point equal to zero.

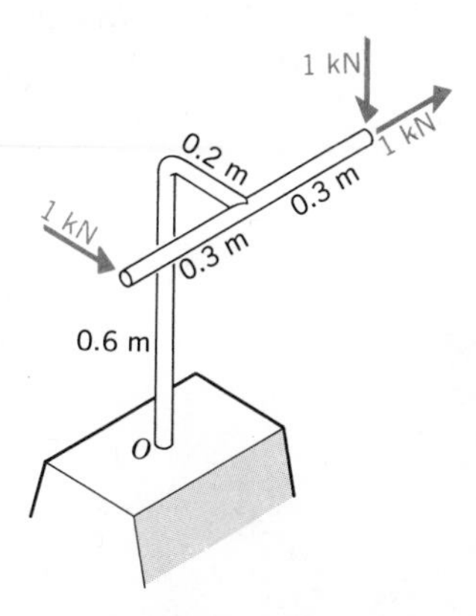

3-104 Specify the wrench equivalent to the forces of the previous problem.

3-105 Specify the wrench equivalent to the force system of Problem 3-97.

3-106 Referring to Problem 3-54 (p. 74), let **A**, **B**, and **C** represent forces of magnitudes 2 kN, 3 kN, and 4 kN, respectively, and let the numerical values on the sketch be distances in meters. Find the resultant force, the location of the line of action of the equivalent wrench, and the moment of the associated couple.

3-107 Work the previous problem with reference to Problem 3-55 instead of 3-54.

3-108 Referring to the example on page 91, what are the coordinates (x_0, y_0) of the point where the line of action of the wrench intersects the plane $z = 0$?

3-109 (a) Evaluate the equivalent force-couple system acting at the origin.
(b) Evaluate the equivalent wrench. Point P: (2 ft, 3 ft, 6 ft)

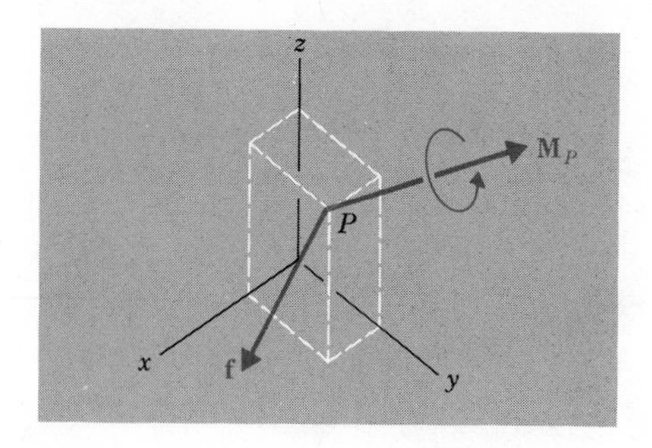

$f = 13$ lb; $f_x = 12$ lb; $f_y = 5$ lb.
$\mathbf{M}_P = -(18 \text{ lb} \cdot \text{ft})\, \mathbf{u}_x + (9 \text{ lb} \cdot \text{ft})\, \mathbf{u}_y + (6 \text{ lb} \cdot \text{ft})\, \mathbf{u}_z$

3-110 During a wind gust, the airplane is subjected to aerodynamic loads equivalent to the four forces shown. Points P_1, P_2, and P_3 are each 5 m from the origin, and P_3 is 1.5 m from the y-axis. Determine the equivalent wrench.

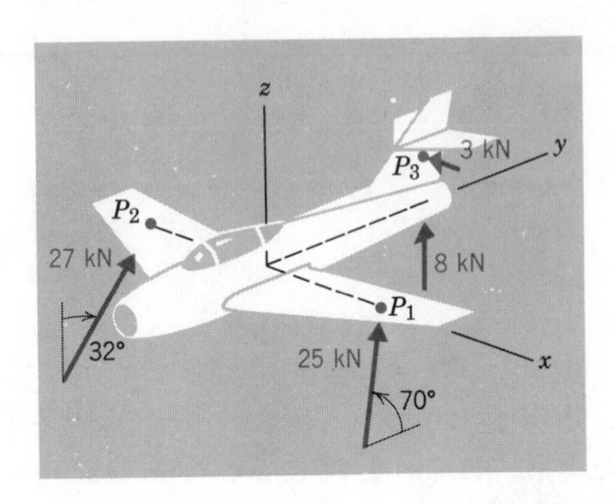

STATIC EQUILIBRIUM

Problems in statics entail the determination of unknown forces that are induced by known forces acting on bodies at rest. For the solution of these problems there are two fundamental laws: that of force equilibrium and that of moment equilibrium.

Force equilibrium is the law that the sum of all forces acting on a body at rest is zero:

$$\boxed{\mathbf{f} = \mathbf{0}} \tag{4-1a}$$

This law is a special case of Newton's second law of motion, $\mathbf{f} = m\,\mathbf{a}$.

Moment equilibrium is the law that *the resultant moment about any point of all forces acting on a body at rest is zero*:

$$\boxed{\mathbf{M} = \mathbf{0}} \tag{4-2a}$$

This law is a special case of a law of rotational motion, which we will derive from Newton's second and third laws.

Each of the laws, (4-1a) and (4-2a), is a vector equation in three dimensions, implying equality of components in three different directions. Resolving all forces into a set of mutually perpendicular directions often provides the basis for problem solution. After these directions are selected, the equation of force equilibrium may be written in the form

$$\left(\sum f_x\right)\mathbf{u}_x + \left(\sum f_y\right)\mathbf{u}_y + \left(\sum f_z\right)\mathbf{u}_z = \mathbf{0}$$

in which Σf_x, Σf_y, and Σf_z are the summations of force components in the three directions. This vector equation implies the three component equations

$$\boxed{\begin{aligned}\sum f_x &= 0\\ \sum f_y &= 0\\ \sum f_z &= 0\end{aligned}} \qquad \textbf{(4-1b)}$$

Similar resolution of the moment resultant about a selected point O leads to the three equations

$$\boxed{\begin{aligned}\sum M_{Ox} &= 0\\ \sum M_{Oy} &= 0\\ \sum M_{Oz} &= 0\end{aligned}} \qquad \textbf{(4-2b)}$$

in which ΣM_{Ox}, ΣM_{Oy}, and ΣM_{Oz} are the summations of components of moments about O in the three directions.

In general, then, we can expect to have *six* independent equations of equilibrium for each body considered.

In cases in which all the forces act in one plane, one of the three equations from (4-1) becomes trivial ($0 = 0$). Also, the resultant moment of the forces about any point in the plane will be perpendicular to this plane, so that (4-2) yields only one nontrivial equation. Thus, in planar systems, we can expect to have *three* equations of equilibrium for each body.

4-1 FREE-BODY DIAGRAMS

To apply (4-1) and (4-2) it is necessary to identify carefully a specific body (which may be a part of a larger structural system) and consider *all* the forces acting on the body. Experience has shown that to consistently do this correctly, it is necessary to draw a *free-body diagram*.

To illustrate this procedure, let us consider the device shown in Figure 4-1. A number of different free-bodies are possibly useful, so we will construct several of them.

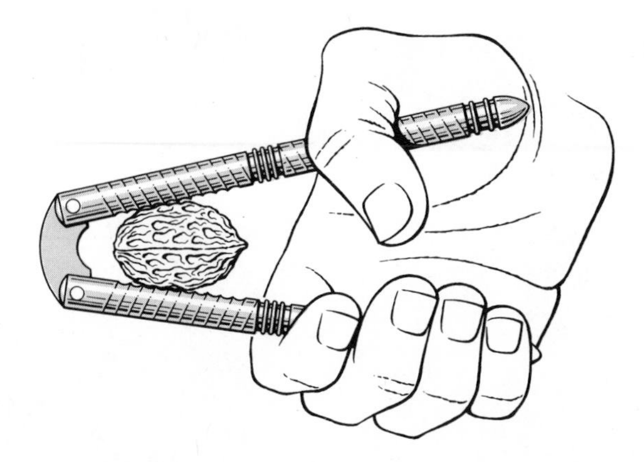

FIGURE 4-1

What To Exclude and Include

First, consider the system consisting of the nutcracker together with the walnut. This system is shown isolated from all other objects, with arrows depicting the forces that come from objects *external* to the system. Assuming forces of gravity to be negligible, the only external body that exerts forces on this system is the hand. Therefore, the appropriate arrows are those shown in Figure 4-2*a*. The interaction between the nut and the cracker is *not* shown on this free-body, since this is internal to this system.

To expose the force tending to break the nut, we might consider free-bodies of the nut and of the nutcracker (Figure 4-2*b* and *c*). Now, external to the nutcracker are the hand and the walnut. Therefore, arrows depicting the forces from the nut as well as from the hand are included on the free-body of the nutcracker.

Another possibly useful free-body would be the upper handle. Objects external to this are the hand, the walnut, and the connecting pin. This free-body appears in Figure 4-2*d*.

To summarize: First, a sketch must be made showing clearly the system to be considered for equilibrium. The system boundary is normally chosen so that it passes through a point where a force interaction of particular interest occurs. Next, *all* forces acting on the system, from bodies *external* to the system, must be properly represented. Force interactions between bodies *within* the system are *not* shown.

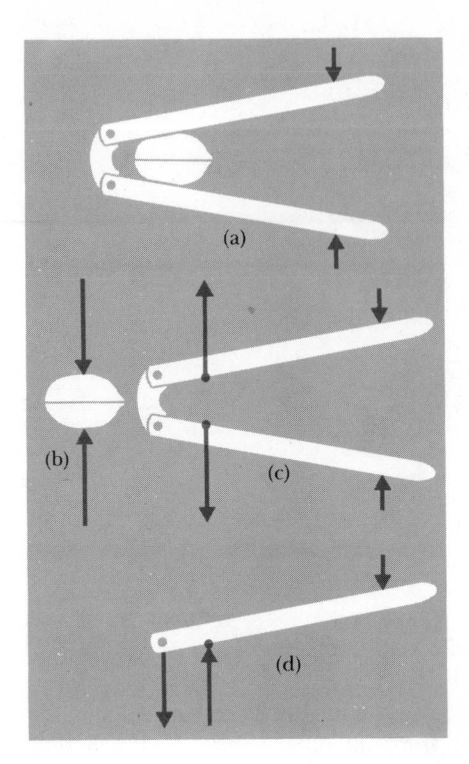

FIGURE 4-2

Labeling

To show clearly the physical significance of quantities in written equations, magnitudes of forces must be properly identified on the free-body diagrams. The implication of the letter placed beside an arrow for this purpose must be precisely understood. The letter represents a *scalar* multiplier of a unit vector in the direction of the arrow. Thus, since this quantity can take on positive or negative values, a force in the same or opposite direction as indicated by the arrow can be represented.

Figure 4-3 indicates appropriate labeling of our free-bodies for analysis of the nut-cracking system. If an analysis leads to the values, say, $P = 80$ N and $R = 400$ N, this would imply that the forces acting on the upper handle are 80 newtons *downward* on the right-hand end and 400 newtons *upward* where the walnut makes contact. If different values led to, say, $P = -15$ N and $R = -75$ N, these values would imply that the forces acting on the upper handle are 15 newtons *upward* on the right-hand end and 75 newtons *downward* where the walnut makes contact. (A little adhesive between the walnut and the nutcracker would make this possible.)

Observe that the forces on the walnut have the same labels as their

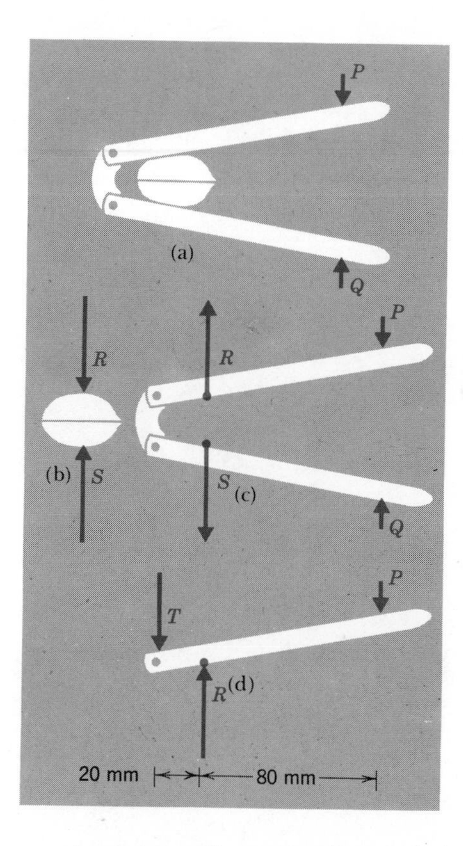

FIGURE 4-3

counterparts on the nutcracker, and the arrows have opposite directions. We have in this way implied satisfaction of Newton's third law without further fuss. Other relatively simple aspects of force analysis can be taken care of as the free bodies are constructed; for example, unless the walnut is to accelerate, it is evident from a glance at its free-body that $R = S$. Writing this equation can be circumvented by simply labeling both arrows with the same letter. Can you find another free-body in Figure 4-3 in which some analysis can be taken care of as the labeling is done?

Occasionally, a force depicted on a free-body diagram is labeled with a *boldface* letter. This, of course, implies a *vector* value, which can take on any direction and magnitude. In this case the labeling appears differently, as a comparison of (a) and (b) in Figure 4-4 shows. Notice the necessity of indicating the base directions with $\mathbf{u}_x$ and $\mathbf{u}_y$ before a specific value can be identified with the boldface labeling. Also, note that the consistent manner of labeling the oppositely-directed reaction $-\mathbf{R}$ incorporates a minus sign, whereas in Figure 4-4*a* the force components acting on the incline have positive signs on their labels.

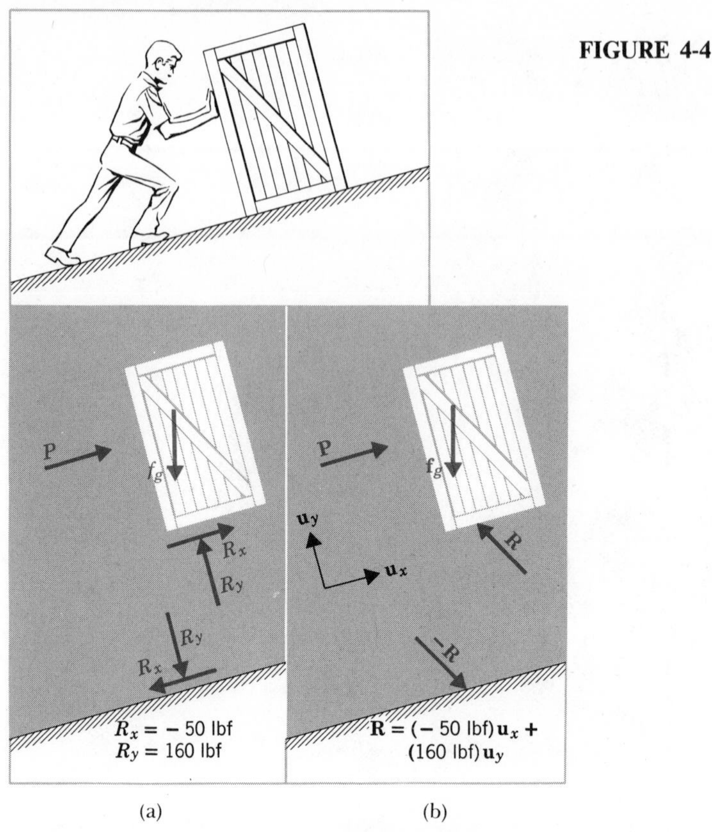

(a) (b)

FIGURE 4-4

Connections

In the above examples, we have represented by a single force the action at each place where an object makes contact with the isolated system. This single force is the *equivalent* (in the sense discussed in Section 2-5 and on pages 86–87) of an actual set of forces that are distributed over the area in which there is contact. Often it is necessary to introduce a couple in addition to the force to properly represent the set of forces in the region of contact. For example, the reaction at the support A of the cantilever beam in Figure 4-5a is normally depicted as a single force at A and a couple, as in the free-body diagram b, although it can also be represented by a pair of forces as indicated in the free-body diagram c. Neither of these representations shows the details of the actual distribution of forces over the contact surface, but each is equivalent.

The force and couple components that can be transmitted to the free body depend on the nature of the mechanical connection at the place of connection. For example, the pin that transmits the force T shown in Figure 4-3d forms a hinge at that point, and this hinge is incapable of transmitting a moment about

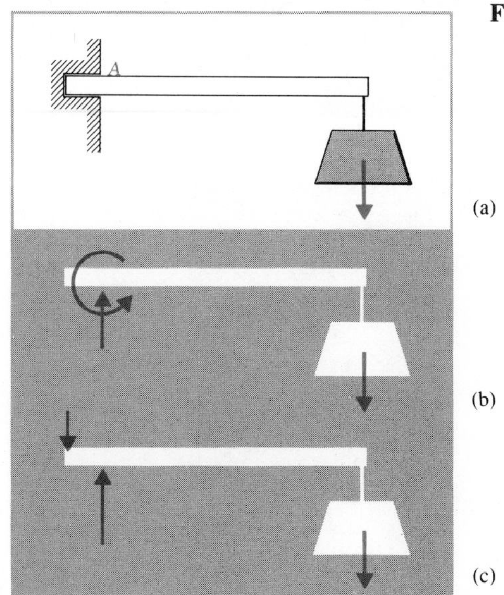

FIGURE 4-5

(a)

(b)

(c)

this point to the handle. If, instead of this construction, the handles were welded to the connecting plate, the free-body diagram would show a couple in addition to the force T at this point. Examples of several possible types of connections are shown in Problem 4-1.

PROBLEMS

4-1 The sketches depict a number of the types of connections between structural members. On a free-body diagram of the member like that indicated on the right, show the components of force and couple that the connection is capable of transmitting. (The free-body diagram will be incomplete, inasmuch as the reaction from the upper part of the member need not be shown for this exercise.)

(a)

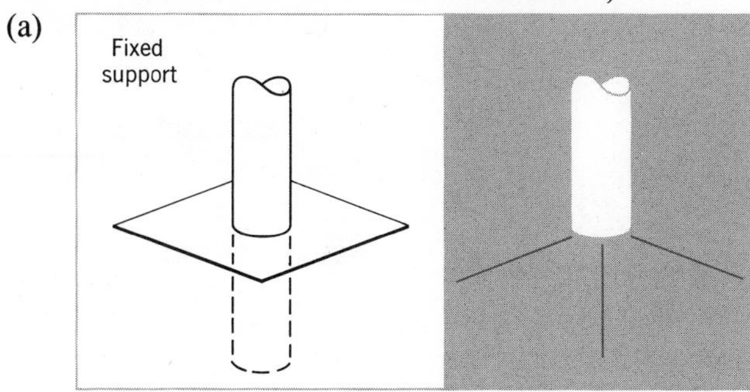

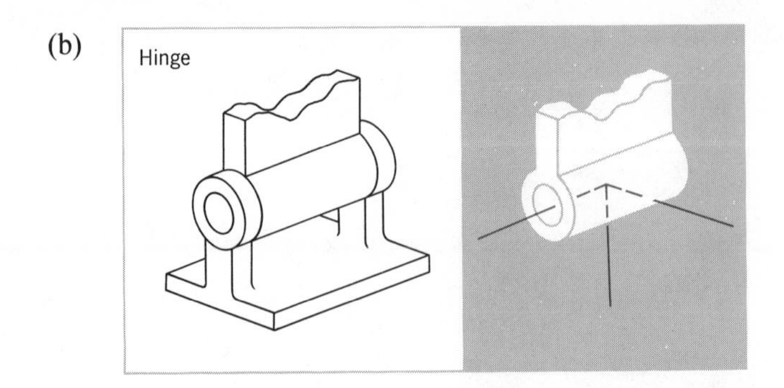
(b)
Hinge

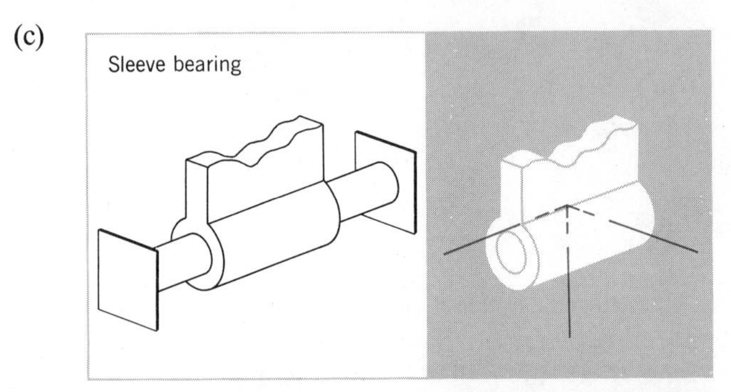
(c)
Sleeve bearing

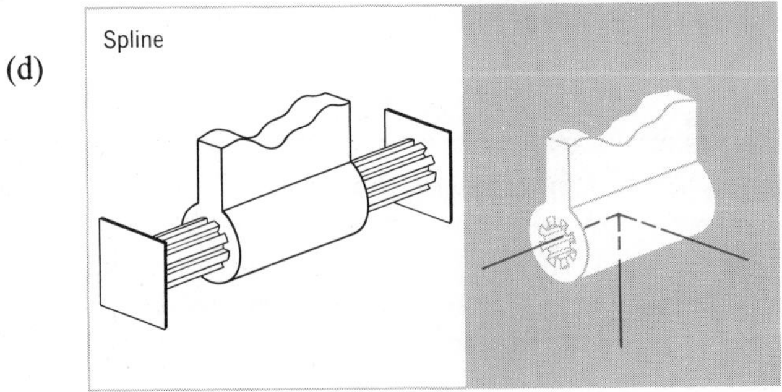
(d)
Spline

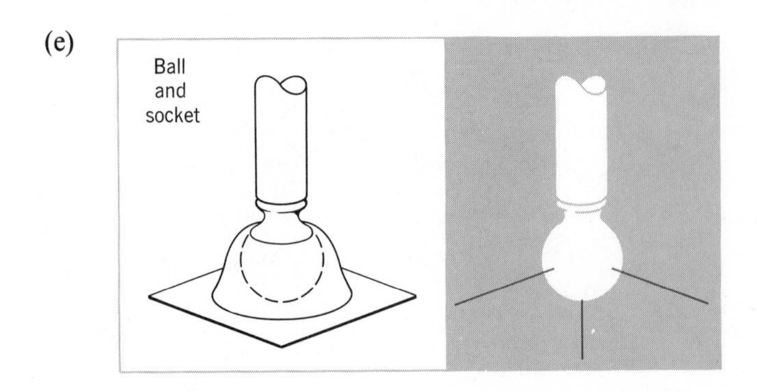
(e)
Ball
and
socket

(f)

(g)

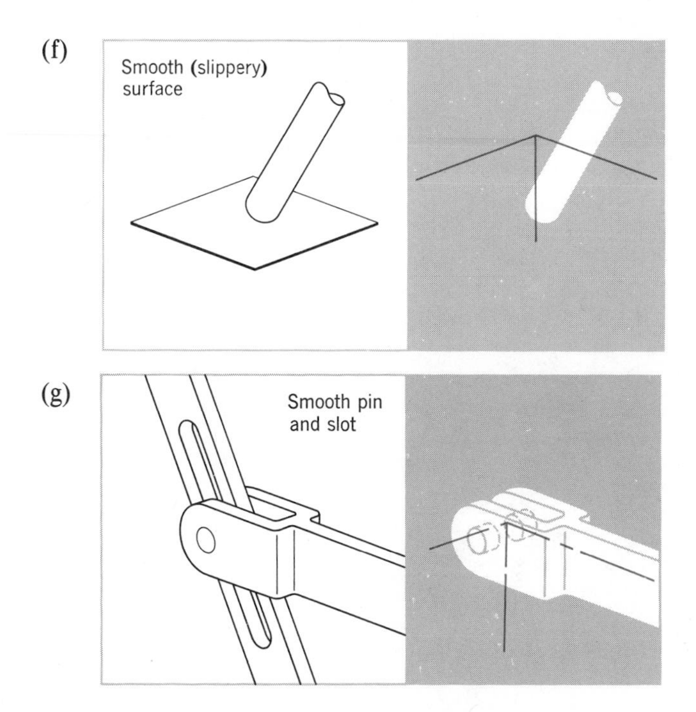

For Problems 4-2 through 4-7, indicate which of the free-body diagrams are correct and indicate changes that must be made to correct those that are incorrect.

4-2 A force of 200 N is applied to the handles of the bolt cutters as shown. Gravity forces are negligible.

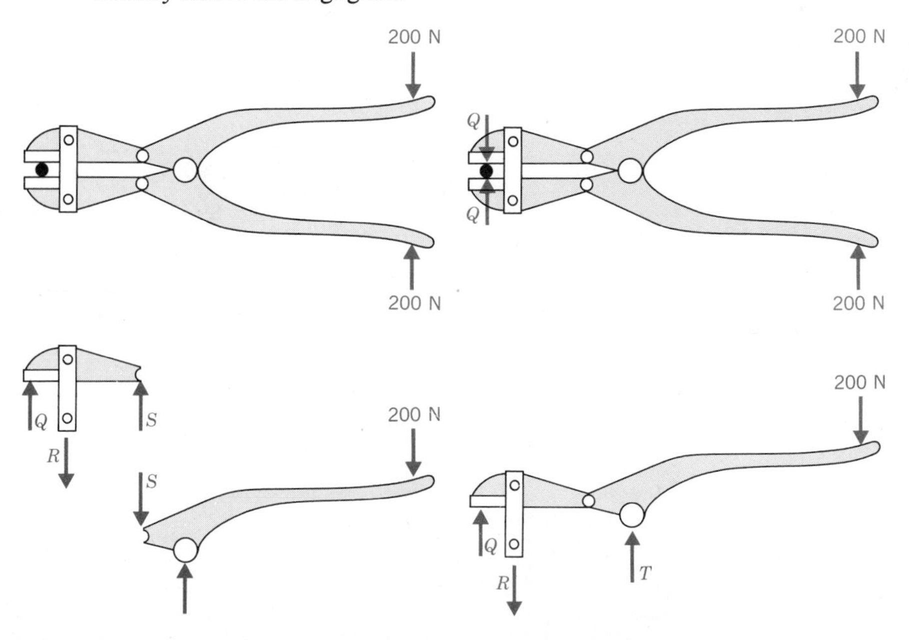

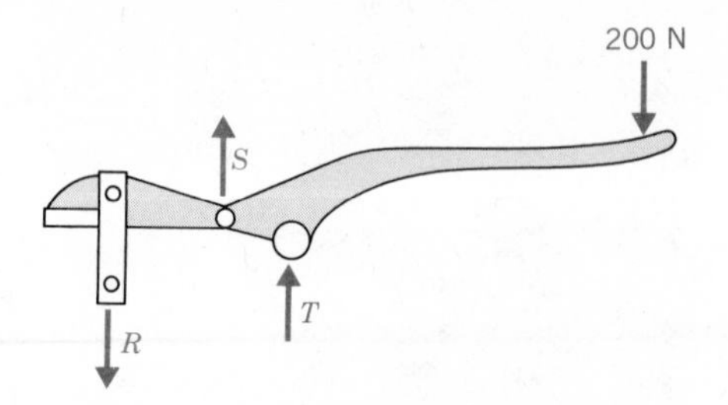

4-3 The pipe wrench is being used to tighten the pipe joint. Gravity forces are negligible.

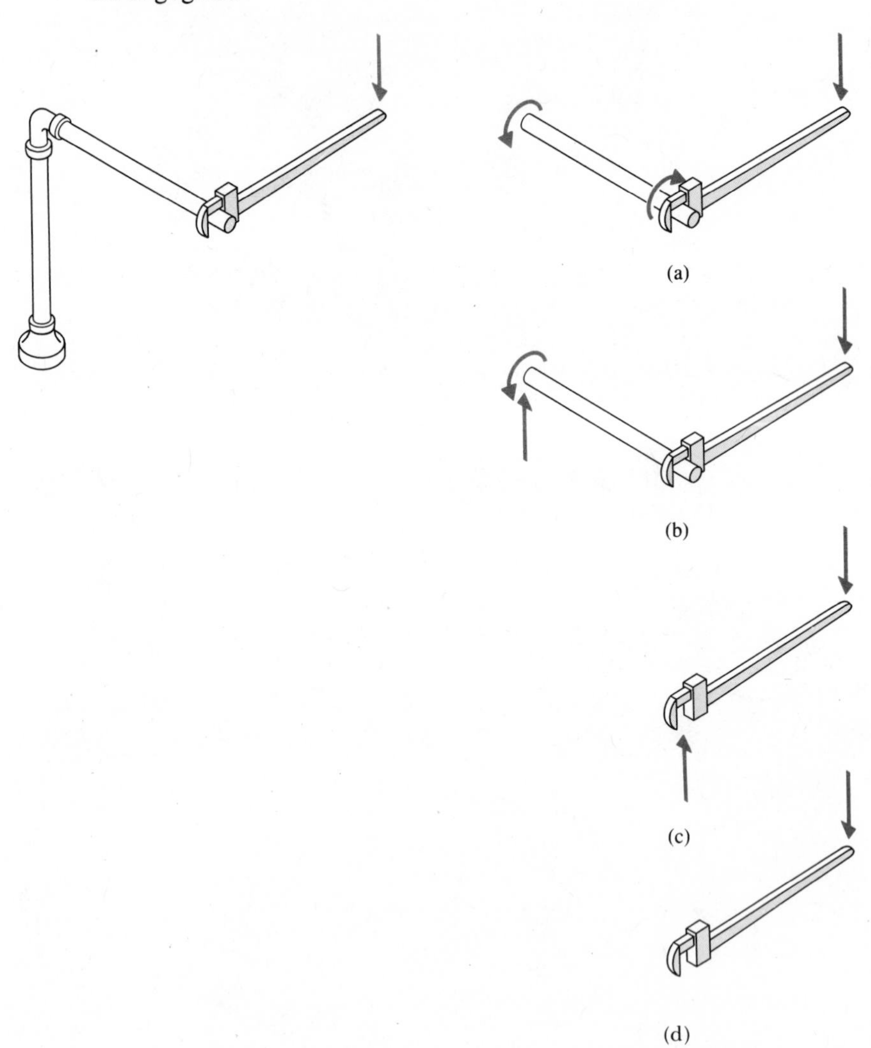

4-4 The crowbar is being used to extract the nail. Gravity forces are negligible.

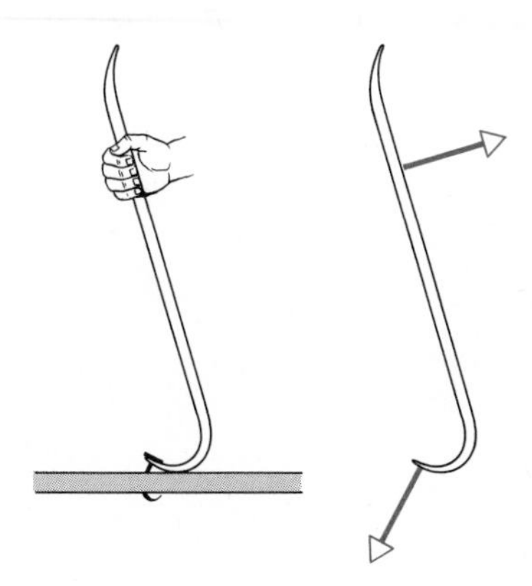

4-5 The "nippers" are being used to cut a piece of wire. Gravity forces are negligible.

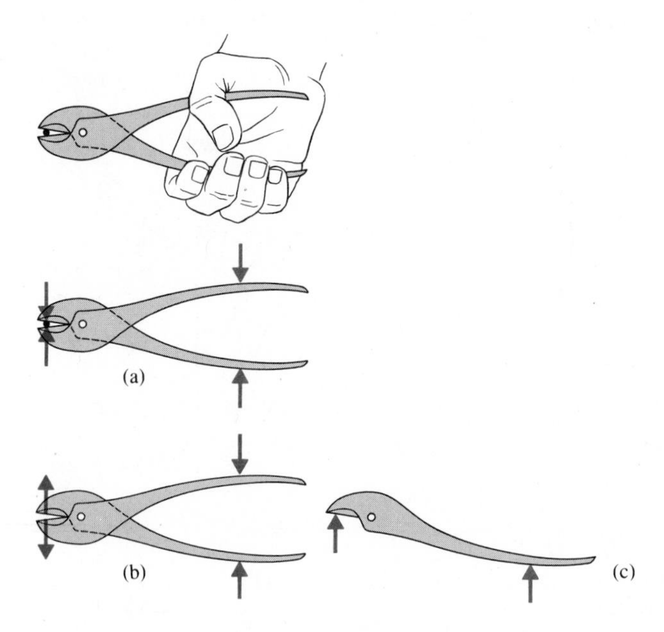

4-6 The motor at A is being used to hoist the load at B. Gravity forces other than that on the load are negligible.

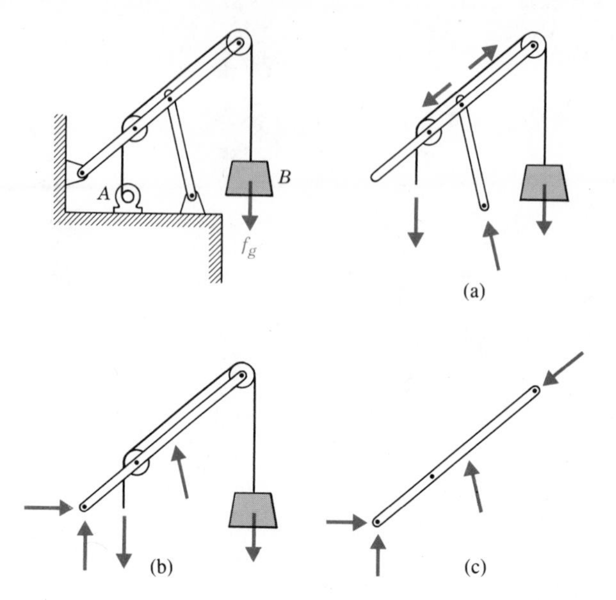

4-7 The pole is supported with a ball joint at O and three guy wires under tension.

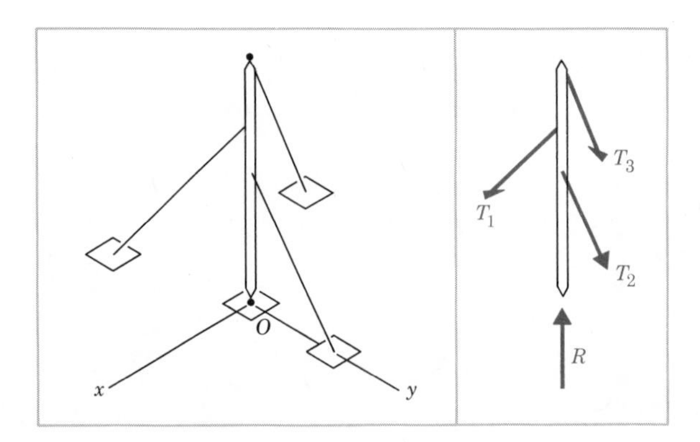

4-8 Draw a free-body diagram of:

(a) The pulley.

(b) The frame.

(c) The system consisting of the frame, pulley, line, block, and man.

4-9 Draw a free-body diagram of:

(a) Member A.
(b) Member B.
(c) Member C.
(d) The block M.
(e) The system consisting of A, B, C, and M.

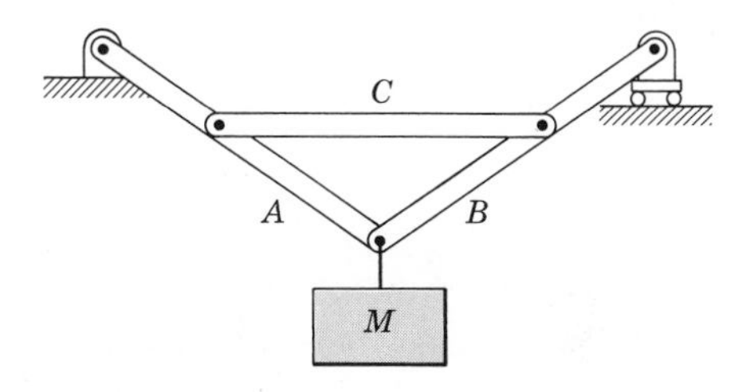

4-10 Draw a free-body diagram of:

(a) The rod A.
(b) The lifeboat davit B.
(c) The lifeboat C.
(d) The system consisting of A, B, and C.

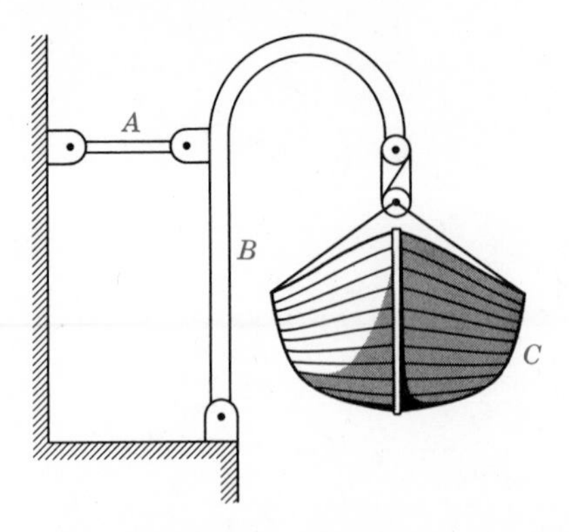

4-11 Draw a free-body diagram of the flagpole and flag.

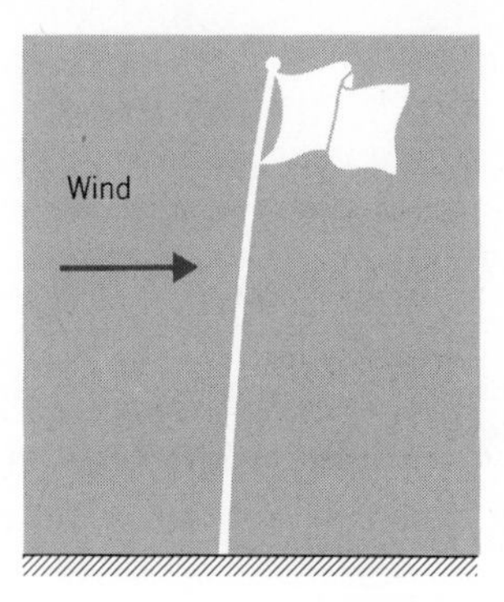

4-12 Draw a free-body diagram of each of the logs in the pile.

4-13 Draw a free-body diagram of the hand of Figure 1-1.

4-14 Draw a free-body diagram of each of the two large parts of the channel-lock pliers.

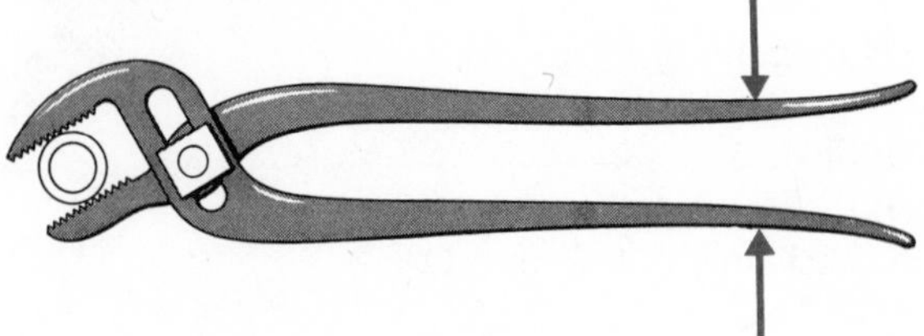

4-15 Draw a free-body diagram of the wheel being pulled over the curb.

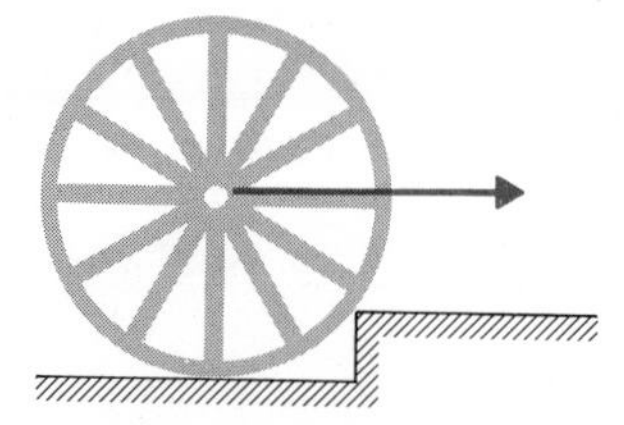

4-16 Draw a free-body diagram of the heavy plate. The hinge at *B* supports all of the force acting along the horizontal axis *AB*.

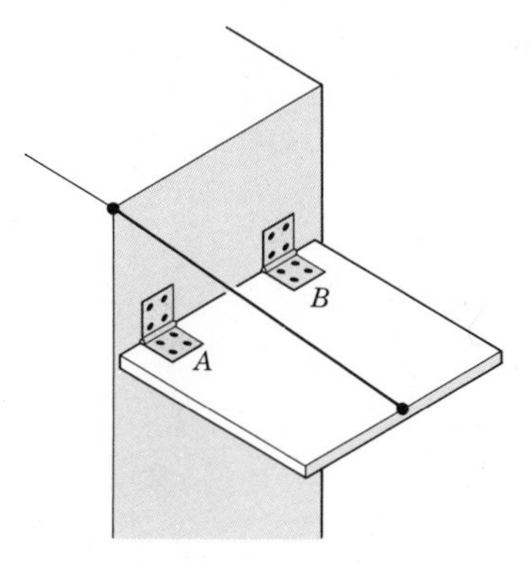

4-17 Draw a free-body diagram of the system consisting of the frame, pulleys, line, and load.

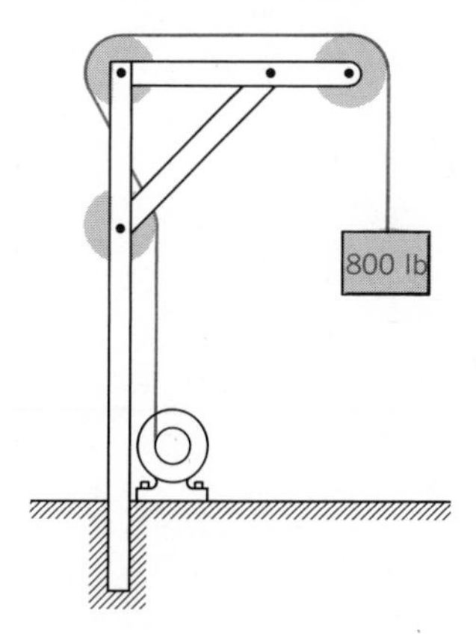

4-18 Draw free-body diagrams of the tractor, of one rear wheel, and of the tractor without its rear wheels.

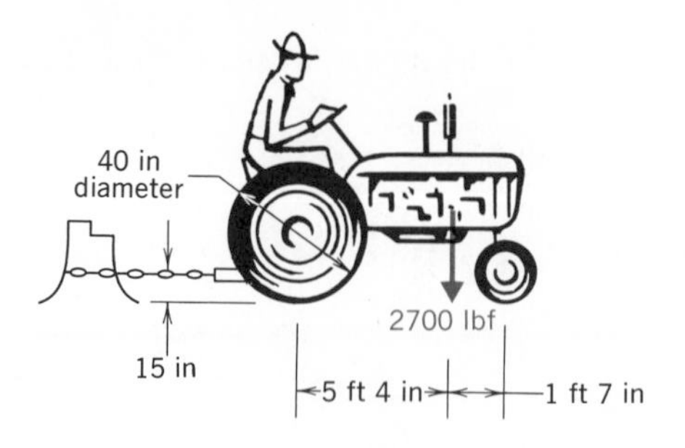

4-2 EQUATIONS OF EQUILIBRIUM

With properly labeled free-body diagrams drawn, equations of equilibrium can be written for any of the chosen bodies. For example, considering the vertical direction for force equilibrium of the free-body of Figure 4-3 (repeated from page 105), we can write

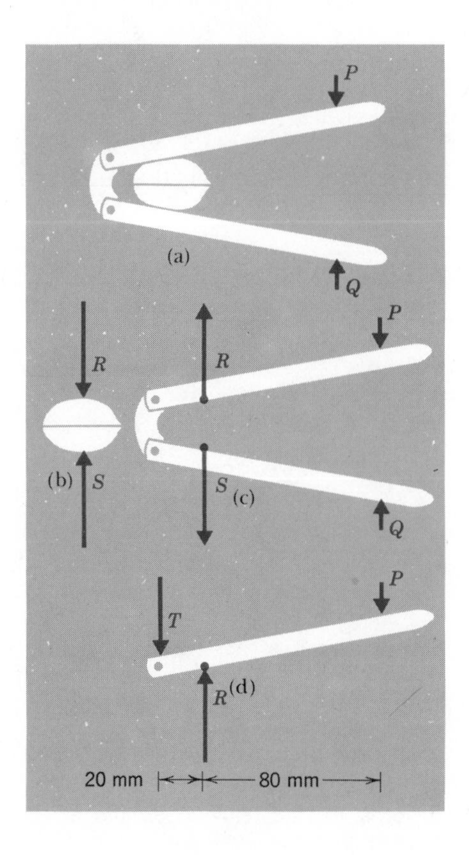

FIGURE 4-3 (repeated from page 105)

$$Q - P = 0$$

Each of the other two equations of force equilibrium, involving components in the horizontal direction and components perpendicular to the page, is the trivial $0 = 0$. Account of moment equilibrium tells us that the lines of action of the two forces must coincide, a fact that has already been incorporated in the free-body diagram. Similar analysis of the free-bodies of Figure 4-3*b* and *c* leads to

$$S - R = 0$$

$$Q - P + R - S = 0$$

Now suppose that the reason for this analysis is to obtain an estimate of how hard one must squeeze in order to crack a nut, given that cracking requires 245 N applied to the nut. Then

$$R = 245 \text{ N}$$

and the three equations above contain the three unknowns P, Q, and S. Unfortunately, attempts to solve these for P or Q will fail, because the three equations are not independent. (The last equation can be deduced from the first two by addition.) Equilibrium of still another body will then have to be considered to obtain an independent equation. The free-body of the handle in Figure 4-3*d* can provide two more equations, the vertical component of force equilibrium,

$$T + P = 245 \text{ N} \qquad \textbf{(a)}$$

and an equation of moment equilibrium. Summing moments about the point of contact with the walnut leads to

$$(20 \text{ mm})T - (80 \text{ mm})P = 0 \qquad \textbf{(b)}$$

To solve for P, we can multiply equation (a) by 20 mm and subtract equation (b) from the result, yielding

$$(100 \text{ mm})P = (20 \text{ mm})(245 \text{ N}) \qquad \textbf{(c)}$$

This gives

$$P = 49\text{N}$$

A more direct analysis stems from considering moments about a different point on the handle, resulting in (c) as the first equation written.

The following examples provide further illustration of the use of the basic considerations of static equilibrium.

EXAMPLE

Neglecting gravity forces except that on the 700-lb load, determine the forces in the cable and in the boom shown in Figure 4-6.

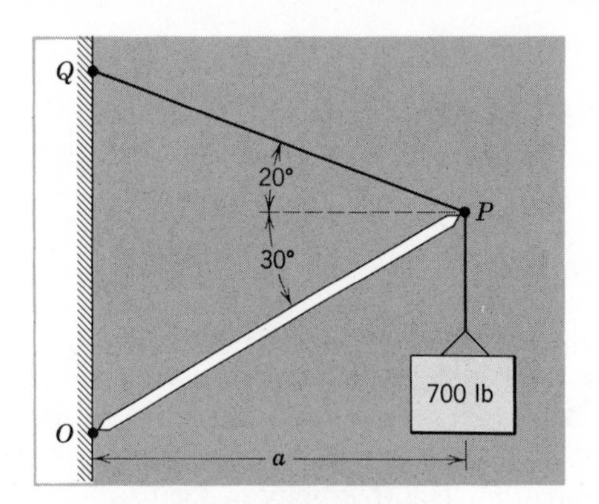

FIGURE 4-6

Free-body diagrams of the boom and of the suspended block are shown in Figure 4-7. First, equilibrium of the block requires that

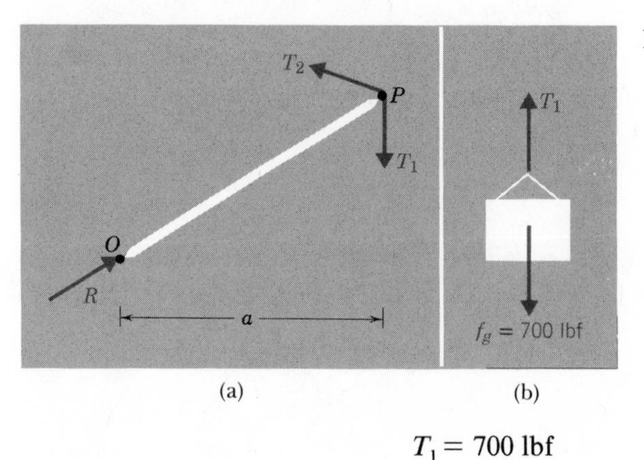

FIGURE 4-7

$$T_1 = 700 \text{ lbf}$$

Considering moment equilibrium about point O of the boom, we can write

$$(T_2 \sin 20° - T_1)a + (T_2 \cos 20°)(a \tan 30°) = 0$$

from which

$$T_2 = \frac{T_1}{\sin 20° + \cos 20° \tan 30°}$$

$$= 791 \text{ lbf}$$

Considering moment equilibrium about point P of the boom, we see that the reaction R must be directed along the boom. Equilibrium of horizontal forces then gives

$$R \cos 30° - T_2 \cos 20° = 0$$

from which

$$R = \frac{T_2 \cos 20°}{\cos 30°}$$

$$= 859 \text{ lbf}$$

EXAMPLE

Neglecting gravity forces except that on the 2-ton load, determine the tension in the cable AB holding up the crane boom in Figure 4-8.

Figure 4-9 shows a free-body diagram of the boom and pulley, the load, and portions of the cables. The tension in the upper cable is readily found (see Problem 4-31) to be 2 tons. To avoid introducing the unknown reaction at O into the equilibrium equations, let us consider moments of forces about this point. Letting r denote radius of the pulley, we write the equation of moment equilibrium about point O as

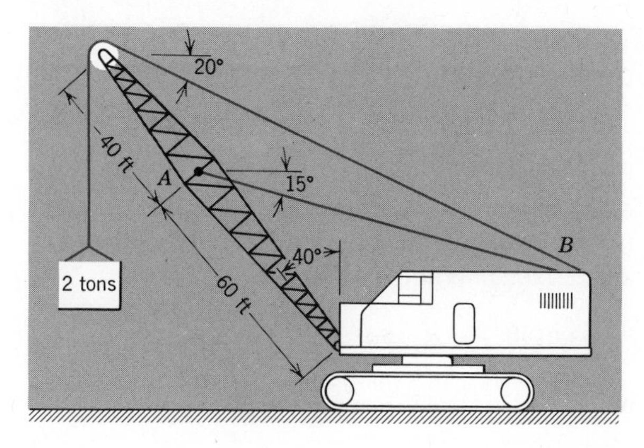

FIGURE 4-8

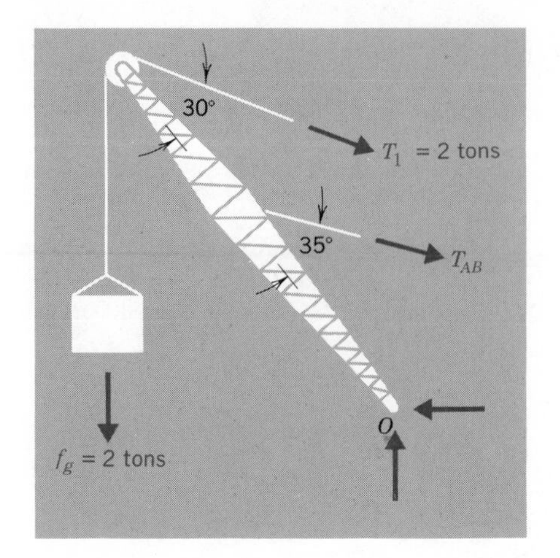

FIGURE 4-9

$$f_g[(100 \text{ ft}) \sin 40° + r] - T_1[(100 \text{ ft}) \sin 30° + r] - T_{AB}(60 \text{ ft}) \sin 35° = 0$$

With $T_1 = f_g = 2$ tons, this equation is readily solved for the desired tension:

$$T_{AB} = \frac{(2 \text{ tons})(100 \text{ ft})(\sin 40° - \sin 30°)}{(60 \text{ ft}) \sin 35°}$$

$$= 0.83 \text{ tons}$$

Note that, because the force of gravity on the boom has been neglected, the actual tension will be somewhat higher.

EXAMPLE

Gravity forces on the structural members are negligible compared with P and Q. Evaluate all the forces acting on each of the three members in the A-frame in Figure 4-10.

Figure 4-11 shows free-body diagrams of the overall structure and of the individual members. The roller support at D means that no horizontal force can be transmitted to the ground at that point. From the free-body (a) we can write equations of equilibrium of moments about point E:

$$Q(3a) + P(3a \tan 30°) - R_D(6a \tan 30°) = 0$$

horizontal force equilibrium:

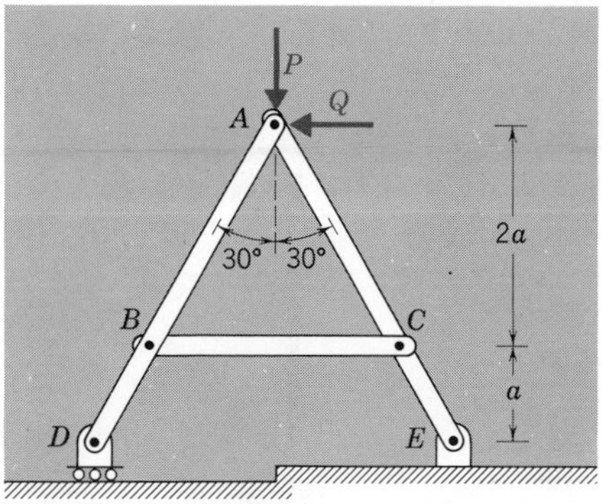

P
Q
A
B
C
D
E
30° 30°
2a
a

FIGURE 4-10

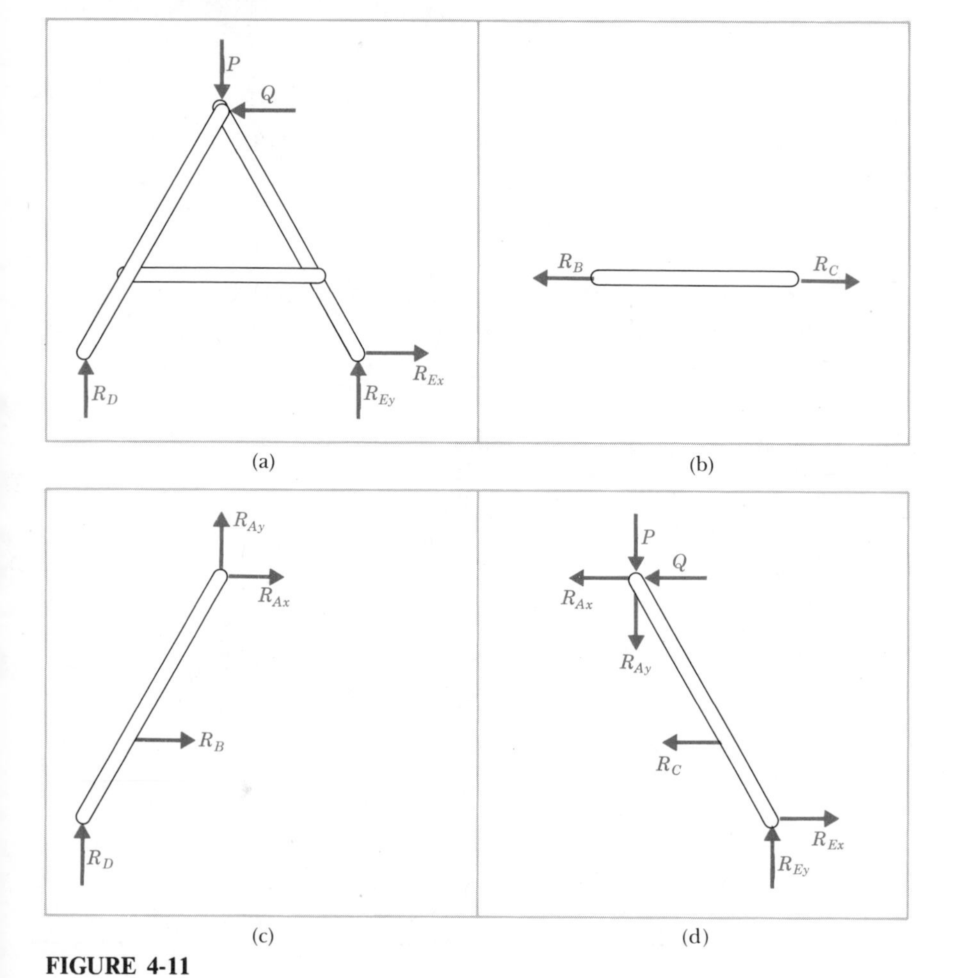

FIGURE 4-11

$$R_{Ex} - Q = 0$$

and equilibrium of moments about point D:

$$R_{Ey}(6a \tan 30°) + Q(3a) - P(3a \tan 30°) = 0$$

These equations can then be solved for the support reactions:

$$R_D = \frac{1}{2}P + \frac{\text{ctn } 30°}{2}Q$$

$$R_{Ex} = Q$$

$$R_{Ey} = \frac{1}{2}P - \frac{\text{ctn } 30°}{2}Q$$

As a check, you might examine vertical force equilibrium.

Moment equilibrium of the free-body (b) implies that the lines of action of R_B and R_C are along the bar. Its horizontal force equilibrium gives us

$$R_C - R_B = 0$$

Now turning to the free-body (c), we can write equations of moment equilibrium about point A:

$$R_B(2a) - R_D(3a \tan 30°) = 0$$

and horizontal and vertical force equilibrium:

$$R_{Ax} + R_B = 0$$

$$R_{Ay} + R_D = 0$$

With values of R_D, R_{Ex}, and R_{Ey} above, these equations give the following values of the remaining unknown reactions.

$$R_B = R_C = \frac{3 \tan 30°}{4}P + \frac{3}{4}Q$$

$$R_{Ax} = -\frac{3 \tan 30°}{4}P - \frac{3}{4}Q$$

$$R_{Ay} = -\frac{1}{2}P - \frac{\text{ctn } 30°}{2}Q$$

EXAMPLE

The cables OA, OB, and OC support the suspended block shown in Figure 4-12. Determine the tension in each cable in terms of the gravitational force f_g.

The desired tensions, P, Q, and R, are shown on the free-body of the portion of the structure in the neighborhood of point O. To resolve the forces in the directions of $\mathbf{u}_x$, $\mathbf{u}_y$, and $\mathbf{u}_z$, the direction cosines between OA, OB, and OC and these directions must be evaluated. This is done, as in the example on page 72, by dividing the projection of the cable onto the axis by the length of the cable; for instance, the direction cosine between OA and $\mathbf{u}_x$ is

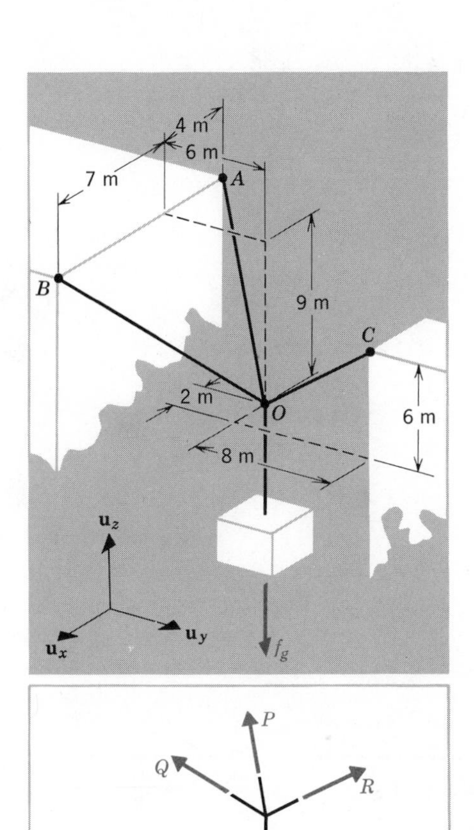

FIGURE 4-12

$$\cos \measuredangle_{OA}^{\mathbf{u}_x} = \frac{-4 \text{ m}}{\sqrt{(4 \text{ m})^2 + (6 \text{ m})^2 + (9 \text{ m})^2}} = -0.346\,84$$

The results are summarized in the following:

$$\mathbf{u}_{OA} = -0.346\,84\,\mathbf{u}_x - 0.520\,27\,\mathbf{u}_y + 0.780\,40\,\mathbf{u}_z$$

$$\mathbf{u}_{OB} = 0.543\,31\,\mathbf{u}_x - 0.465\,69\,\mathbf{u}_y + 0.698\,54\,\mathbf{u}_z$$

$$\mathbf{u}_{OC} = 0.196\,12\,\mathbf{u}_x + 0.784\,46\,\mathbf{u}_y + 0.588\,35\,\mathbf{u}_z$$

Force equilibrium requires that

$$P\mathbf{u}_{OA} + Q\mathbf{u}_{OB} + R\mathbf{u}_{OC} - f_g\mathbf{u}_z = \mathbf{0}$$

With the above resolutions inserted, this vector equation implies the three component equations:

$$\sum f_x = 0.346\,84P + 0.543\,31Q + 0.196\,12R = 0$$

$$\sum f_y = -0.520\,27P - 0.465\,69Q + 0.784\,46R = 0$$

$$\sum f_z = 0.780\,40P + 0.698\,54Q + 0.588\,35R = f_g$$

These three equations can be solved simultaneously for the three unknown forces. The results are

$$P = 0.6601f_g$$

$$Q = 0.2169f_g$$

$$R = 0.5666f_g$$

EXAMPLE

Determine the forces in the boom OC and the cables CA and CB, in terms of the gravitational force f_g, in Figure 4-13.

Considering moments about point C of the forces acting on the free-body of the boom, we see that the force **R** must be directed along the axis of the boom. With the directions of **P**, **Q**, and **R** known, we could, as illustrated in the previous example, write $\Sigma f_x = 0$, $\Sigma f_y = 0$ and $\Sigma f_z = 0$, and solve the

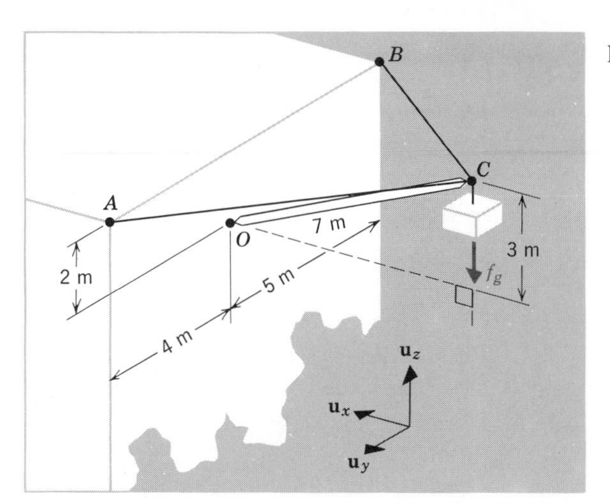

FIGURE 4-13

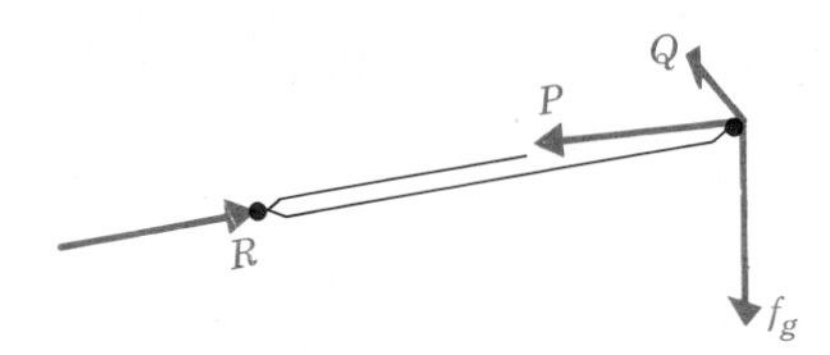

resulting simultaneous equations for P, Q, and R. As an alternative, let us consider the moments about three different axes that we select in such a way as to eliminate some of the unknown forces from each resulting equation.

Consider first the axis OB. Since the lines of action of both $\mathbf{Q}$ and $\mathbf{R}$ pass through this axis, the only forces that will contribute to the moment about OB are $\mathbf{P}$ and $\mathbf{f}_g$. Denoting by $\mathbf{r}_{C/O}$ the position vector to C from O and by $\mathbf{u}_{B/O}$ the unit vector in the direction of B from O, we write the moment about the axis OB as (see Equation 3-28a, p. 79.)

$$\mathbf{M}_{OB} = \left[\mathbf{u}_{B/O} \cdot (\mathbf{r}_{C/O} \times \mathbf{P}) + \mathbf{u}_{B/O} \cdot (\mathbf{r}_{C/O} \times \mathbf{f}_g)\right] \mathbf{u}_{B/O}$$

Since the boom is in equilibrium, the coefficient in the brackets must be equal to zero. The vectors appearing in this coefficient have the following values.

$$\mathbf{u}_{B/O} = \frac{(-5 \text{ m})\,\mathbf{u}_y + (2 \text{ m})\,\mathbf{u}_z}{\sqrt{(5 \text{ m})^2 + (2 \text{ m})^2}}$$

$$= -0.928\,477\,\mathbf{u}_y + 0.371\,391\,\mathbf{u}_z$$

$$\mathbf{r}_{C/O} = \left(-2\sqrt{10}\ \text{m}\right)\mathbf{u}_x + (3\ \text{m})\,\mathbf{u}_z$$

$$\mathbf{P} = P\frac{\left(2\sqrt{10}\ \text{m}\right)\mathbf{u}_x + (4\ \text{m})\,\mathbf{u}_y + (-1\ \text{m})\,\mathbf{u}_z}{\sqrt{\left(2\sqrt{10}\ \text{m}\right)^2 + (4\ \text{m})^2 + (1\ \text{m})^2}}$$

$$= (0.837\,708P)\,\mathbf{u}_x + (0.529\,813P)\,\mathbf{u}_y + (-0.132\,453P)\,\mathbf{u}_z$$

$$\mathbf{f}_g = -f_g\mathbf{u}_z$$

Now, referring to the formula 3-26 (p. 73) we evaluate the triple scalar products as

$$\mathbf{u}_{B/O}\cdot\left(\mathbf{r}_{C/O}\times\mathbf{P}\right) = \begin{vmatrix} 0 & -0.928\,477 & 0.371\,391 \\ -6.324\,555\ \text{m} & 0 & 3\ \text{m} \\ 0.837\,708P & 0.529\,813P & -0.132\,453P \end{vmatrix}$$

$$= (-2.800\,05\ \text{m})P$$

$$\mathbf{u}_{B/O}\cdot\left(\mathbf{r}_{C/O}\times\mathbf{f}_g\right) = \begin{vmatrix} 0 & -0.928\,477 & 0.371\,391 \\ -6.324\,555\ \text{m} & 0 & 3\ \text{m} \\ 0 & 0 & -f_g \end{vmatrix}$$

$$= (5.872\,20\ \text{m})f_g$$

With these values, the vanishing of moment about the axis OB is written as

$$-(2.800\,05\ \text{m})P + (5.872\,20\ \text{m})f_g = 0$$

which yields

$$P = 2.097f_g$$

A similar calculation of moments about the axis OA leads to

$$(3.133\,40\ \text{m})Q - (5.656\,85\ \text{m})f_g = 0$$

which yields

$$Q = 1.805f_g$$

Finally, we can calculate R by summing moments about the axis AB. However, the geometry for this calculation is somewhat simpler than for the previous two axes, so that we can use Equation 3-28c instead of Equation

3-28a. This requires that we find the perpendicular distances from the lines of action of **R** and $\mathbf{f}_g$ to the axis AB. Once this is done, calculation requires much less arithmetic than the previous two moment sums. Consider the view along the axis AB. The distance d may be determined from proportionality between the two similar triangles:

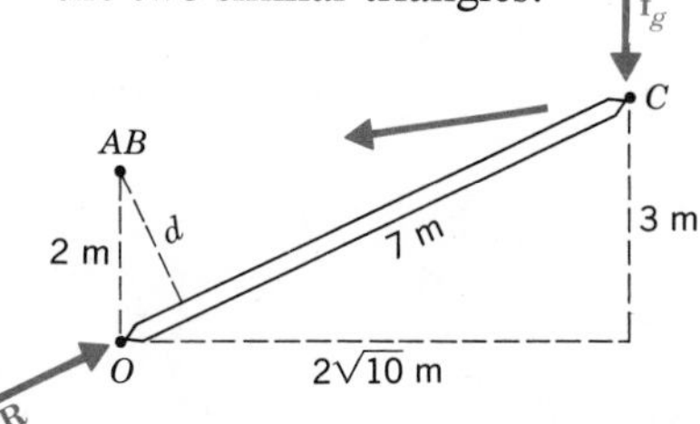

$$d = \frac{2\sqrt{10}}{7}(2\text{ m}) = \frac{4\sqrt{10}}{7}\text{ m}$$

Summation of moments about AB is then written as

$$R\left(\frac{4\sqrt{10}}{7}\text{ m}\right) - f_g\left(2\sqrt{10}\text{ m}\right) = 0$$

from which

$$R = 3.5f_g$$

Two- and Three-Force Members

In several of the preceding examples, we have dealt with bodies on which only two forces act. Such a body is called a *two-force member*. Force equilibrium requires that the forces acting on a two-force member have equal magnitudes and opposite directions. In addition, the forces must have a common line of action; for, otherwise, they would form a couple with a nonzero moment.

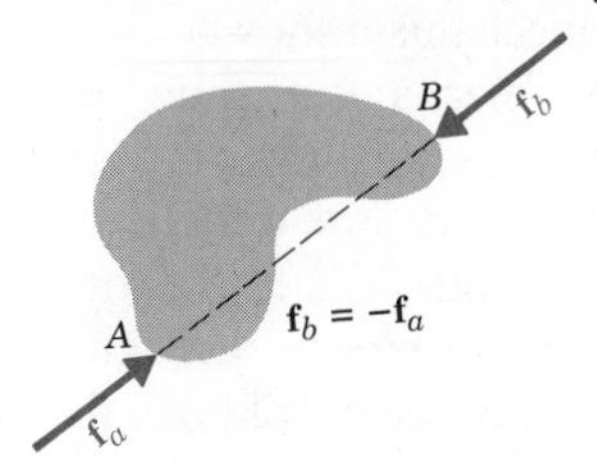

A *three-force member* is a body on which exactly three forces act. For such a body to be in equilibrium, *the lines of action of the forces must lie in a common plane and intersect at a common point*. For verification of this statement, consider the equilibrium of the three-force member shown in Figure 4-14. The resultant moment about point C must vanish:

$$\mathbf{M}_C = \mathbf{r}_a \times \mathbf{f}_a + \mathbf{r}_b \times \mathbf{f}_b = \mathbf{0}$$

so that

$$\mathbf{r}_b \times \mathbf{f}_b = -\mathbf{r}_a \times \mathbf{f}_a$$

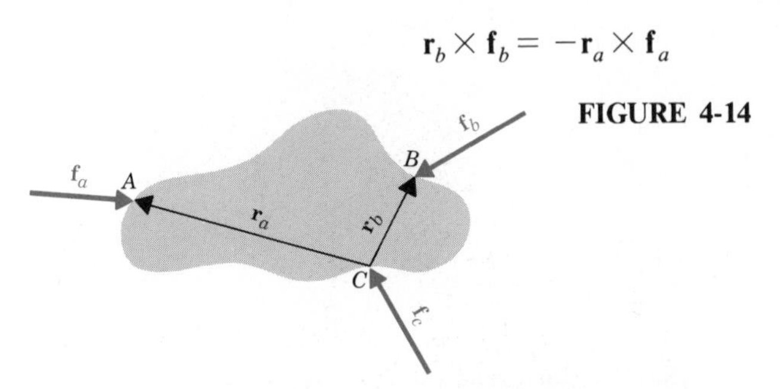

FIGURE 4-14

This relationship implies that the plane containing $\mathbf{r}_b$ and $\mathbf{f}_b$ must be parallel to the plane containing $\mathbf{r}_a$ and $\mathbf{f}_a$, because the equation states that vectors normal to each of these planes are equal. Furthermore, the planes must coincide, because each contains the point C. But the force $\mathbf{f}_c$ must also lie in this plane, since otherwise there would be a nonzero component of the resultant force perpendicular to this plane. Thus, we see that the lines of action of all three forces must be in the plane of points A, B, and C. Next, consider the resultant moment about the point of intersection of any two of the lines of action. If the line of action of the third force does not pass through this point, the resultant moment about this point cannot be zero. Thus, the three lines of action must pass through a common point. In the special case in which the three forces are parallel, we can consider the intersection point to be "at infinity."

EXAMPLE

In terms of P, evaluate the clamping force at E in the mechanism shown.

Free-body diagrams of the three parts of the mechanism show the reactions at the pin connections. Considering first the two-force member AB, we see that the force Q must be directed as shown, that is, its line of action is the line AB. Turning next to the handle, we see that this is a three-force member, so that the lines of action of P, Q, and R must pass through a common point. In terms of the directions of P and Q, which we now know, this

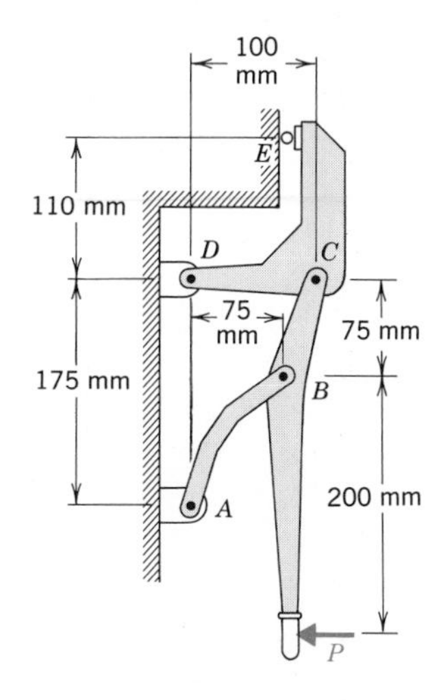

point of intersection is that shown as O. This determines the line of action of R, the location of which we can determine from the sketch showing the relative positions of O, A, B, and C (next page):
The distance from O to B is

$$OB = \sqrt{(200 \text{ mm})^2 + (150 \text{ mm})^2} = 250 \text{ mm}$$

and the angle BOC has the value

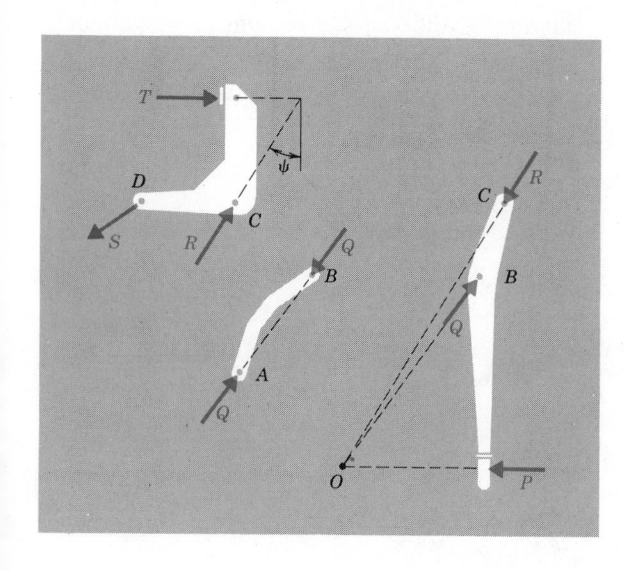

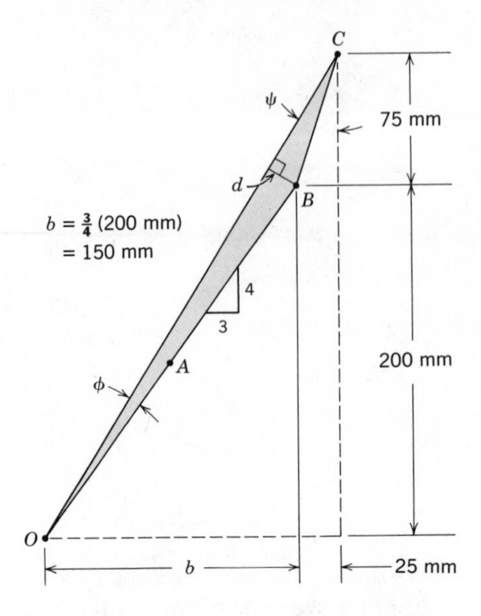

$$\phi = \tan^{-1}\left(\frac{275}{175}\right) - \tan^{-1}\left(\frac{4}{3}\right) = 4.3987°$$

so that the perpendicular distance from B to the line of action of R is

$$d = OB \sin\phi = 19.1741 \text{ mm}$$

Returning to the free-body diagram of the handle, we sum moments about point B:

$$M_B = R(19.1741 \text{ mm}) - P(200 \text{ mm}) = 0$$

which yields the magnitude of the reaction at C:

$$R = 10.4307P$$

Finally, we consider the equilibrium of the clamping arm DCE. Summing moments about D, we write

$$M_D = R\cos\psi(100 \text{ mm}) - T(110 \text{ mm}) = 0$$

which yields

$$T = \frac{10}{11} R\cos\psi$$

$$= \frac{10}{11}(10.4307P)\cos\left[\tan^{-1}\left(\frac{175}{275}\right)\right]$$

$$= 8.00P$$

Note the steps, each carefully taken, that are common to all the problem solutions of this section: Free-body diagrams are drawn, known and unknown forces are identified and labeled, and equations of equilibrium for each free-body are written. These steps reduce the problem to one of computation, such as the solution of simultaneous algebraic equations.

The ability to readily apply these steps to new systems comes from *practicing* them; little can be gained at this point by following additional worked-out examples. As you practice these steps, keep the following in mind.

Many different equivalent sets of equations are possible, depending on which sets of objects are isolated as free-bodies, which directions are selected for vector resolutions, and which points or axes are selected to compute moments. Therefore, if an analysis leads to a large set of equations with many unknowns, it is worth considering a different free-body, a different component resolution, or a different point or axis for moment summation. For example, we could write three equations of equilibrium for the free body in Figure 4-9, containing the unknown force T_{AB} and the two reaction components at O. However, selection of point O for moment summation results in an equation that does not contain these last two unknown components, so that T_{AB} is directly solvable without recourse to additional equations.

PROBLEMS

4-19 Show how equation c (p. 117) may be written directly from a single equilibrium consideration of one of the free bodies in Figure 4-3.

4-20 Draw a free-body diagram of the mountain climber (p. 21) and evaluate the force exerted by the mountain on the climber's feet.

4-21 Evaluate the reaction at O for the example on page 119.

4-22 Evaluate the reaction at the wall induced by the 240-lb plumber of Problem 2-62.

4-23 Evaluate the force at the pin B in the device of the example on page 128.

4-24 Evaluate the reaction at D in the device of the example on page 128.

4-25 The surfaces are smooth where the drum makes contact. Determine the reactions at these points. *Ans.* 200 lb; 346 lb.

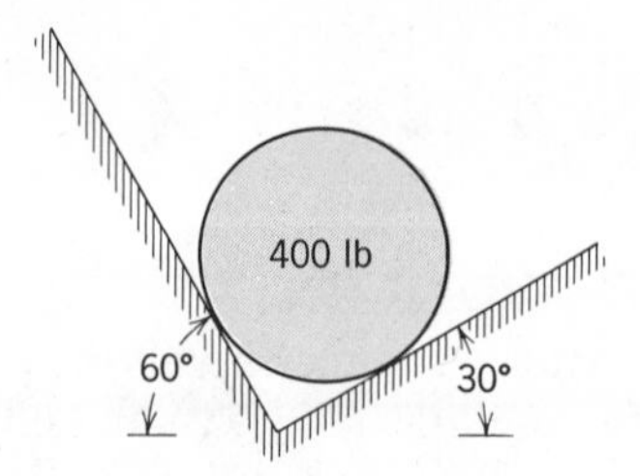

4-26 The surfaces of the bars are smooth where the drum makes contact. Determine the reactions at A, B, and C.

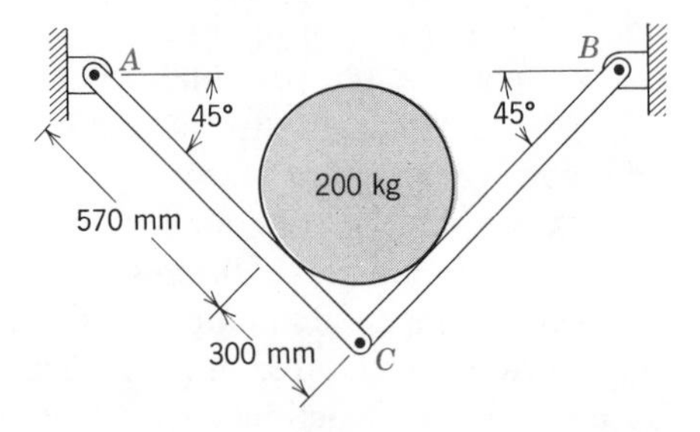

4-27 The mass of the rod AB is negligible compared with m. The rod and walls are smooth at A and B. Evaluate the tension in the line CD and the reactions at A and B.

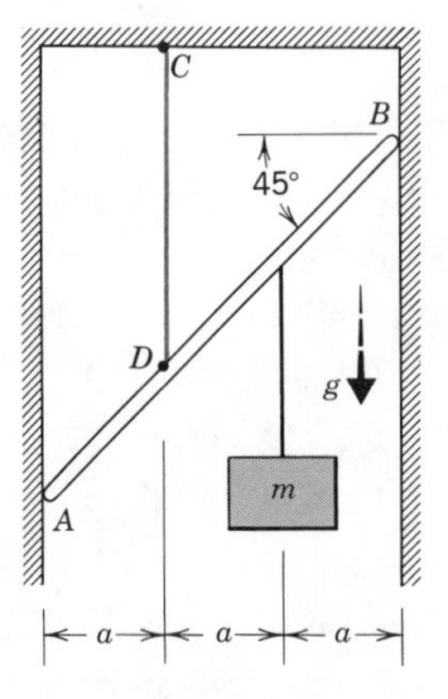

4-28 The surfaces are smooth where the drum makes contact with the bars. The bars form a 90° angle at the pin connection A, and have negligible mass compared with that of the drum. Determine the length l.

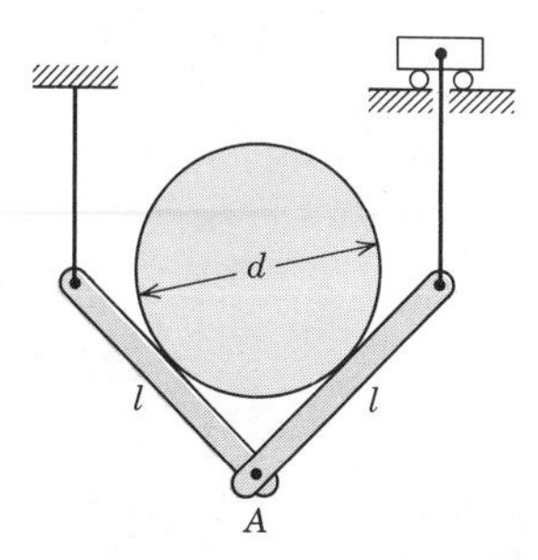

4-29 Two wheels, each of radius a but of different mass, are connected by the rod of length R. The assembly is free to roll in the circular trough. Determine the angle θ for equilibrium.

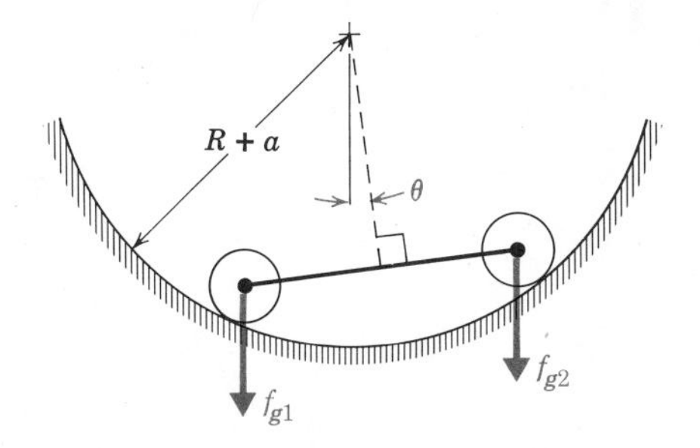

4-30 What force P is necessary to begin pulling the 40-lb, 30-inch diameter wheel over the curb?

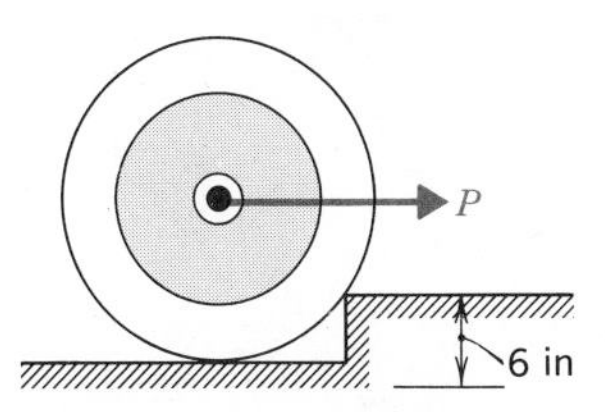

4-31 Show why the tension on either side of the pulley is the same.

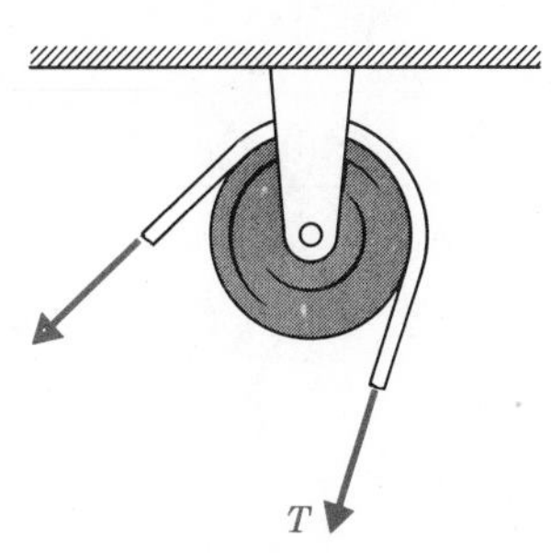

4-32 Determine the tension force T in terms of the gravitational force f_g.

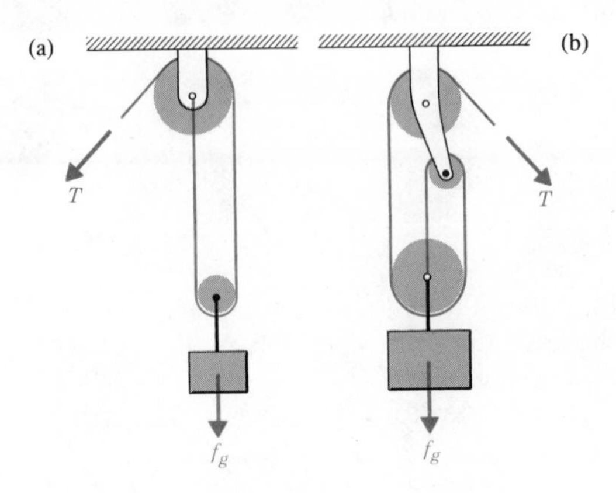

4-33 For a given force T, how much load can be lifted by the block and tackle?

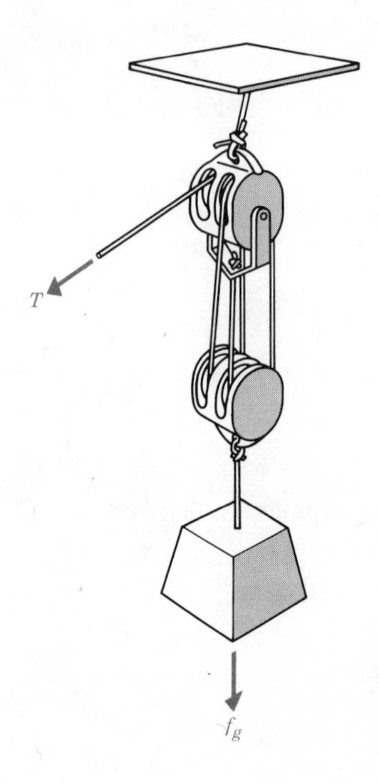

4-34 Determine the force with which the 80-kg man must pull on the rope in order to support himself. *Ans.* 157 N.

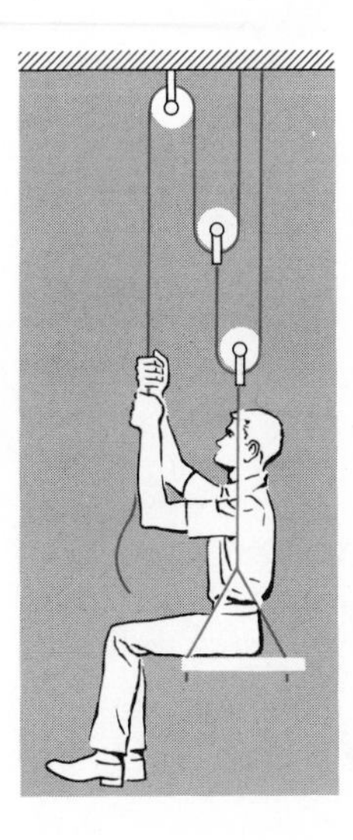

4-35 For equilibrium of the system, what must be the mass m and the tension T? The mass of each suspended pulley is negligible.

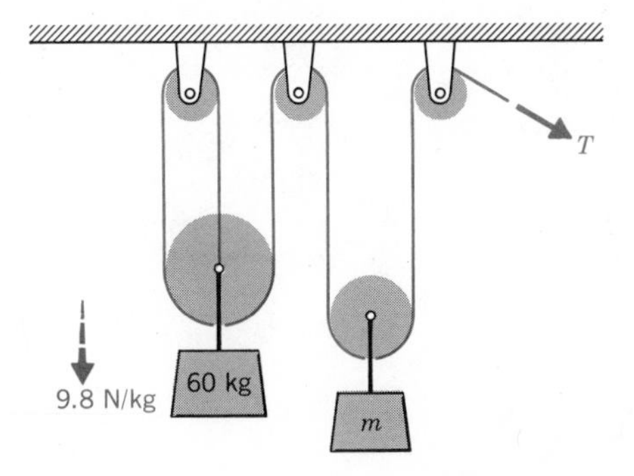

4-36 Determine the angle α and the tension T in terms of f_g, for equilibrium of the system. The mass of the suspended pulley is negligible.

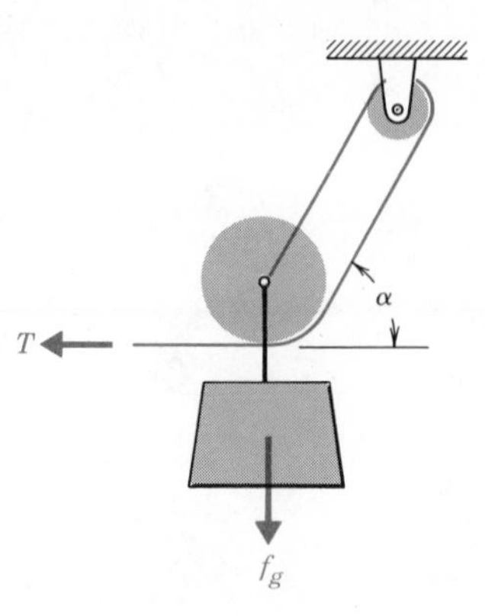

4-37 Each of the tracks in the upper pulley unit is recessed to fit the chain, so as to prevent slipping. The smaller track has a radius equal to 0.9 times that of the larger track. Evaluate the force P necessary to lift the block by means of the differential chain hoist. *Ans.* 0.392 kN.

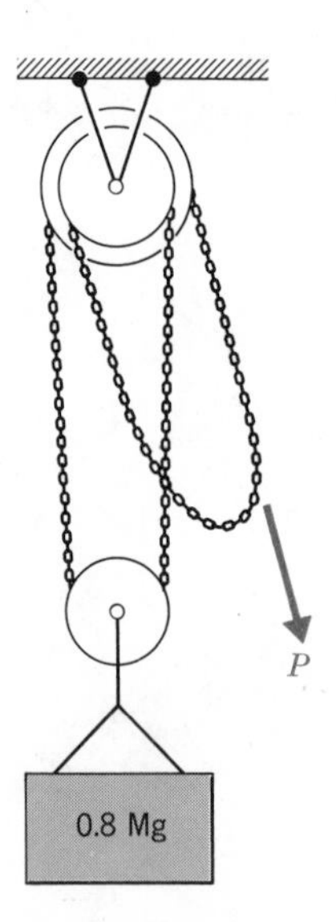

4-38 Neglecting gravity forces on the structural members, evaluate all reactions between bars and at the supports.

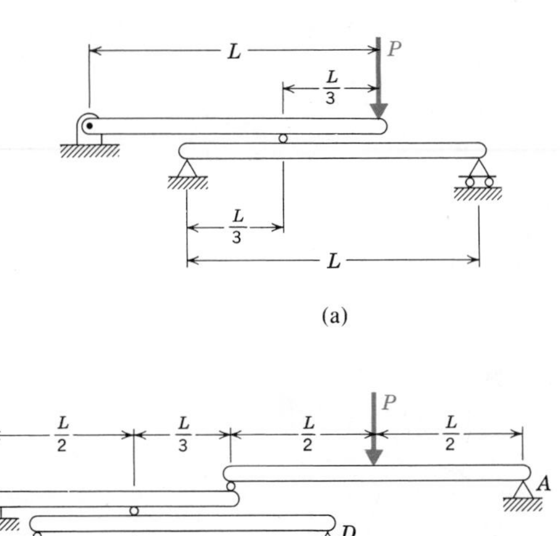

(a)

(b)

4-39 Determine the tension in the cable and the reaction at A. The mass of the structure is negligible compared with that of the 1500-lb load.

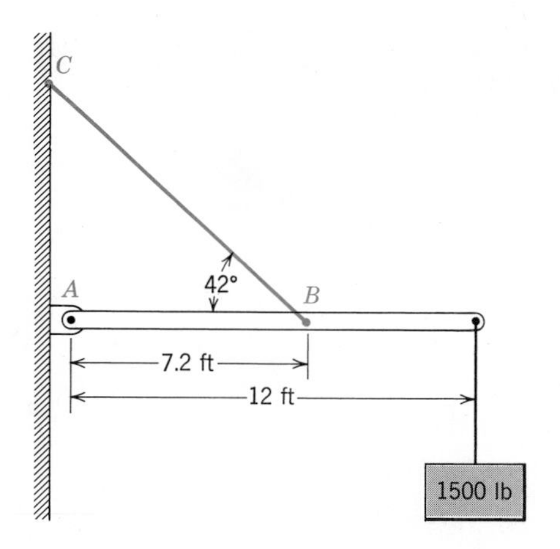

4-40 Neglecting the mass of the structure, evaluate the force in the member A.

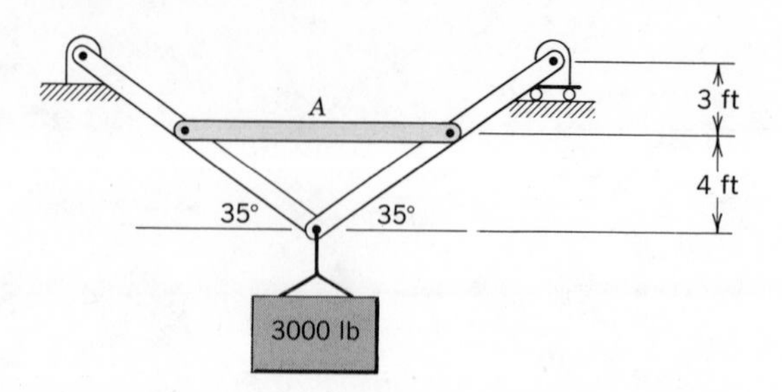

4-41 The mass of the linkage is negligible compared with m. What is the value of P necessary to maintain equilibrium?

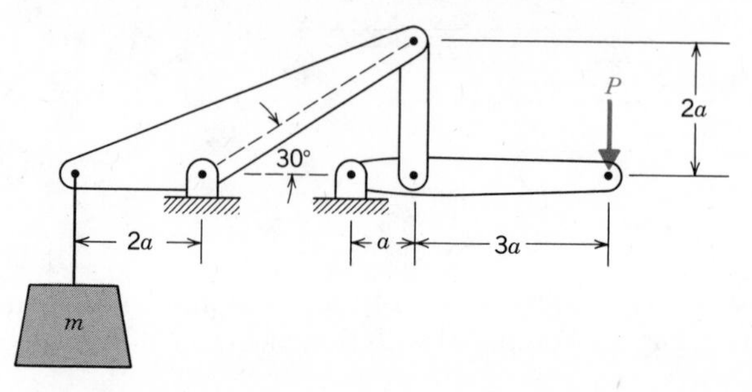

4-42 Three identical uniform bars, each of mass m, are hinged at A, B, and C. The end D is guided to slide in the vertical slot with negligible friction. In terms of θ, m, and g, what are P and Q such that the bar BC is vertical? *Ans.* $\frac{5}{2}\, mg \tan\theta$; $\frac{1}{2}\, mg \tan\theta$.

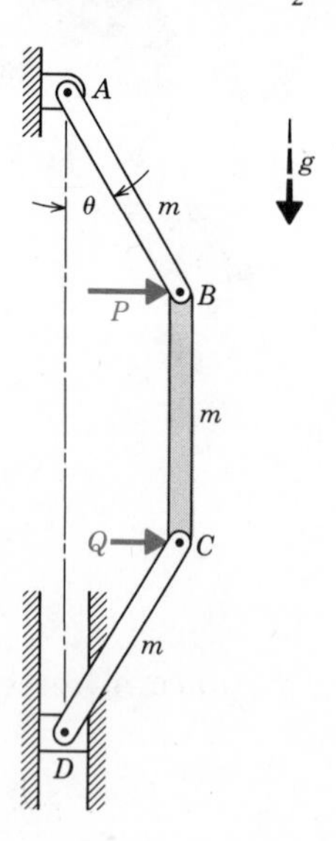

4-43 Neglecting the mass of the structure, determine the force at each pin connection.

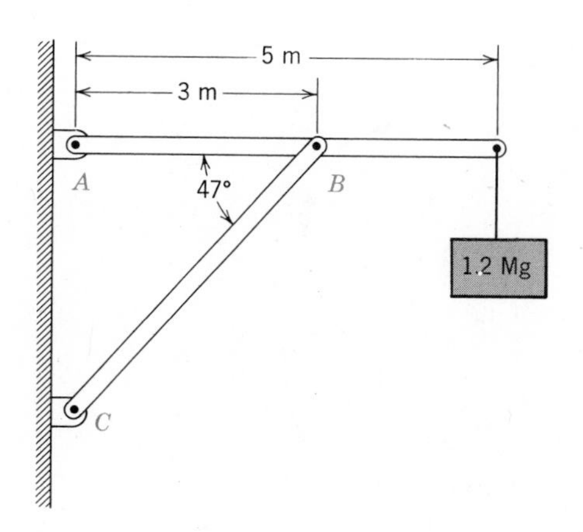

4-44 Neglecting the mass of the structure, determine the reactions at A, B, and O, induced by the 700-lb lifeboat.

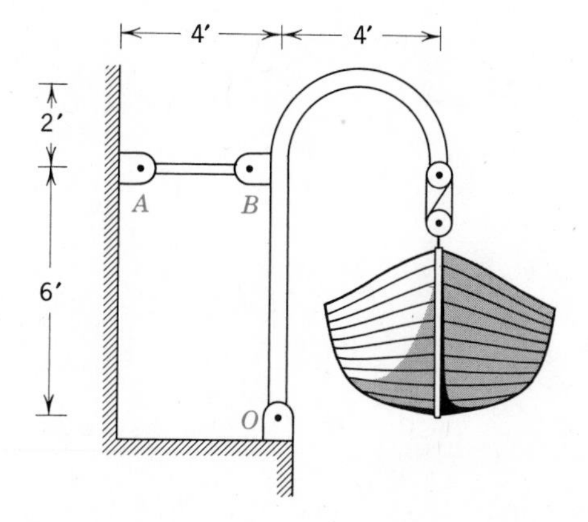

4-45 Neglecting the mass of the structure, evaluate the reactions at the supports for each case.

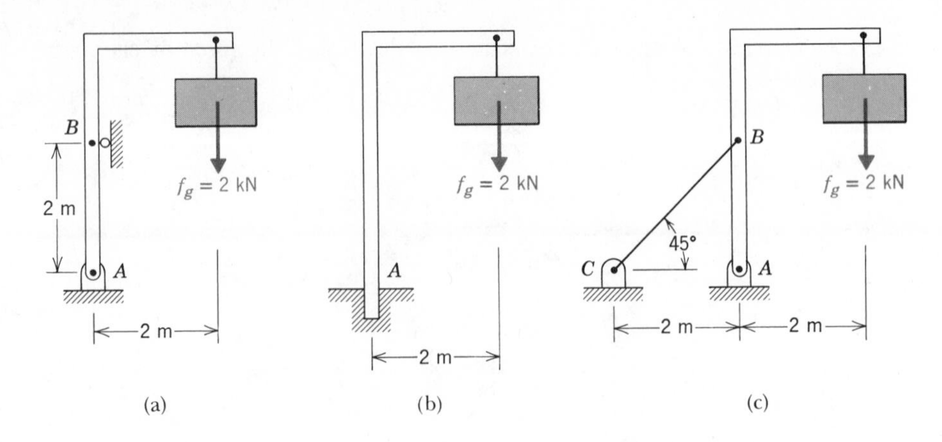

4-46 Determine the tension in the cable AB and the reaction at C. The mass of the boom is negligible compared with that of the 0.5-Mg load.

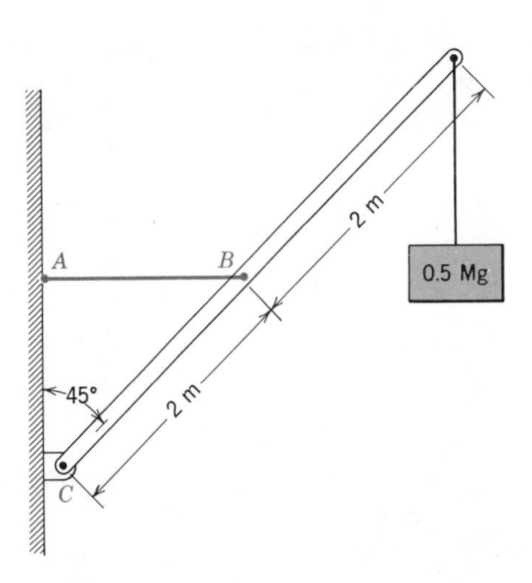

4-47 The frame structure shown supports a 1.2-Mg load, which is suspended by a cable that runs over the 80-kg pulley and is attached at E. Neglecting the masses of the members of the frame, determine the reactions at A, B, C, and D.

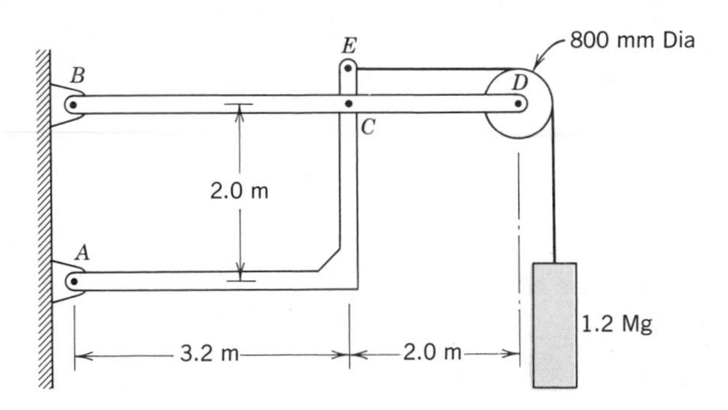

4-48 Evaluate the tension in the cable AB and the reaction at C. *Ans.* 203.5 kN; 368.3 kN.

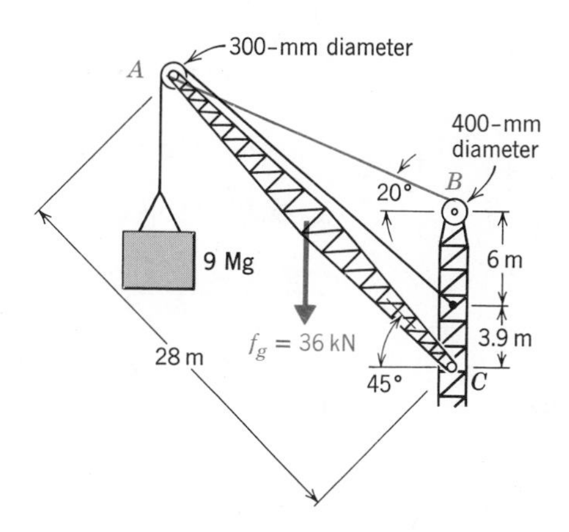

4-49 The maximum load that the dockyard crane must handle is $f_g = 20\ 000$ lbf. The gravitational forces acting on the boom itself are represented by the 800-lb force at mid-span. Evaluate the tension in the cable and the reaction at O, and show how these vary with x.

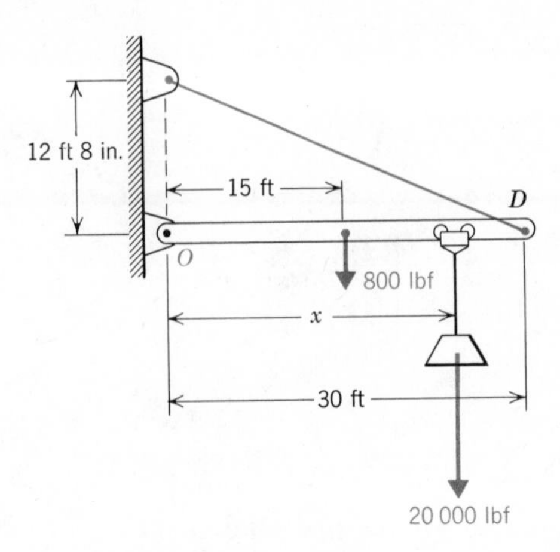

4-50 Show that the moment M of the couple acting on the crank must equal Pe in order for the system to be in static equilibrium.

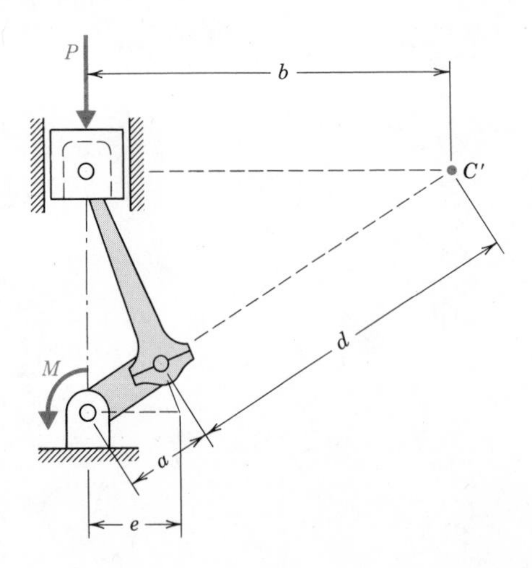

4-51 The piston slides with negligible friction inside a cylinder (not shown). Determine the reactions at the crank bearings, and the moment M of the couple acting on the crankshaft for equilibrium.

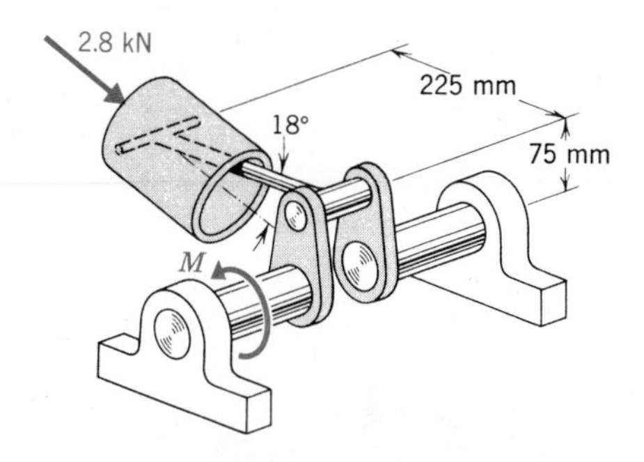

4-52 The gravitational forces acting on the cart and its load of coal are represented by the 600-lb force acting as shown. Determine the tension T required to hold the cart, and the reactions at the wheels.

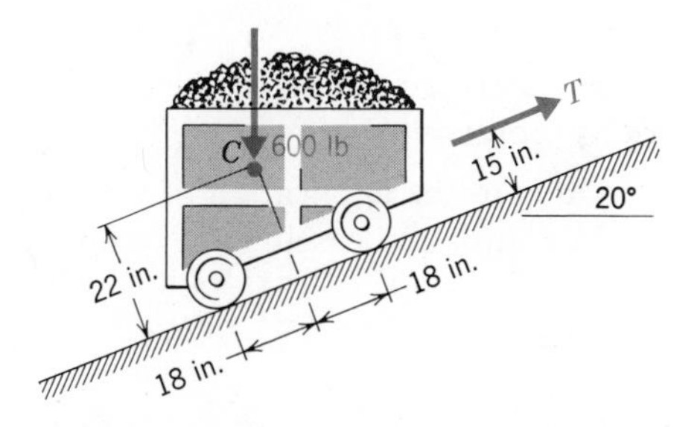

4-53 Evaluate the reactions at the trailer hitch and at each wheel.

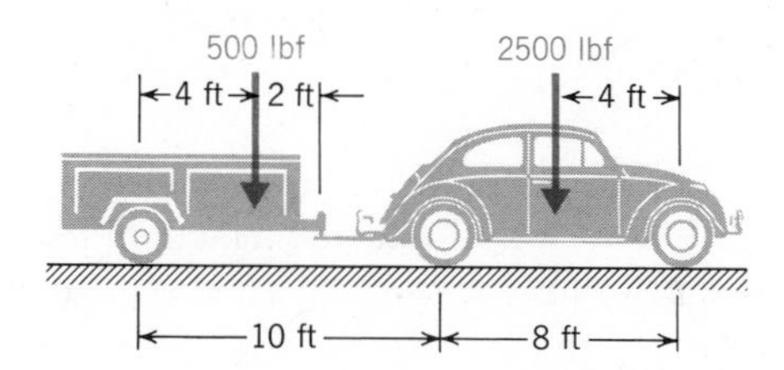

4-54 In the trailer "load-leveler" hitch, the angle bar slips into the cylindrical socket at A, forming a thrust bearing where it bottoms. The end B is then attached by a short chain to the towing vehicle. The pretension in the chain is 1.7 kN. Evaluate the reactions at A, C, D, and E. *Ans.* 1.7 kN, 0.986 kN·m; 5.25 kN; 8.05 kN; 6.20 kN.

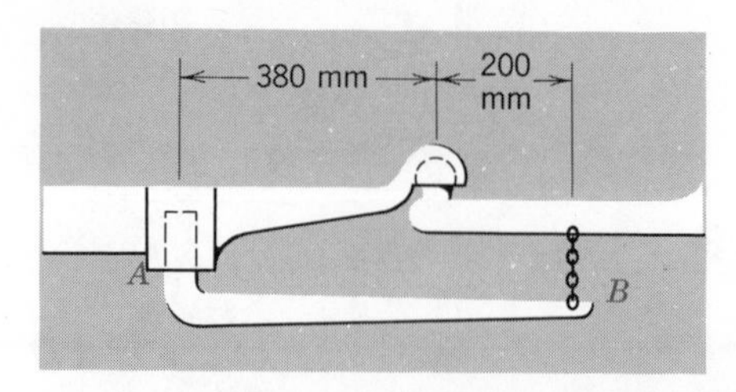

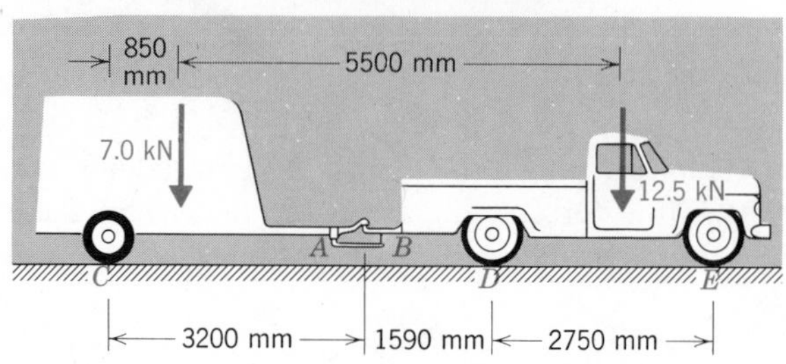

4-55 What changes in reactions at the wheels are caused by the pretensioning of the chain at B of the previous problem?

4-56 Compare the response of the system to the application of the couple at each of the two places. Analyze the situation carefully in terms of basic laws of equilibrium.

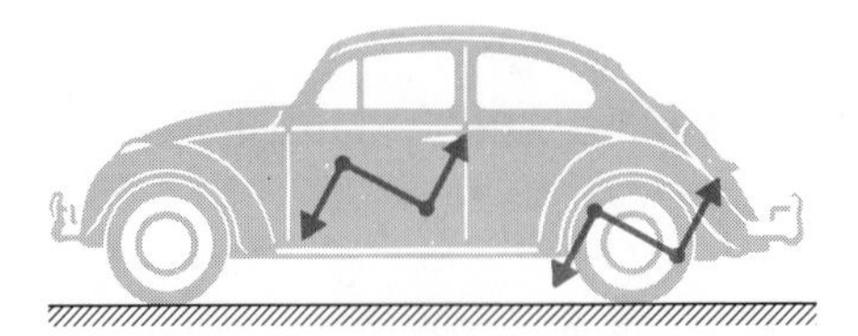

4-57 The tension in the chain of Problem 4-18 is 1200 lbf and the gravity force has a magnitude of 2700 lbf. Evaluate the reaction from the ground on each wheel and the torque transmitted to each rear wheel through the axle.

4-58 Assuming that there is sufficient friction to prevent slipping of the drive wheels of Problem 4-18, how much tension can be transmitted through the chain before the front wheels of the tractor lift off the ground? What would be the corresponding torque transmitted to the rear wheels through the axle?

4-59 The speed reducer consists of the input shaft S_1 and attached small gear A, which drives gear B and its output shaft S_2 with a 2 : 1 reduction.

(The radius of B is twice that of A.) A clockwise input torque of 80 N·m is applied to the shaft S_1 and the output shaft S_2 drives a machine at constant speed. What will be the horizontal components of the reactions at the supports C and D?

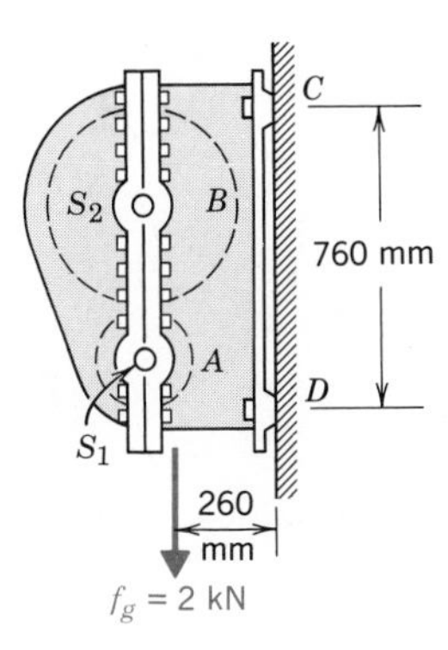

4-60 The long-reach limb trimmer is actuated by a 50-lb pull on the rope. With what force is the blade cutting into the limb at C?

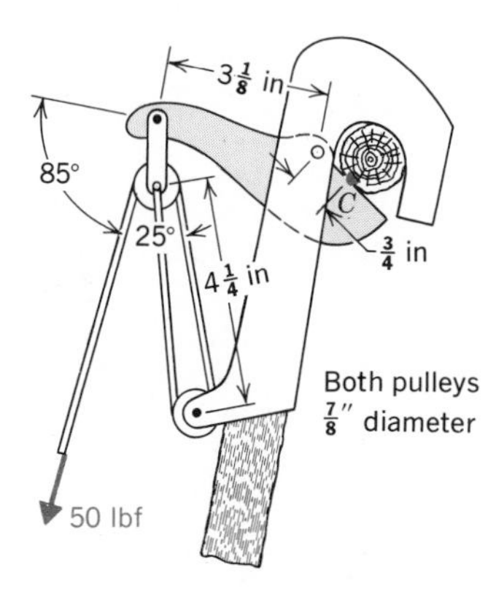

4-61 What is the punching force at C in terms of the force P on the handles?

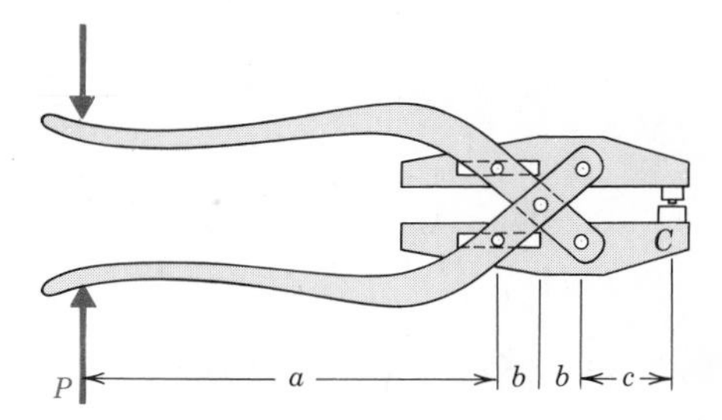

4-62 Evaluate the cutting force at C in terms of the force P on the handles of the compound snips.

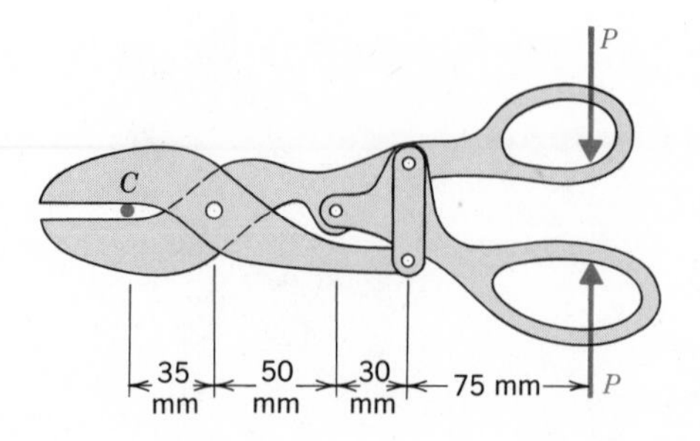

4-63 The reloading press consists of the ram, which is guided to slide inside the frame, with the linkage connecting the ram to the handle. What is the ratio of the force that drives the shell case into the die to the force P at the handle? Sketch a curve showing how this ratio varies with θ.

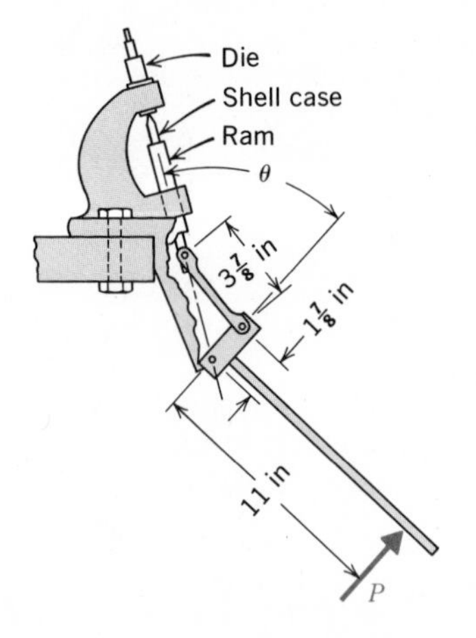

Ans. $$\frac{R}{P} = \frac{5.867}{\sin\theta\left(1 + \dfrac{\cos\theta}{\sqrt{4.271 - \sin^2\theta}}\right)}.$$

4-64 What is the force on the nut at C exerted by the jaws of the vise-grip wrench?

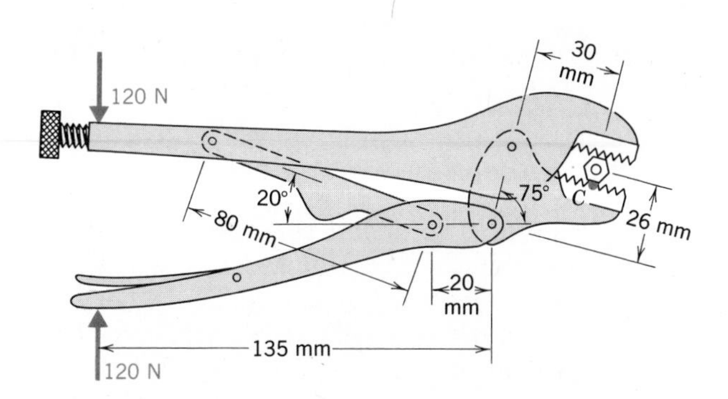

4-65 The loader mechanism is operated by two identical hydraulic cylinders A and two identical hydraulic cylinders B. Evaluate the force transmitted by each cylinder.

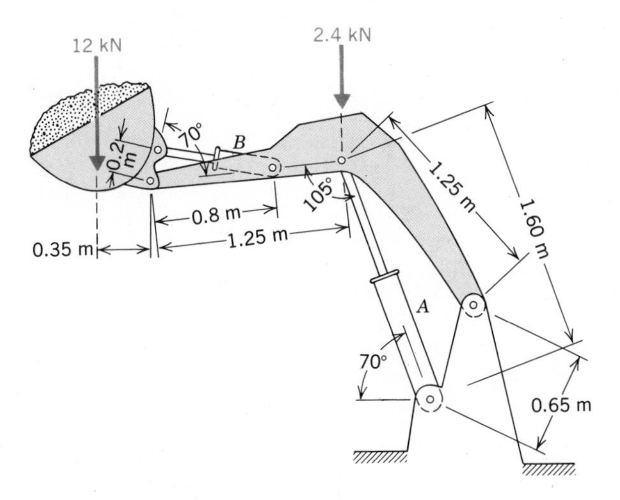

4-66 Determine the reaction at each bearing and the moment of the couple transmitted by the shaft at section A.

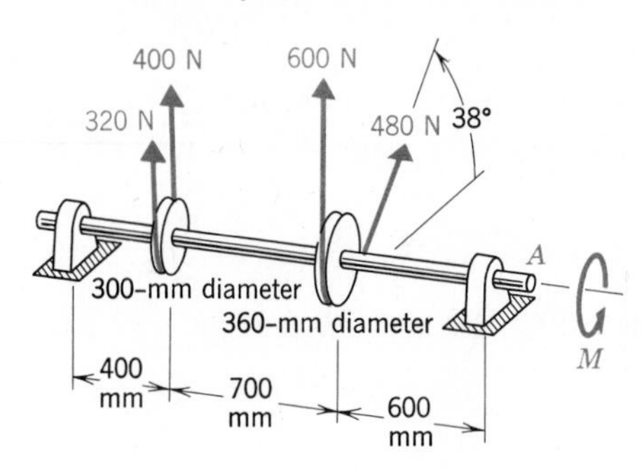

4-67 The tension in cable $O'R$ in Figure 3-3 is equal to 700 lb. Determine the tensions in the cables $O'P$ and $O'Q$.

4-68 The slider A has a mass of 4 kg and is constrained to slide without friction along the fixed vertical rod. The mass of the wire AB is negligible. The slider B is constrained to slide along the horizontal rod without friction. What must be the magnitude F of the force applied to the slider B to maintain equilibrium?

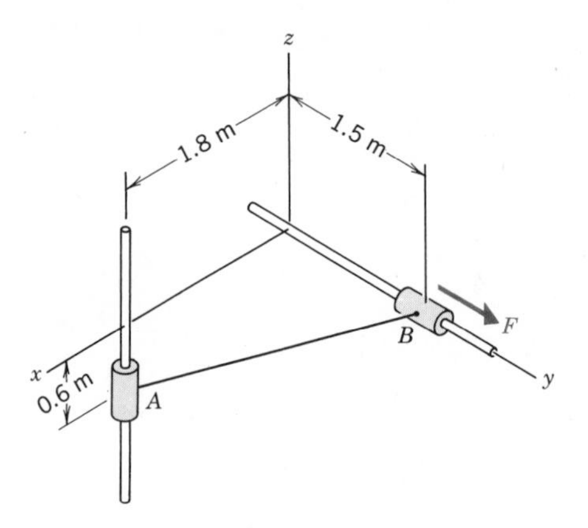

4-69 Write out the equations $f_x = 0$, $f_y = 0$ and $f_z = 0$ for the example on page 124. Verify that these are satisfied by the values of P, Q, and R computed in the example.

4-70 Calculate the force in cable OA by using equilibrium about the axis BC for the example on page 123.

4-71 Evaluate the tensions in the cables AD and BD and the force in the rod CD.

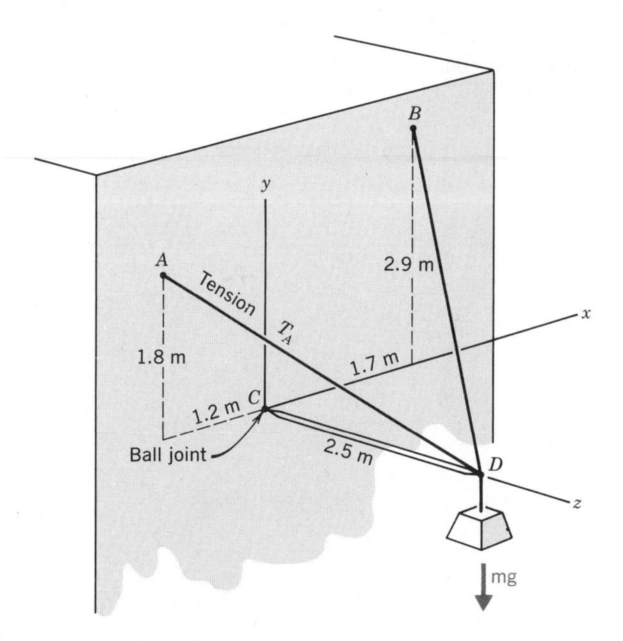

4-72 The turnbuckle in the guy line AE is to be tightened such that the vertical component of the reaction at O is 800 N. What must be the tension in AE?

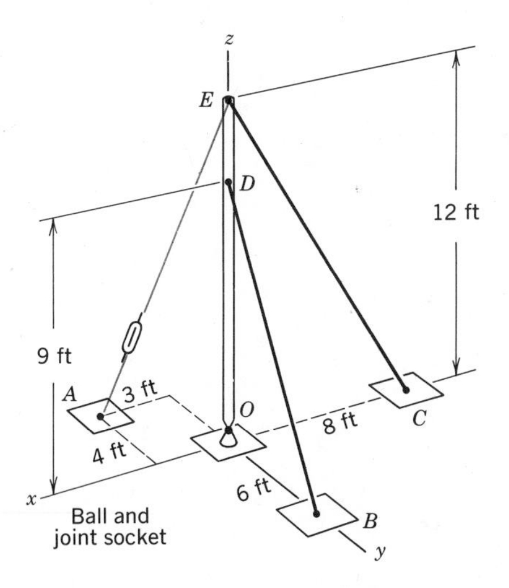

4-73 Referring to the guyed pole of the previous problem, if the tension in the cable AE is 400 N, what will be the reaction at O and the tensions in the other two cables?

4-74 Evaluate the tensions in the guy lines AP and BQ, supporting the mast.
Ans. 0.228 kN; 0.709 kN.

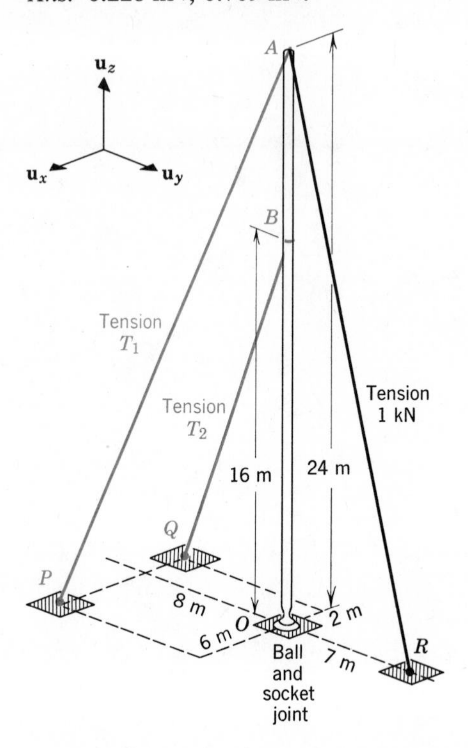

4-75 Determine the tensions in the cables AB and CD.

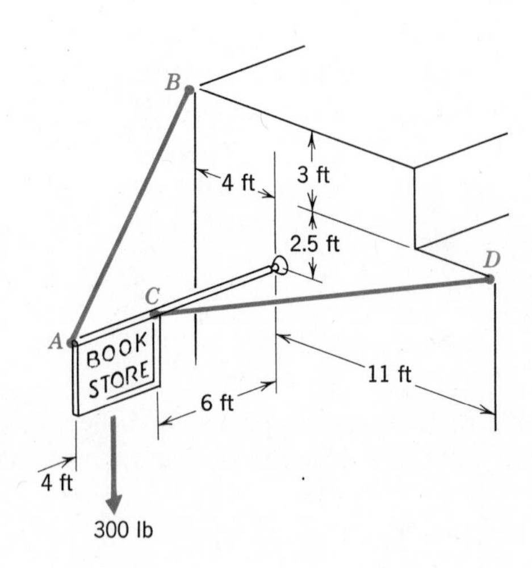

4-76 Evaluate the tensions in each of the cables supporting the boom in Problem 3-90.

4-77 The 30-kg trapdoor is lifted by the corner as the illustration shows. Evaluate the reactions at the hinges at A and B. f_g acts through the geometric center of the plate. *Ans.* -36.8 N; 183.9 N.

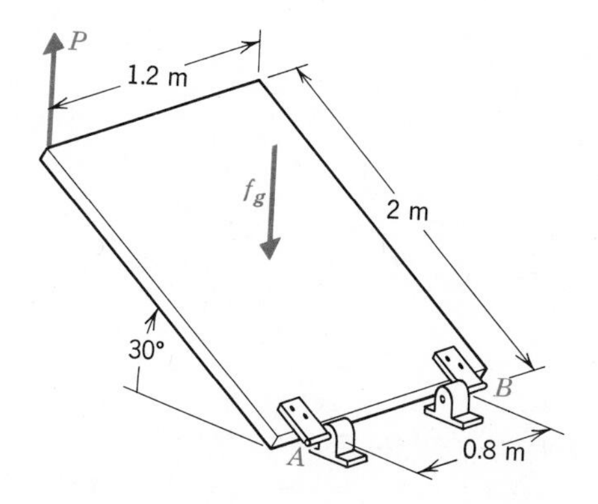

4-78 Determine the reactions at the hinge supports at A and B. The hinge at B supports all of the force acting along the horizontal axis AB. f_g acts through the geometric center of the plate.

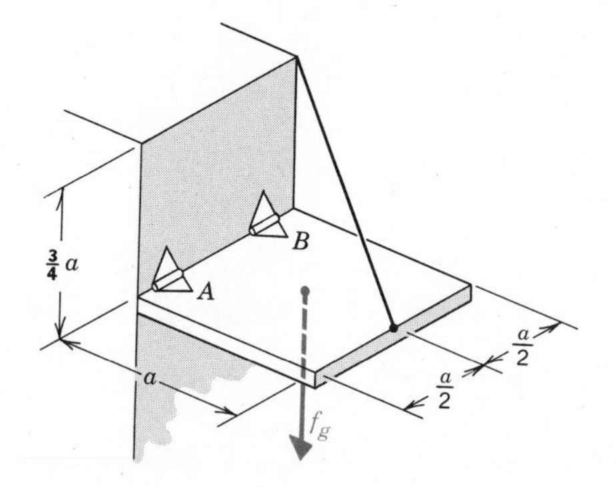

4-79 The 40-kg rectangular plate is held by hinges along its edge OA and by the wire BD. The gravity force f_g acts through the geometric center of the plate. What is the tension in the wire?

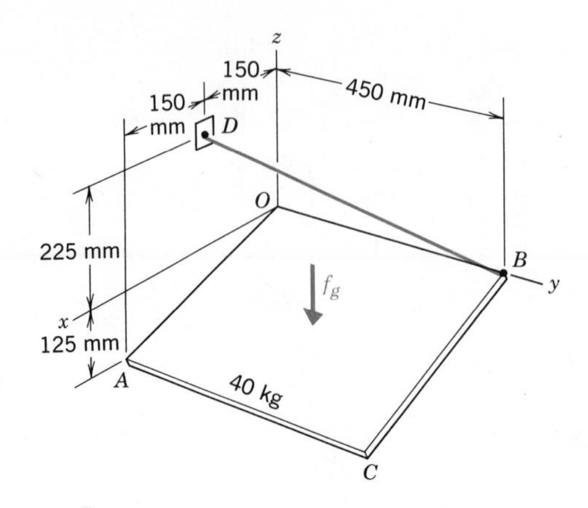

4-80 Each connection is a ball joint. Neglecting gravity forces on the structural members, evaluate the forces at all joints.

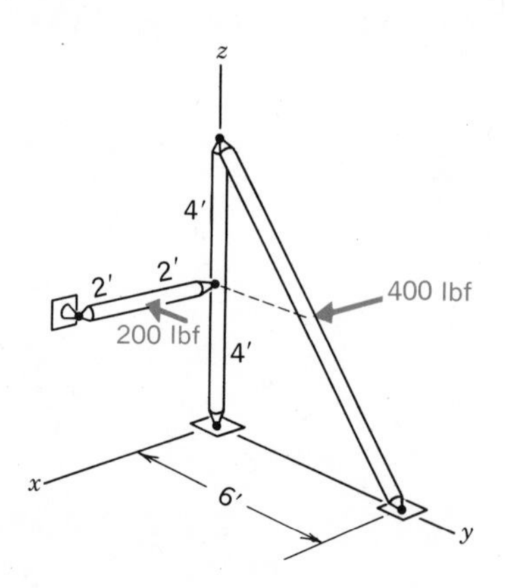

4-81 Each connection is a ball joint. Neglecting forces of gravity on the bars, evaluate the forces at all joints.

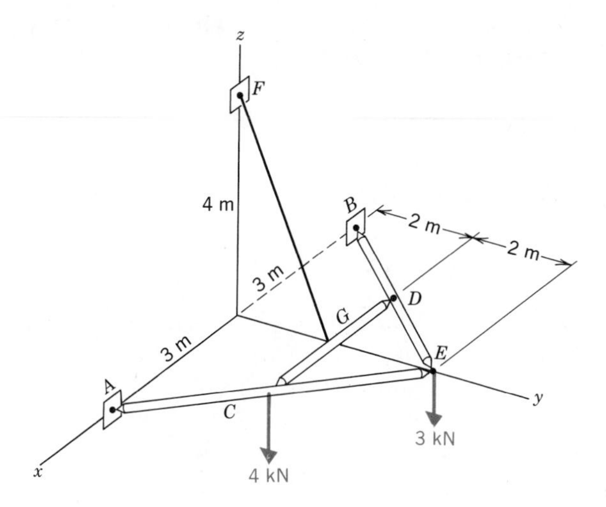

4-82 The force P acts perpendicular to the arm CD and to the drum axis AB. Evaluate the magnitude of P and the reactions at the bearing supports A and B.

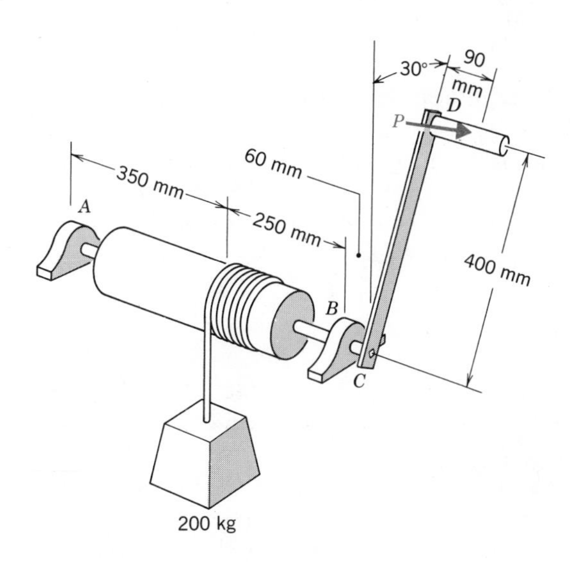

4-3 STATICALLY INDETERMINATE AND IMPROPER SUPPORT SYSTEMS

Each of the bodies considered in the preceding examples is connected with its surroundings in such a way that it is possible to deduce the forces at these connections from equations of equilibrium.

Redundant Supports

If the degree of support is increased, this will generally increase the number of possible reaction components above the number of independent statics equations that can be written. The support system is then said to be *statically indeterminate*. For example, if the roller support at D in Figure 4-10 (p. 121) were replaced by the pinned type shown at E, two reaction components would be possible there in addition to the two at E. Since planar equilibrium of the free-body in Figure 4-11a can yield only three independent equations, the four reaction components could not be determined from statics alone.

To determine the forces acting on a body supported by a statically indeterminate system, *it is necessary to abandon the rigid body idealization and account for deformations of the body*. Such deformations are a primary focus of studies in the mechanics of deformable bodies (or "strength of materials") and will not be considered here.

Deficient Supports

If the support system is incapable of preventing motion when certain loads are applied, it is said to be *improper*. For example, if the support at E in Figure 4-10 were replaced by a roller type as at D, supports would be incapable of preventing horizontal motion.

Criteria

Usually the best procedure for determining whether a support system is indeterminate or improper or both is to consider the following tests.

(a) The support system is statically indeterminate if some degree of support can be removed, leaving the body still capable of withstanding any applied loads; or, equivalently, if reactions from the supports can exist in the absence of other applied forces.

(b) If the support system is incapable of preventing motion induced by all applied forces, additional support must be added to make the system proper.

The answers to these tests are nearly always obvious after simple inspection. However, there are situations where a more formal procedure may be needed. It can be shown that necessary and sufficient conditions for a proper support system for a rigid body are that it be capable of transmitting six forces with lines of action that cannot be intersected by a straight line.* In the planar case, the support system must be capable of transmitting three forces with lines of action that do not intersect at a common point. In dealing with clamped supports, where the reaction may be represented by a force and a couple, the criteria can be applied by considering the force-couple combination as an equivalent pair of forces as indicated in Fig. 4-5 (See p. 107).

Any supports in addition to those that meet these requirements are redundant, and their presence means that the support system is statically indeterminate.

PROBLEMS

4-83 Is the support system for the A-frame in Figure 4-10 improper?

4-84 Do the cables and ball joint at the deck form a statically determinate support system for the mast in Figure 3-3?

4-85 Are the supports for the trailer and for the car of Problem 4-53 statically determinate? Are the supports for the trailer and for the car of Problem 4-54 statically determinate?

4-86 Are the supports for the pole of Problem 4-72 statically determinate?

4-87 Which of the following support systems are statically indeterminate, and which are improper? Consider only forces in the plane of the diagram.

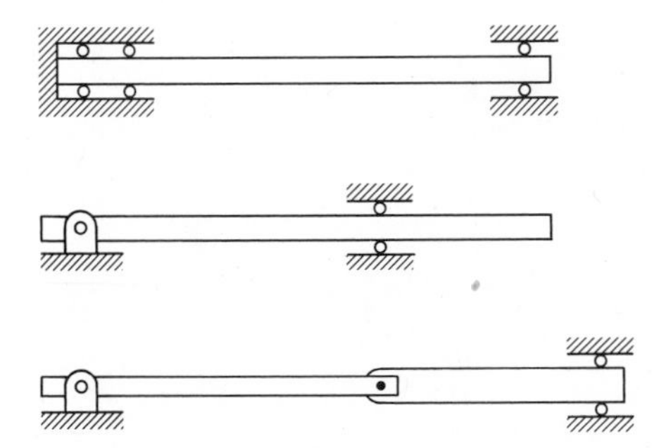

*G. W. Housner and D. E. Hudson, *Applied Mechanics Statics*, Second Edition, D. Van Nostrand Co., New York, 1961.

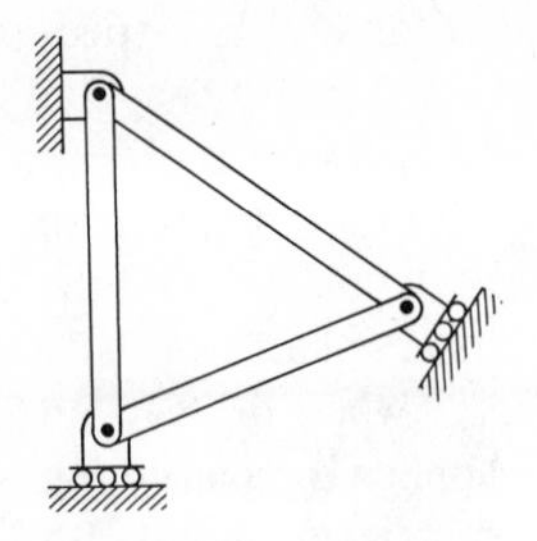

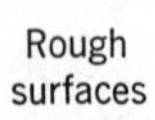
Rough
surfaces

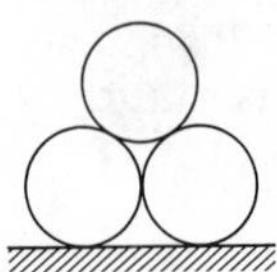

5

SOME SPECIAL APPLICATIONS

Most mechanical devices, structures, and machines have features that lead naturally to classification of some kind. To group them accordingly as our study proceeds is appropriate, because many of these special features are worthy of concentrated attention. However, important as they are, the special topics discussed in this and the next chapter are secondary to the basic equilibrium principles. Therefore, the most important reason for considering them now is to provide a further reinforcement of skills for drawing and labeling of free-body diagrams and for writing and solving equations of equilibrium.

5-1 TRUSSES

Trusslike structures are used to support bridge decks, roofs, power lines, and many other loads.

A truss is a structure that is built up with interconnected *axial force members*. Such a member is a straight rod that can transmit force along its axis but it does not carry a couple. The limitation to axial force results from the fact that the interconnections within a truss are of the ball and socket type; that is, they constrain the end points of the connected members against relative position change, but allow the members complete freedom to rotate about the connection point. Also, external forces are applied only at these joints.

Each member, then, is subjected to only two forces, one at each end. From moment equilibrium it follows that the lines of action of the two forces must coincide with the line connecting the points of application.

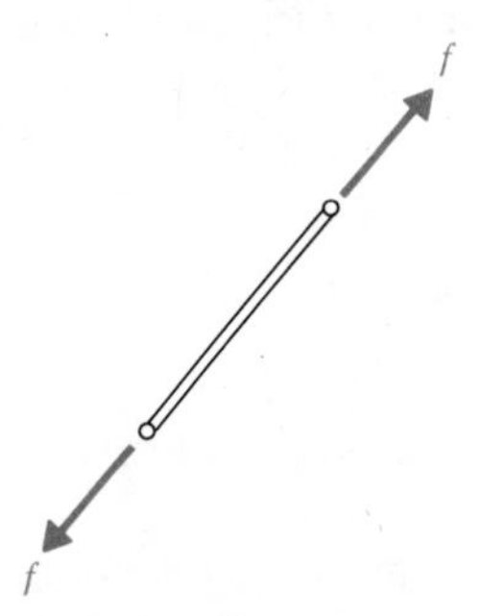

Real trusslike structures normally have joints that can actually transmit couples and transverse components of forces to the members, as indicated by the dashed arrows in the sketch on the next page. However, when the members are slender and the loads are applied only near the joints, the idealization as a truss usually leads to very good estimates of the internal forces. And because the idealization results in significant simplification of analysis, it is usually worthwhile.

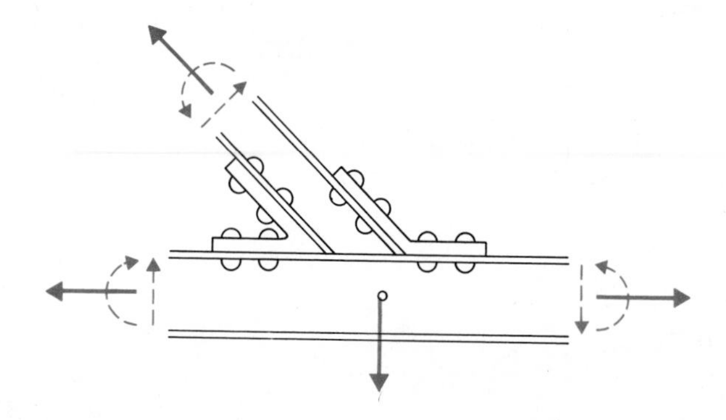

Notation

In analyzing trusses, it is convenient to adopt a uniform notation for specifying the forces within the structure. Here we specify the various joints with letters, A, B, C, and so forth, and the *tensile* force in the member connecting the joints I and J as T_{IJ} (I, $J = A, B, C, \ldots$). Thus if the value of T_{IJ} is positive, the bar is in tension, whereas if the value of T_{IJ} is negative, the bar is in compression. This convention must be kept in mind as free-body diagrams are labeled, and as results are interpreted.

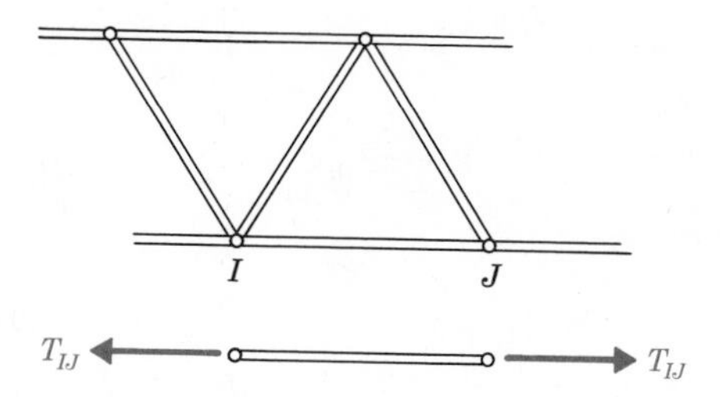

Equations from Joints

One approach to determining the forces in the individual members within a truss is to isolate as a free-body the portion of the structure in the neighborhood of each joint, and to write equations of force equilibrium for each of them. This procedure is illustrated in terms of the plane truss depicted in Figure 5-1. The free-body diagrams and corresponding equations of horizontal and vertical force equilibrium are as follows. The direction cosines used to evaluate horizontal and vertical components of the forces in the members have been evaluated from the dimensions given in Figure 5-1.

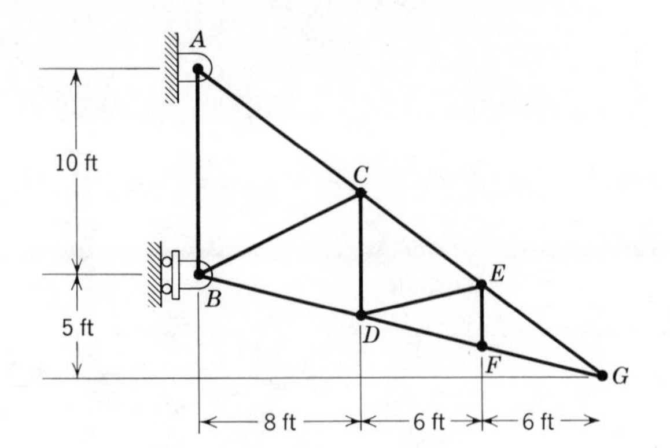

FIGURE 5-1

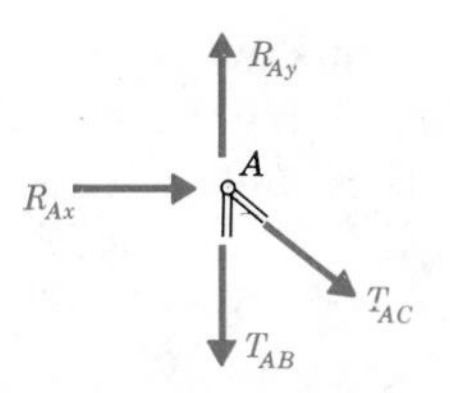

$$R_{Ax} + \frac{4}{5} T_{AC} = 0$$

$$R_{Ay} - T_{AB} - \frac{3}{5} T_{AC} = 0$$

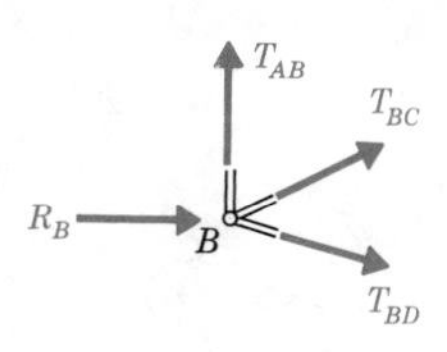

$$R_B + \frac{2}{\sqrt{5}} T_{BC} + \frac{4}{\sqrt{17}} T_{BD} = 0$$

$$T_{AB} + \frac{1}{\sqrt{5}} T_{BC} - \frac{1}{\sqrt{17}} T_{BD} = 0$$

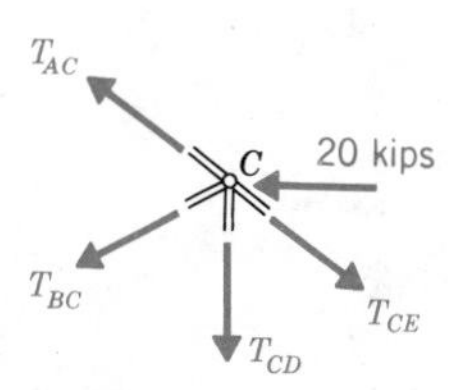

$$-\frac{4}{5} T_{AC} - \frac{2}{\sqrt{5}} T_{BC} + \frac{4}{5} T_{CE} = 20 \text{ kips}$$

$$\frac{3}{5} T_{AC} - \frac{1}{\sqrt{5}} T_{BC} - T_{CD} - \frac{3}{5} T_{CE} = 0$$

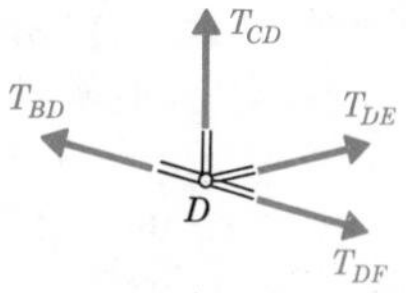

T_{CD}, T_{BD}, T_{DE}, D, T_{DF}

$$-\frac{4}{\sqrt{17}} T_{BD} + \frac{4}{\sqrt{17}} T_{DE} + \frac{4}{\sqrt{17}} T_{DF} = 0$$

$$\frac{1}{\sqrt{17}} T_{BD} + T_{CD} + \frac{1}{\sqrt{17}} T_{DE} - \frac{1}{\sqrt{17}} T_{DF} = 0$$

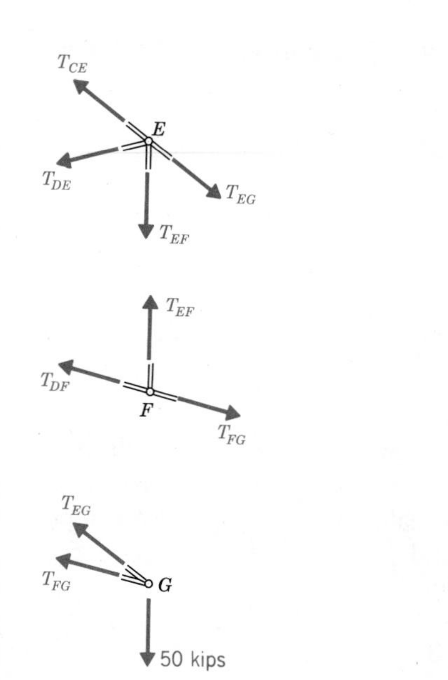

$$-\frac{4}{\sqrt{17}}T_{DE}-\frac{4}{5}T_{CE}+\frac{4}{5}T_{EG}=0$$

$$-\frac{1}{\sqrt{17}}T_{DE}+\frac{3}{5}T_{CE}-T_{EF}-\frac{3}{5}T_{EG}=0$$

$$-\frac{4}{\sqrt{17}}T_{DF}+\frac{4}{\sqrt{17}}T_{FG}=0$$

$$\frac{1}{\sqrt{17}}T_{DF}+T_{EF}-\frac{1}{\sqrt{17}}T_{FG}=0$$

$$-\frac{4}{5}T_{EG}-\frac{4}{\sqrt{17}}T_{FG}=0$$

$$\frac{1}{\sqrt{17}}T_{FG}+\frac{3}{5}T_{EG}=50\text{ kips}$$

The three support reaction components and 11 forces in the members can be determined from the above 14 equations of equilibrium. Although these equations could be solved by an appropriately programmed computer, the internal forces and reactions can be determined by less powerful means. The last two equations contain only two of the unknown forces, so that they can be easily solved simultaneously for T_{EG} and T_{FG}, with the result

$$T_{EG}=125\text{ kips}$$

$$T_{FG}=-25\sqrt{17}\text{ kips}$$

Or, if we prefer to deal with even fewer than two simultaneous equations, we can resolve the forces at joint G in directions perpendicular and parallel to FG

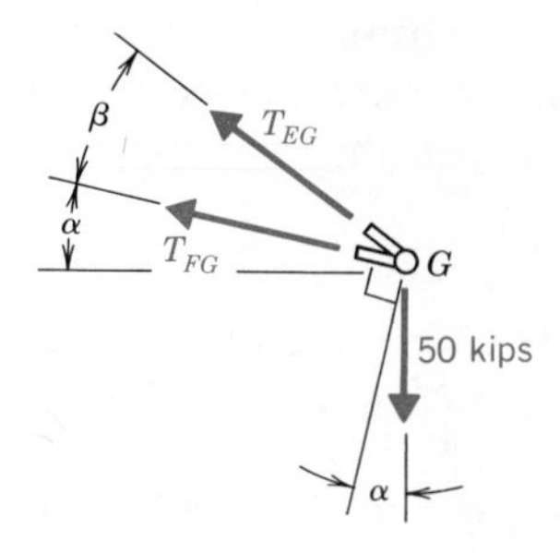

rather than in horizontal and vertical directions. Equilibrium of components perpendicular to FG is expressed by the equation

$$T_{EG} \sin \beta = (50 \text{ kips}) \cos \alpha$$

The values of α and β may be determined from the dimensions given in Figure 5-1:

$$\alpha = \tan^{-1}\left(\frac{5 \text{ ft}}{20 \text{ ft}}\right) = 14.04° \qquad \beta = \tan^{-1}\left(\frac{15 \text{ ft}}{20 \text{ ft}}\right) - \alpha = 22.83°$$

and used in the above equilibrium equation to evaluate T_{EG}.

Turning next to joint F, we see that with T_{FG} now known, the two equations of equilibrium of this joint can be solved for T_{DF} and T_{EF}. However, rather than dealing with the equations of vertical and horizontal force components, consider components perpendicular and parallel to DG. The equation expressing equilibrium perpendicular to DG is

$$T_{EF} \cos \alpha = 0$$

which yields

$$T_{EF} = 0$$

The situation at joint F arises often enough to warrant special attention. Two of the three members connected at this joint are oriented in the same direction; thus, by considering the equilibrium in the direction perpendicular to these two members, we can find the force in the third member by considering force components perpendicular to the two collinear members. The force can often be inferred by inspection.

We can next proceed to joint E, where T_{CE} and T_{DE} are now the only unknowns. Once these forces are evaluated, the equilibrium equations for joint D contain only two unknowns. Proceeding in this manner, we can evaluate the remainder of the internal forces and the reactions from the supports.

By proper selection of the order in which joints of the truss are considered, it is normally possible to work through a planar truss in the manner indicated above. If the values of all the forces are not required, however, the use of the equations from joints may be much less efficient than the approach explained in the following discussion.

Equations from Sections

The forces external to *any* portion of a system in equilibrium have zero resultants of force and moment. Often we can use this fact to obtain an equivalent, but much simpler set of equations.

To illustrate, reconsider the truss of Figure 5-1. A free-body diagram of the entire unit ABG is shown in Figure 5-2a. Summation of moments about point A is written as

$$R_B(10 \text{ ft}) - (20 \text{ kips})(6 \text{ ft}) - (50 \text{ kips})(20 \text{ ft}) = 0$$

from which

$$R_B = 112 \text{ kips} = 498 \text{ kN}$$

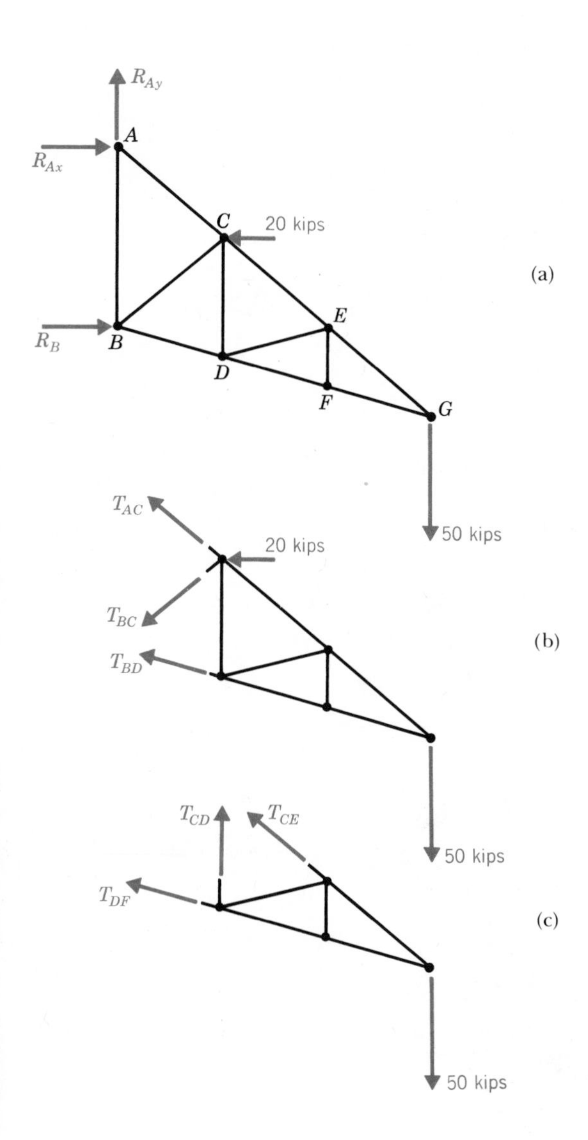

(a)

(b)

(c)

FIGURE 5-2

Then, summation of horizontal force components gives

$$R_{Ax} = -92 \text{ kips} = -409 \text{ kN}$$

and summation of vertical force components gives

$$R_{Ay} = 50 \text{ kips} = 222 \text{ kN}$$

Next, for the free-body shown in Figure 5-2*b*, we can sum moments about point *B*:

$$\frac{4}{5} T_{AC}(10 \text{ ft}) + (20 \text{ kips})(4 \text{ ft}) - (50 \text{ kips})(20 \text{ ft}) = 0$$

to obtain

$$T_{AC} = 115 \text{ kips} = 512 \text{ kN}$$

Then, we can sum moments about point *G*:

$$\frac{1}{\sqrt{5}} T_{BC}(12 \text{ ft}) + \left(\frac{2}{\sqrt{5}} T_{BC} + 20 \text{ kips}\right)(9 \text{ ft}) = 0$$

to obtain

$$T_{BC} = -6\sqrt{5} \text{ kips} = 59.7 \text{ kN}$$

and sum moments about point *C*:

$$-\frac{4}{\sqrt{17}} T_{BD}(6 \text{ ft}) - (50 \text{ kips})(12 \text{ ft}) = 0$$

to obtain

$$T_{BD} = -25\sqrt{17} \text{ kips} = 459 \text{ kN}$$

Now, considering the free-body in Figure 5-2*c*, we can obtain T_{CD} by summing moments about point *G*:

$$T_{CD} = 0$$

By continuing in this fashion, the remaining forces can be obtained without dealing with sets of simultaneous equations.

$$T_{AB} = -19 \text{ kips} = -84.5 \text{ kN}$$

$$T_{EG} = T_{CE} = 125 \text{ kips} = 556 \text{ kN}$$

$$T_{DE} = T_{EF} = 0$$

$$T_{FG} = T_{DF} = -25\sqrt{17} \text{ kips} = -459 \text{ kN}$$

The strategy in this method is to isolate a portion of the structure, with the separation dividing the member in which a force is sought; then, with the free-body completed, to find a direction for force reckoning or an axis for moment reckoning that will yield an equation with as few unknown forces as possible.

For further illustration, consider the truss shown in Figure 5-3*a*. The vertical triangles ABC and DEF are both equilateral, and the rectangle $ABED$ is in a horizontal plane. The force in each of the nine members is to be determined.

Consider first the free-body of the portion of the truss shown in Figure

FIGURE 5-3

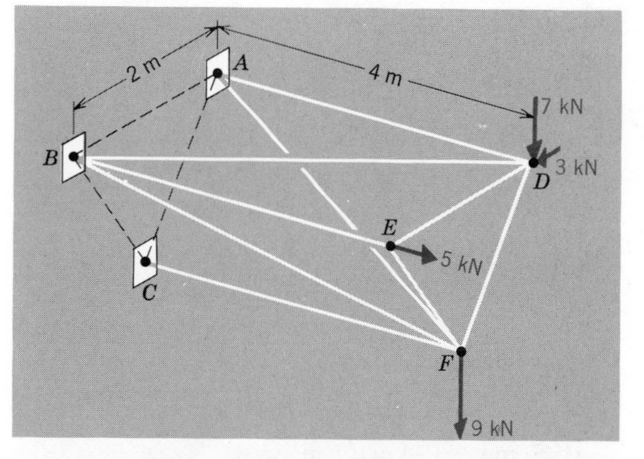

(a)

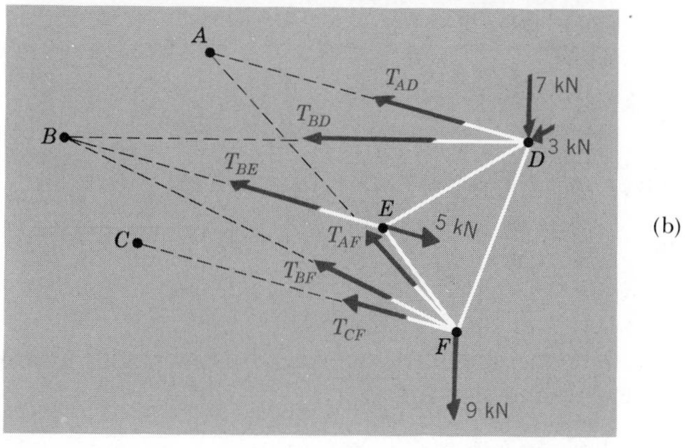

(b)

5-3*b*. To analyze the equilibrium of this section, we select various axes, about which moments involve only one unknown force. The moment about each axis is evaluated most readily by resolving each force as indicated in the illustration accompanying Equation 3-28*c* (p. 79). Recall that the moment of a force about an axis is zero if the line of action intersects the axis.

$$M_{DF} = (T_{BE} - 5\text{ kN})(\sqrt{3}\text{ m}) = 0$$

$$T_{BE} = 5\text{ kN}$$

$$M_{CF} = \left(\frac{2}{\sqrt{20}}T_{BD} + 3\text{ kN}\right)(\sqrt{3}\text{ m}) - (7\text{ kN})(1\text{ m}) = 0$$

$$T_{BD} = \sqrt{5}\left(\frac{7}{\sqrt{3}} - 3\right)\text{ kN} = 2.33\text{ kN}$$

$$M_{FE} = \left(T_{AD} + \frac{4}{\sqrt{20}}T_{BD}\right)(\sqrt{3}\text{ m}) = 0$$

$$T_{AD} = -2\left(\frac{7}{\sqrt{3}} - 3\right)\text{ kN} = -2.08\text{ kN}$$

$$M_{BE} = \left(\frac{2}{\sqrt{20}}T_{AF}\right)(\sqrt{3}\text{ m}) - (9\text{ kN})(1\text{ m}) - (7\text{ kN})(2\text{ m}) = 0$$

$$T_{AF} = 23\sqrt{\frac{5}{3}}\text{ kN} = 29.69\text{ kN}$$

$$M_{DA} = \left(\frac{2}{\sqrt{20}}T_{BF}\right)(\sqrt{3}\text{ m}) - (9\text{ kN})(1\text{ m}) = 0$$

$$T_{BF} = 9\sqrt{\frac{5}{3}}\text{ kN} = 11.62\text{ kN}$$

$$M_{BA} = (T_{CF})(\sqrt{3}\text{ m}) + (7\text{ kN})(4\text{ m}) + (9\text{ kN})(4\text{ m}) = 0$$

$$T_{CF} = -\frac{64}{\sqrt{3}}\text{ kN} = -36.95\text{ kN}$$

To determine the forces in the remaining three members, it is a straightforward matter to isolate joints *D* and *E* and sum forces.

$$T_{DF} = -\frac{14}{\sqrt{3}}\text{ kN} = -8.08\text{ kN}$$

$$T_{DE} = T_{EF} = 0$$

As with planar trusses, equilibrium of three-dimensional trusses can be analyzed by writing equations of force equilibrium for each joint, or by considering a larger portion of the truss. The procedure for writing equilibrium equations for a joint is illustrated on page 123, and, as in that example, usually leads to a set of simultaneous equations for the unknown forces. Often, as in the preceding example, we can avoid the task of solving sets of simultaneous equations by considering an entire section of the truss and finding axes about which only one unknown force has a nonzero moment. However, it is not always possible to find such an axis, so that the equations from joints may provide the best vehicle for computation.

Statically Indeterminate and Nonrigid Trusses

As with support systems for rigid bodies (p. 154), a truss can be statically determinate, statically indeterminate, or incapable of supporting some loads. A few simple examples are shown in Figure 5-4. In simple cases, the determination of whether a truss is rigid and statically determinate can be made by inspection: If one or more members can be removed without destroying the rigidity of the truss, it is statically indeterminate. When the situation is not obvious, more formal means of checking for rigidity and determinateness may be necessary. Discussions of such procedures can be found in many books on structural analysis.*

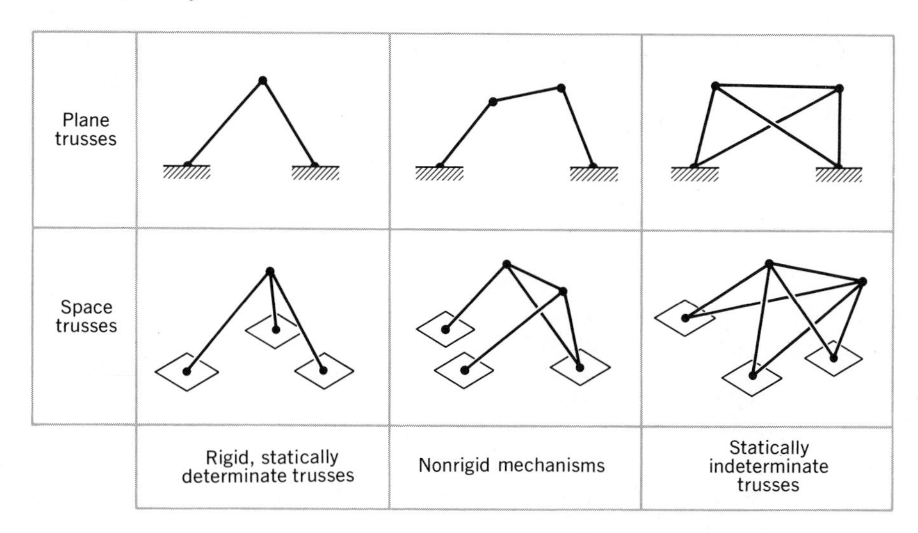

FIGURE 5-4

*See, for example, Timoshenko and Young, *Theory of Structures* (McGraw-Hill, 1945).

PROBLEMS

5-1 For the truss in Figure 5-1, verify that the values of forces obtained from equations of sections satisfy the equations of joints.

5-2 For the truss in Figure 5-3, write the equations of equilibrium of each isolated joint. Verify that the forces obtained in the text satisfy these.

5-3 Reconsider the free-body diagram of joint G, page 161. Instead of resolving forces into components in two mutually perpendicular directions, draw the vector diagram showing equilibrium of $\mathbf{T}_{EG}$, $\mathbf{T}_{FG}$, and the 50-kip force. Solve this triangle for T_{EG} and T_{FG}.

5-4 Using the free-body diagram of joint G on page 161, evaluate T_{FG} by considering moments about point A, and T_{EG} by considering moments about point B.

5-5 Reconsider the free-body diagram of joint E, page 161. Recalling that we found T_{EF} to be zero by inspection of joint F, what does a similar inspection of joint E then tell us about the value of T_{DE}? Applying similar reasoning to the free-body diagram of joint D, what must be the value of T_{CD}?

5-6 Evaluate the force in members EF, DE, and DC.

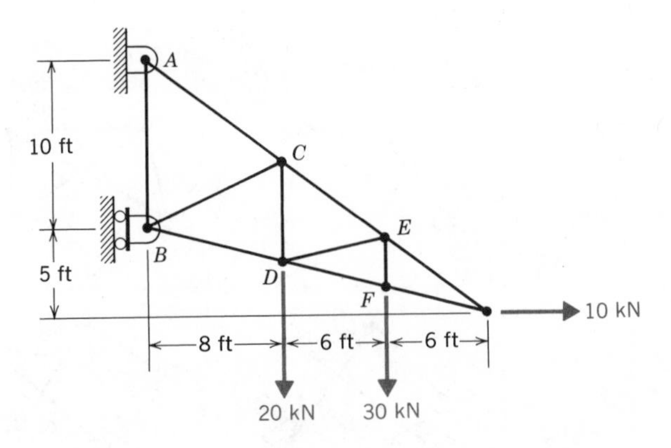

5-7 A snow load transfers the forces shown to each of the upper joints of a Pratt roof truss. Neglecting horizontal components of reactions at the supports, evaluate the forces in members BC, BF, CF, FG, and CG.

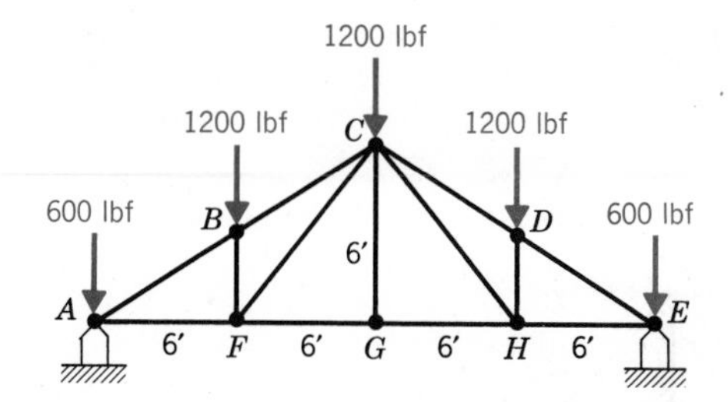

5-8 Evaluate the force in member DB.

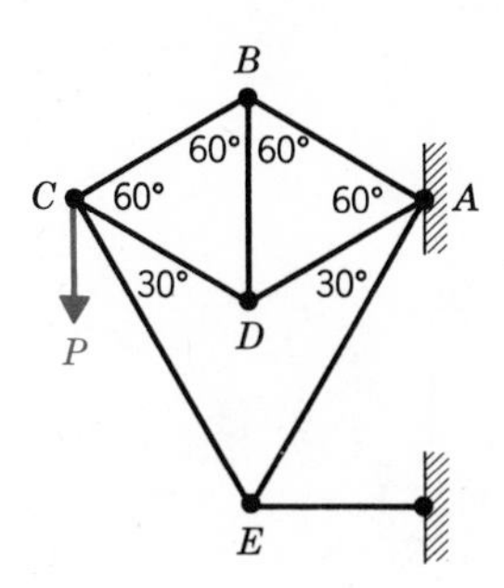

5-9 Evaluate the forces in members CF and CG. *Ans.* -58.9 kN; 0.

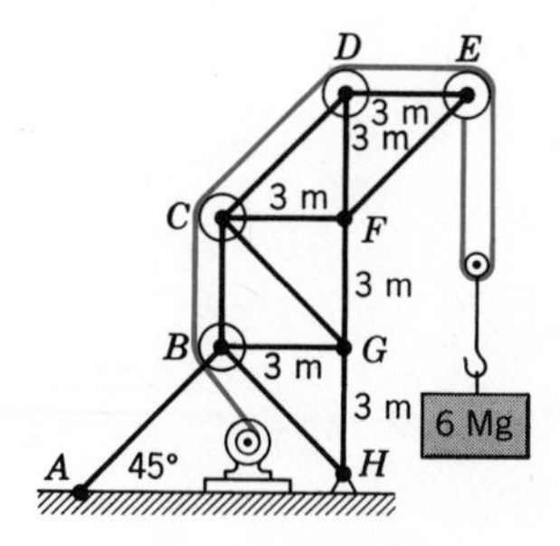

5-10 Evaluate the support reactions and the forces in each member.

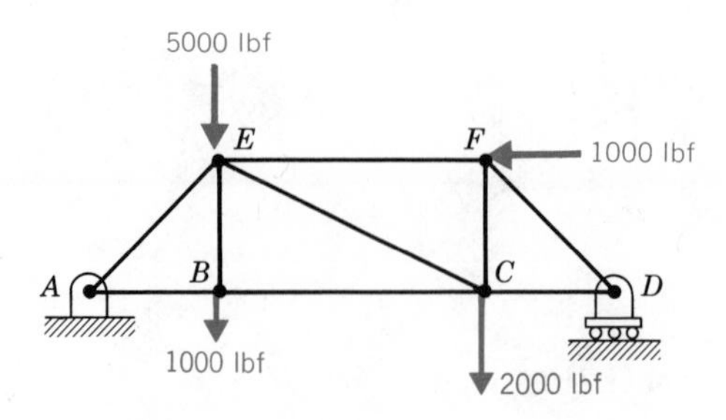

5-11 Evaluate the force in the member *AB*.

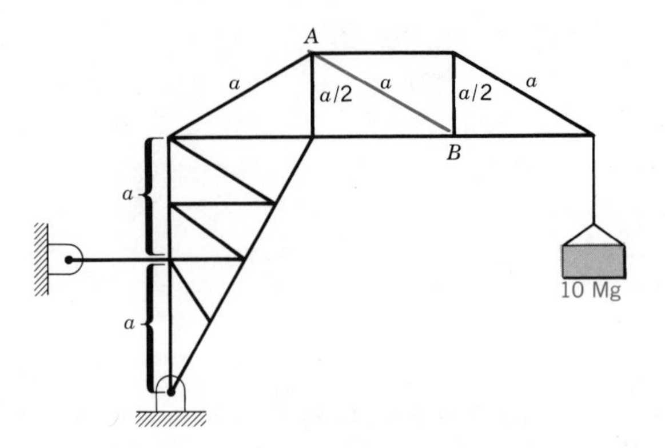

5-12 Evaluate the forces in members *IJ*, *GH*, and *EH*.

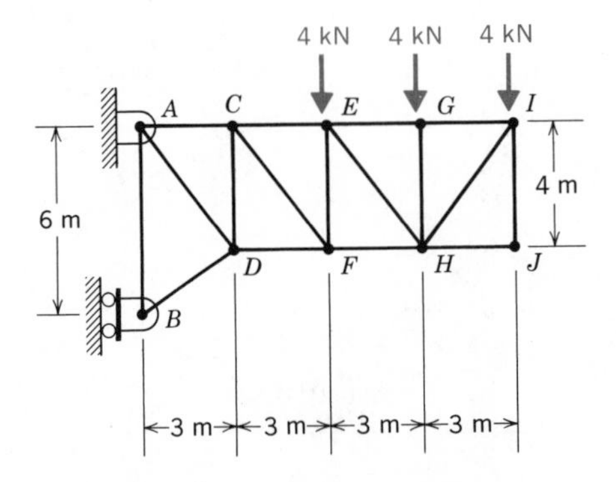

5-13 Evaluate the force in member *ML*.

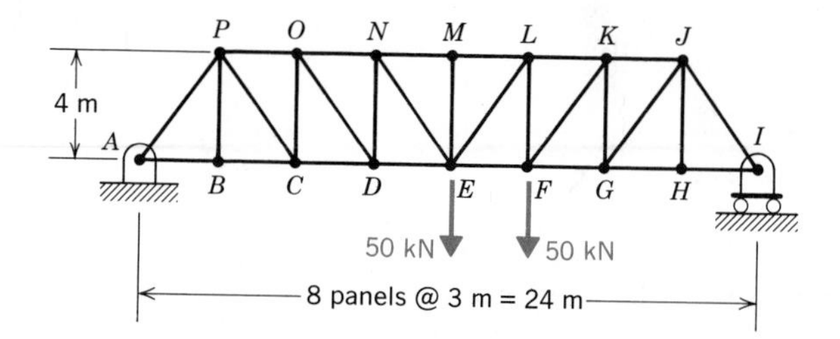

5-14 Evaluate the force in member *CD*. Members are all of equal length. *Ans.* −828.6 kN.

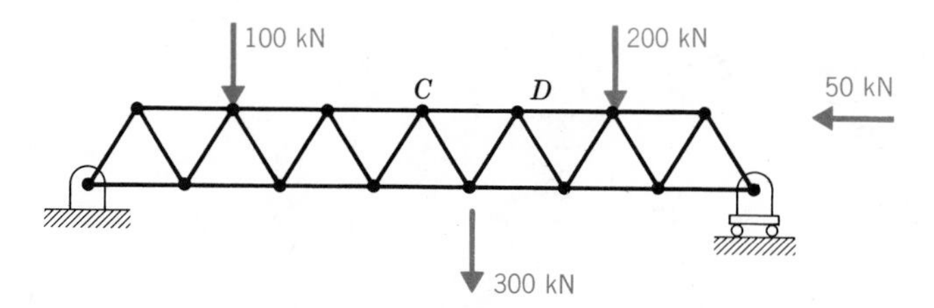

5-15 All angles formed by the members of the truss are 120°, 90°, 60°, or 30°. Determine the force in the member *EP*.

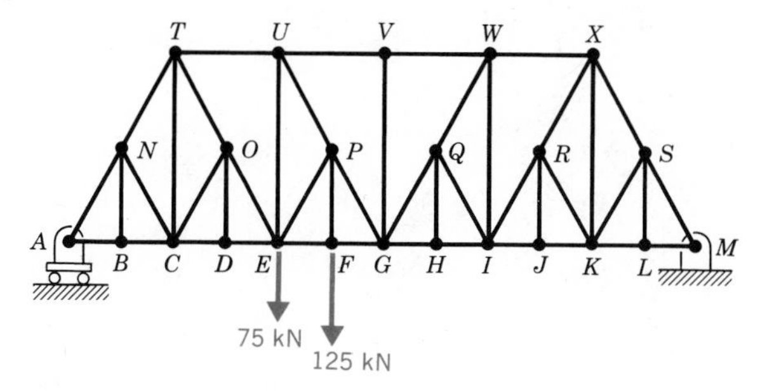

For Problems 5-16 through 5-21, evaluate the forces in those members assigned by your instructor.

5-16

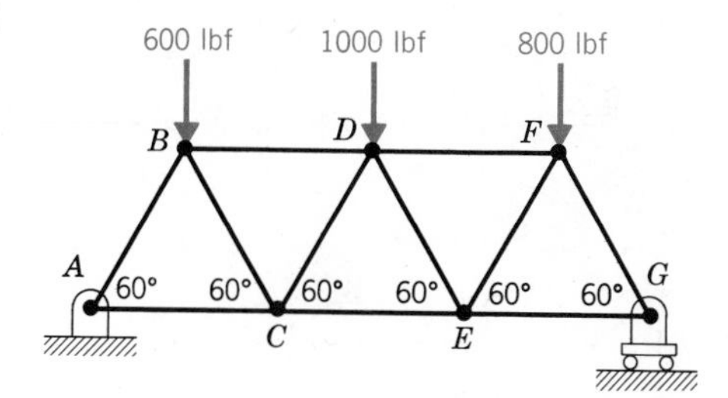

5-17

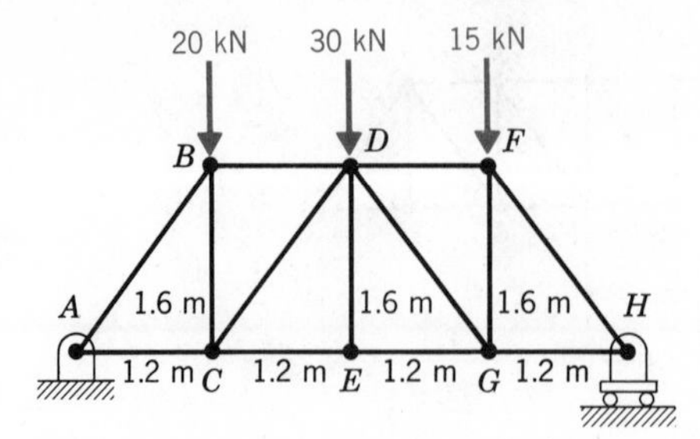

5-18

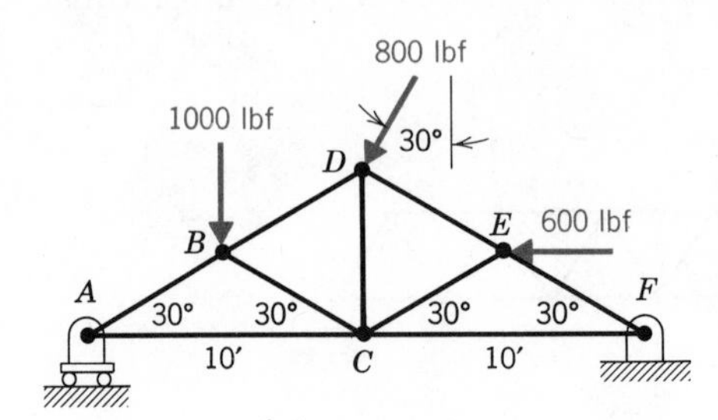

5-19

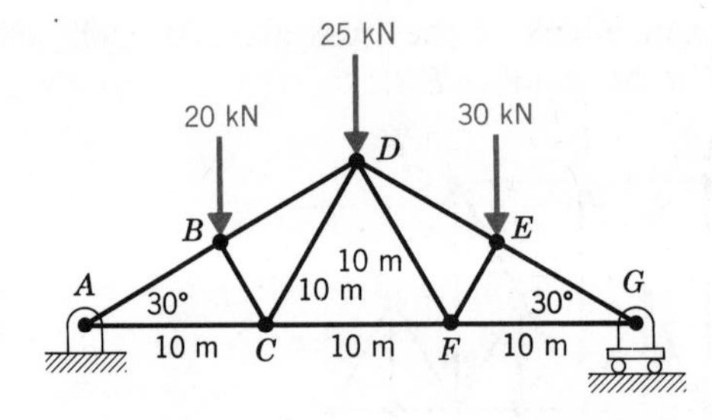

5-20

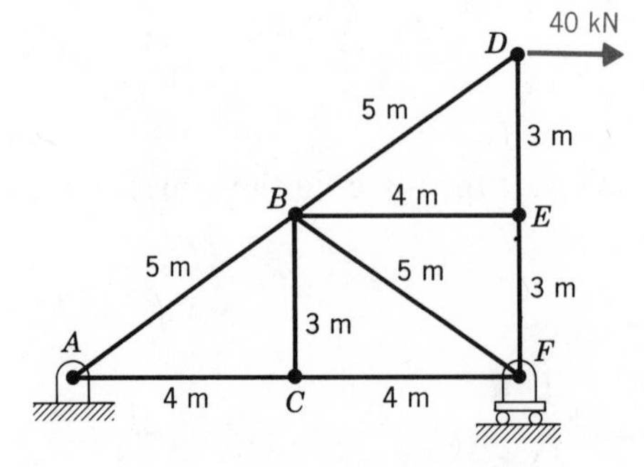

5-21

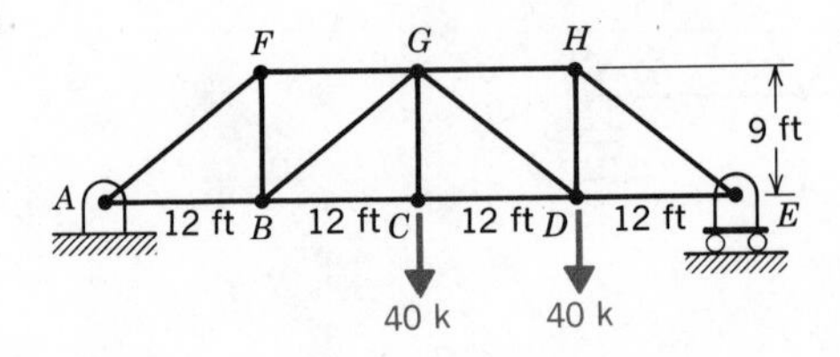

5-22 The crane boom supports the 4-Mg load as shown. Evaluate the force that the pulley bearing transmits to the truss at A. Determine the forces in all the truss members.

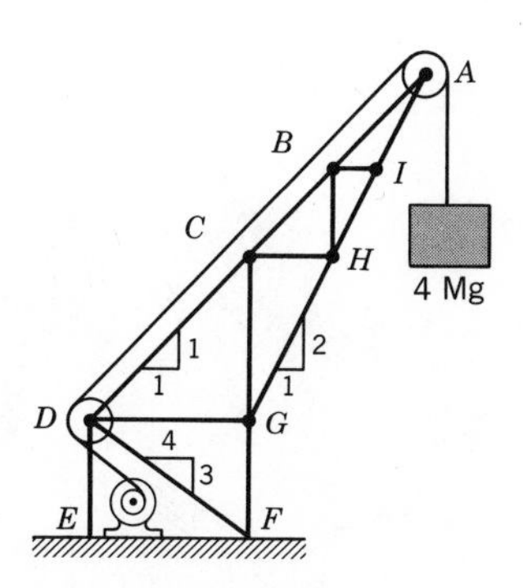

5-23 Determine the forces in members DC and CF, for
(a) $P_1 = 1,\ P_2 = P_3 = 0.$
(b) $P_2 = 1,\ P_3 = P_1 = 0.$
(c) $P_3 = 1,\ P_1 = P_2 = 0.$
(d) Arbitrary $P_1,\ P_2,\ P_3.$

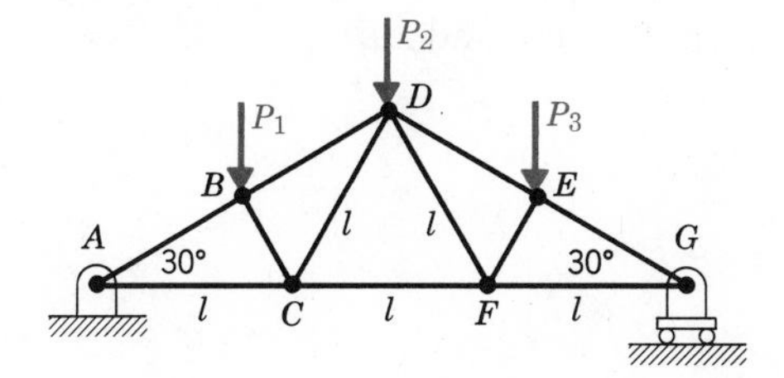

5-24 The members AD, CD, and BD, and the lines AB, BC, and CA are all of equal length. Evaluate the forces in the three members of the truss. *Ans.* $1.1547P$; 0; $-1.1547P$.

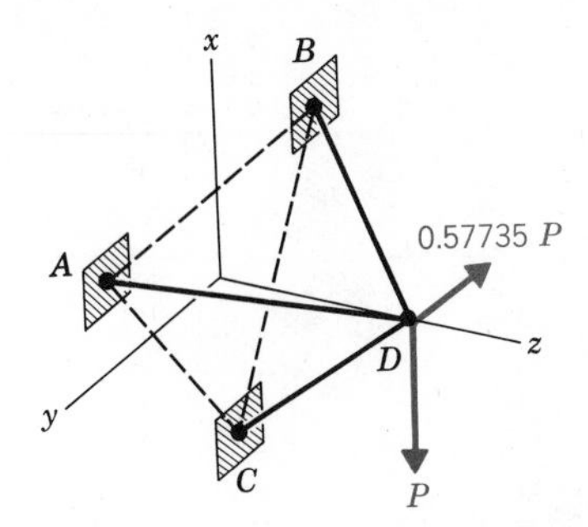

5-25 The member BF, FE, EC, ED, and FD, and the lines AB, BC, and CA are all of equal length, as are the members AF, AE, BD, and DC. BF and CE are perpendicular to ABC. Evaluate the forces in the members of the truss.

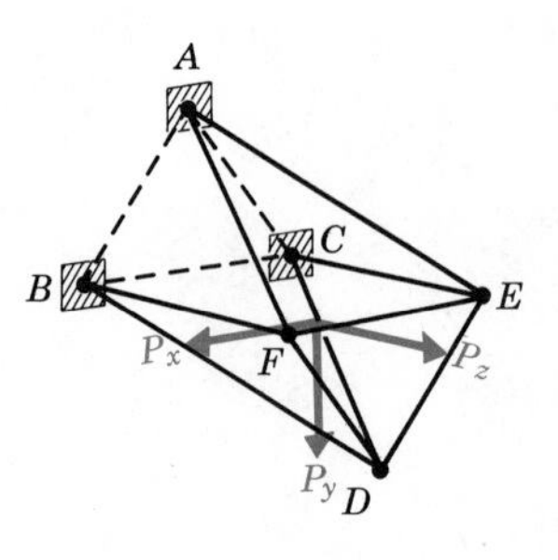

5-26 Members AD, BE, and CF are perpendicular to the plane ABC. Evaluate the forces in members BF, AD, and ED.

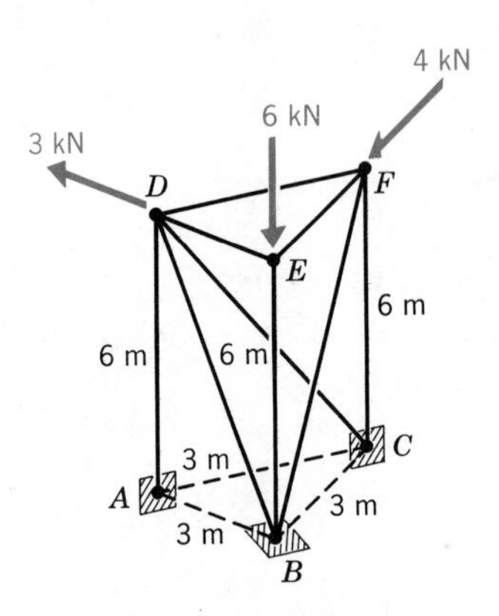

5-27 Evaluate the forces in all the members of the truss.

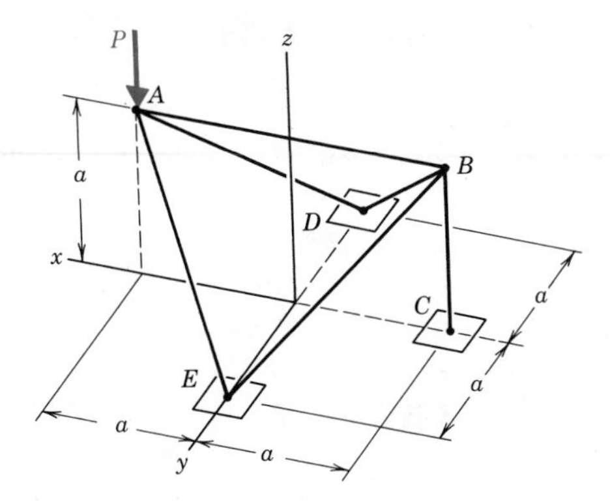

5-28 The truss supports a vertical force P. Assume that the force in member AE is equal to that in BE. Will this be necessarily so? Evaluate the forces in members FE and AE. *Ans.* $1.241P$; $0.886P$.

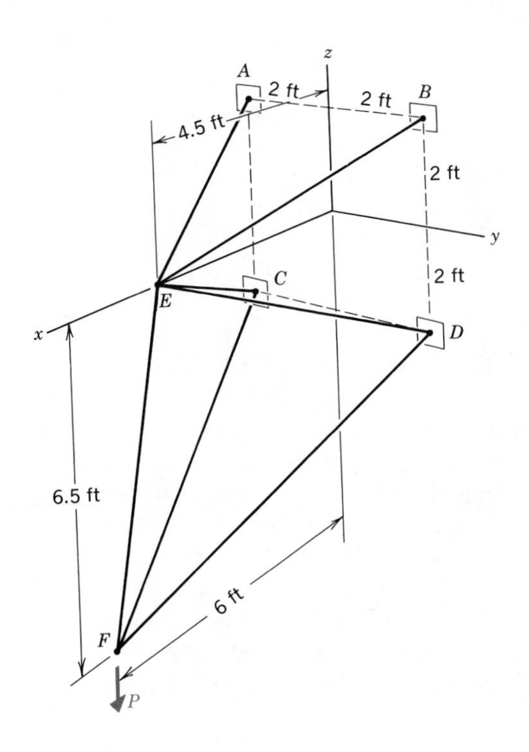

5-29 The support points A, B, and C form an equilateral triangle in the horizontal x-y plane. The length of each of the nine members of the truss is equal to the length of a side of this triangle. The force P acts parallel to EF, the force Q acts perpendicular to P in the horizontal plane, and the force R acts vertically. Evaluate the forces in members AE and AF. *Suggestion*: resolve the reaction at A into a component parallel to EF and a component directed toward the mid-point of EF. Similarly resolve the reactions at B and C. Next, consider moments about axes AB, BC, and CA of forces acting on a free-body of the entire structure. Finally, consider moments about a vertical axis through point O'.

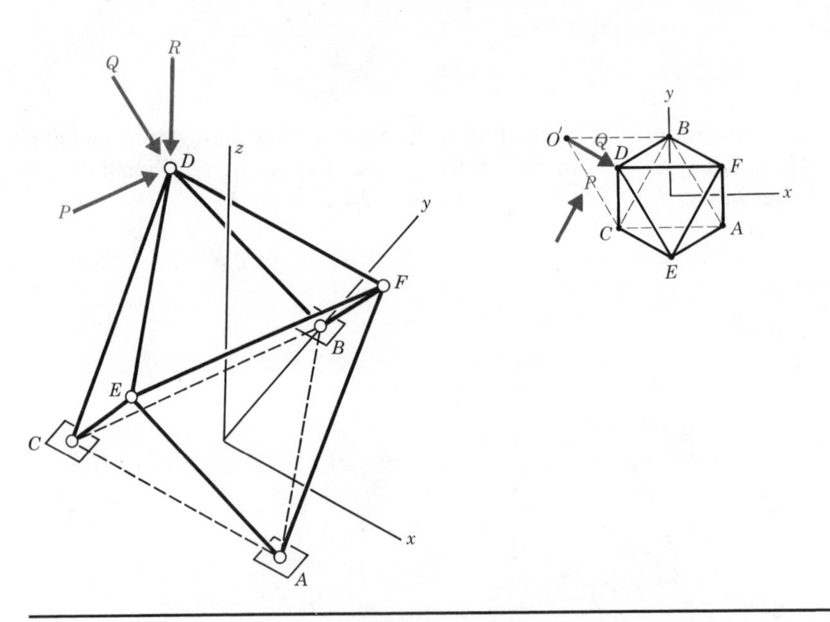

5-2 COUPLE-SUPPORTING MEMBERS

The loads that a stiff bar can carry are not limited to axial forces. If lateral forces are applied, equilibrium requires that couples exist at various sections along the rod, as is demonstrated in Figure 5-5. This figure shows free-body diagrams of the beam and of two separated portions of the beam. Equilibrium of the portion on the left indicates at once that the forces across the plane of separation must form a couple and a lateral force.

The detailed distribution of interaction forces between the two parts of the beam is fairly complicated and cannot be deduced from equilibrium alone. (This is another subject, and is treated in the mechanics of deformable solids.)

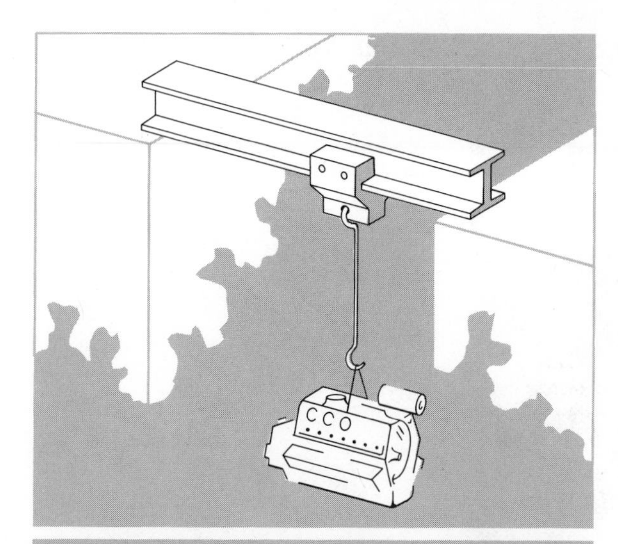

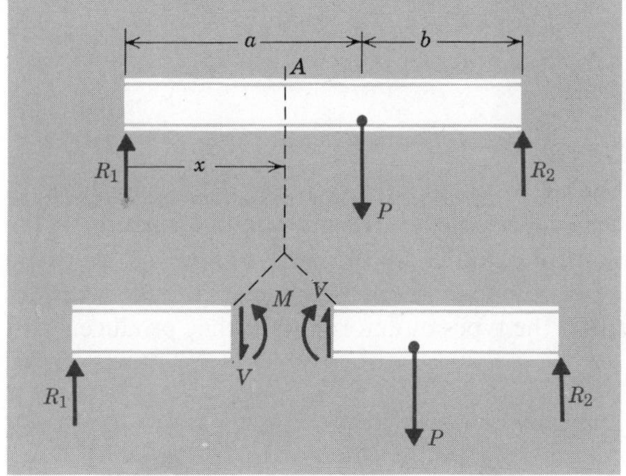

FIGURE 5-5

However, if the support system is statically determinate, it is possible to evaluate the moment of the equivalent couple. Moreover, since structural failure is normally related directly to such moments, their evaluation is usually the most important aspect of the force analysis of a laterally loaded rigid bar.

Twisting and Bending Moments

In general, the moment of the couple at a section can have any direction, depending on how the external loads are applied. For example, the bracket shown in Figure 5-6 has moment components in all three directions induced at section A. The moment vector is normally resolved into a component parallel

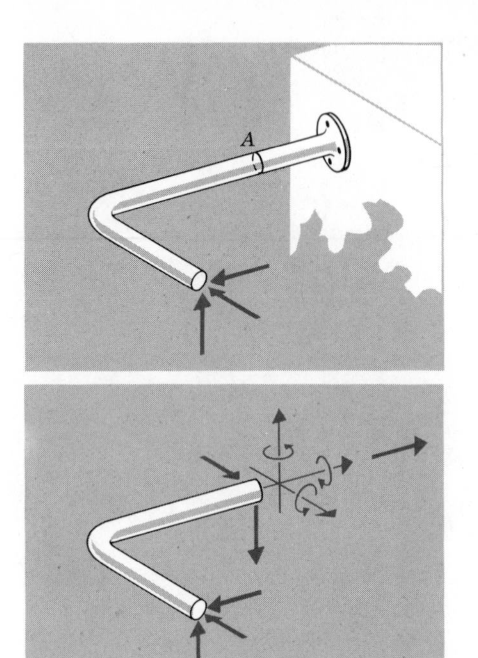

FIGURE 5-6

to the axis of the bar and one or two components perpendicular to the axis. (This will facilitate the further analysis of the strength and deformation of the bar.) The component of moment parallel to the axis of the bar is called the *twisting moment*, and the components perpendicular to the axis are called the *bending moments*, after the types of deformations they produce.

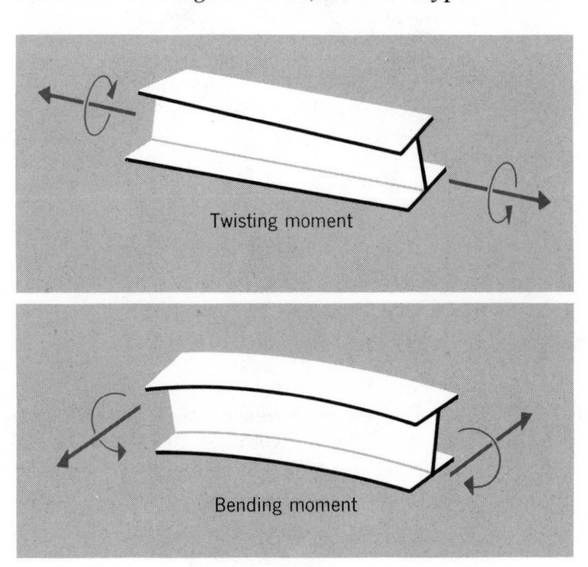

The resultant force acting at a section of a bar is similarly resolved. The component along the axis of the rod is called the *axial force*, and the components perpendicular to the axis are called the *shearing forces*.

Evaluation of these force and moment components is accomplished by using the same basic ideas already examined: a free-body of a portion of the member on either side of the section of interest is isolated and properly labeled, and equations of equilibrium are written and solved.

For example, equilibrium of the portion of the beam to the left of the section A in Figure 5-5 yields

$$V = R_1 \qquad \text{and} \qquad M = R_1 x$$

and equilibrium of the entire beam gives values of the support reactions in terms of the applied load P, as

$$R_1 = \frac{bP}{a+b} \qquad R_2 = \frac{aP}{a+b}$$

Thus, in terms of the applied load, the shear and bending moment are

$$V = \frac{bP}{a+b} \qquad x < a$$

$$M = \frac{bPx}{a+b} \qquad x < a$$

The qualification $x < a$ is necessary because the analysis was done for sections to the left of the applied load. A similar analysis for sections on the other side of the load results in

$$V = \frac{-aP}{a+b} \qquad x > a$$

$$M = \frac{a(a+b-x)P}{a+b} \qquad x > a$$

As another example, let us evaluate the force and moment components at the support O for the automobile torsion bar in Figure 5-7a.

Force equilibrium of the free-body in Figure 5-7b requires that

$$\mathbf{R} + 6.8\ \text{kN}\left(-\frac{8}{17}\mathbf{u}_x + \frac{15}{17}\mathbf{u}_z\right) + 2.5\ \text{kN}\,\mathbf{u}_y = \mathbf{0}$$

Or

$$\mathbf{R} = 3.2\ \text{kN}\,\mathbf{u}_x - 2.5\ \text{kN}\,\mathbf{u}_y - 6.0\ \text{kN}\,\mathbf{u}_z$$

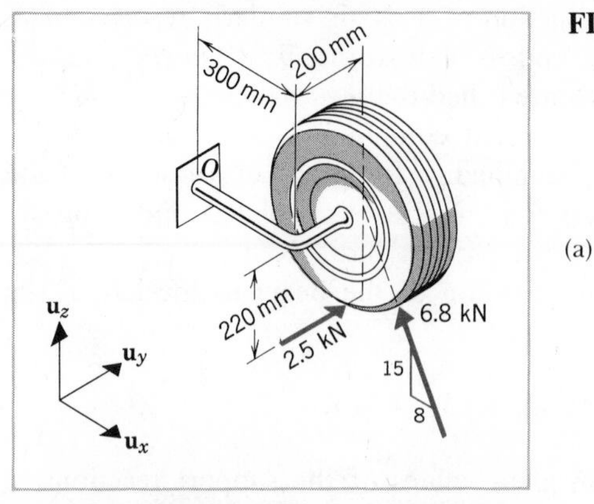

FIGURE 5-7

(a)

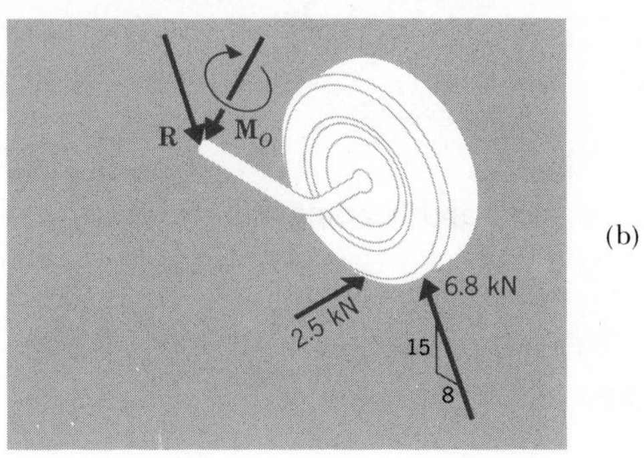

(b)

Therefore, there is a compressive axial force of 3.2 kN and a resultant shearing force of $\sqrt{(2.5)^2 + (6.0)^2}$ kN = 6.5 kN at the section O. Next, moment equilibrium requires that

$$\mathbf{M}_O + [(300\text{ mm})\,\mathbf{u}_x + (200\text{ mm})\,\mathbf{u}_y] \times [(6.0\text{ kN})\,\mathbf{u}_z - (3.2\text{ kN})\,\mathbf{u}_x]$$

$$+ [(300\text{ mm})\,\mathbf{u}_x + (200\text{ mm})\,\mathbf{u}_y - (220\text{ mm})\,\mathbf{u}_z] \times (2.5\text{ kN})\,\mathbf{u}_y = \mathbf{0}$$

Or,

$$\mathbf{M}_O = (-1.75\text{ N}\cdot\text{m})\,\mathbf{u}_x + (1.80\text{ N}\cdot\text{m})\,\mathbf{u}_y - (1.39\text{ N}\cdot\text{m})\,\mathbf{u}_z$$

This indicates a twisting moment,

$$M_t = (1.75)\text{ N}\cdot\text{m}$$

and a bending moment resultant,

$$M_b = \sqrt{(1.80)^2 + (1.39)^2}\ \text{N}\cdot\text{m}$$
$$= 2.27\ \text{N}\cdot\text{m}$$

PROBLEMS

5-30 Write expressions for shear and bending moment in sections to the right of the applied force in Figure 5-5, p. 177.

5-31 Evaluate the shear force, the bending moment, and the twisting moment for each section of pipe shown in Problem 2-62, p. 43. Draw diagrams that show how these vary along the pipe sections.

5-32 Evaluate the axial and shearing forces and the bending moment in the horizontal bar in Problem 4-39, p. 137, and show how these vary along the bar.

5-33 Evaluate the bending moment in the frame of Problem 4-45, p. 139, and sketch its variation along the frame.

5-34 Evaluate the bending and twisting moments in each section of the shaft shown in Problem 4-66, p. 147.

5-35 Evaluate the bending moment in the load leveler bar AB in Problem 4-54, p. 143.

5-36 Evaluate the bending moment along the pole of Problem 4-72, p. 149.

5-37 Evaluate the bending moment along the mast of Problem 4-74, p. 150.

5-38 Plot the bending moment along the entire length of the frame.

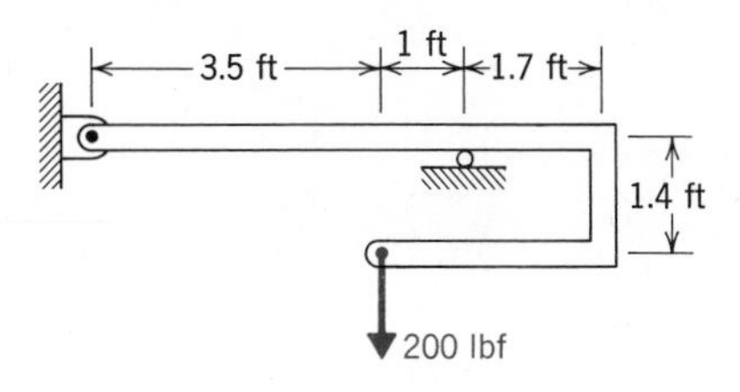

5-39 Plot the bending moment along the horizontal portion of the beam.

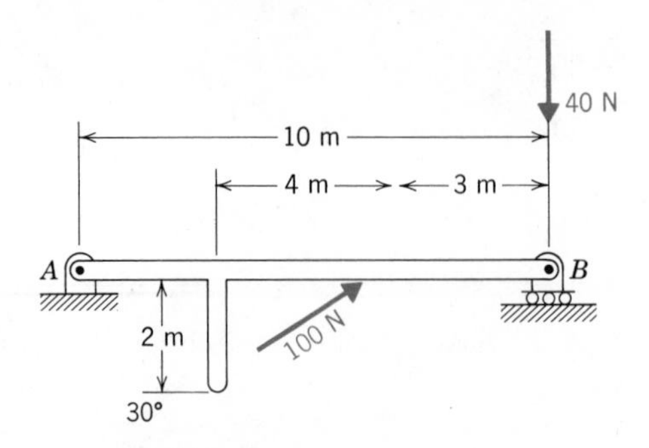

5-40 Plot the bending moment along the entire davit of Problem 4-44, p. 139.

5-41 Plot the bending moment in the lower bar as a function of x, for $d = 0$, $d = \frac{1}{2}b$, and $d = b$.

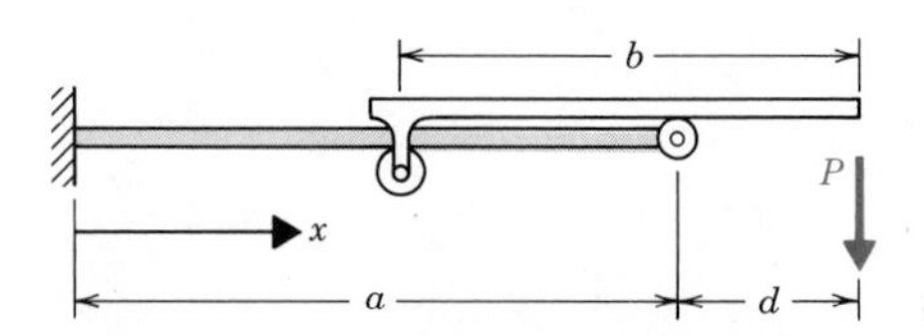

5-42 Evaluate the twisting and bending moments at the position specified by the angle θ. *Ans.* $Pa(1 - \cos\theta)$; $Pa \sin\theta$.

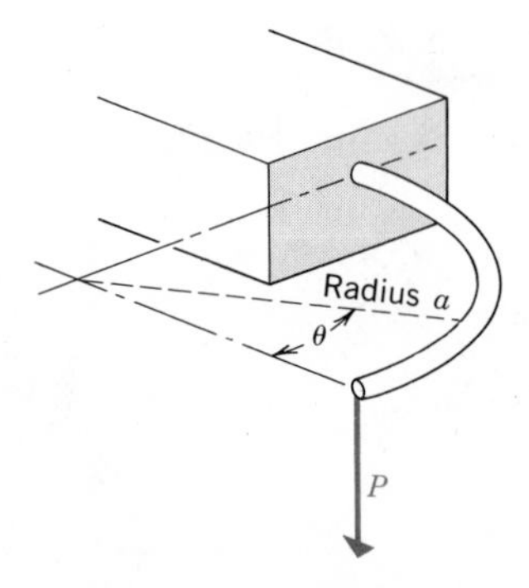

5-43 Evaluate the tensile and shearing forces and the bending moment at the section specified by the angle θ.

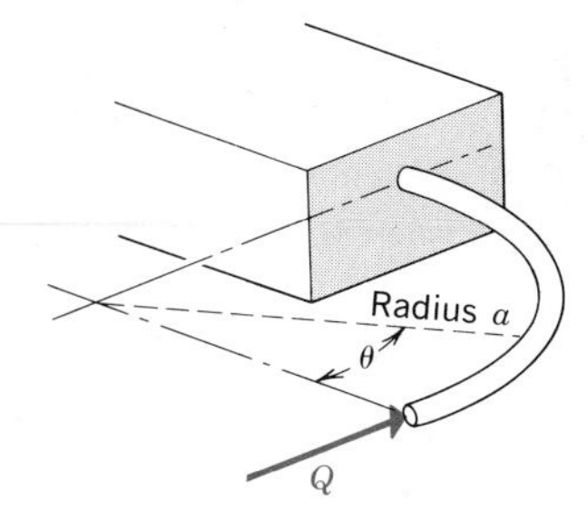

5-44 Evaluate the tensile and shearing forces and the bending moment at the section specified by the angle θ.

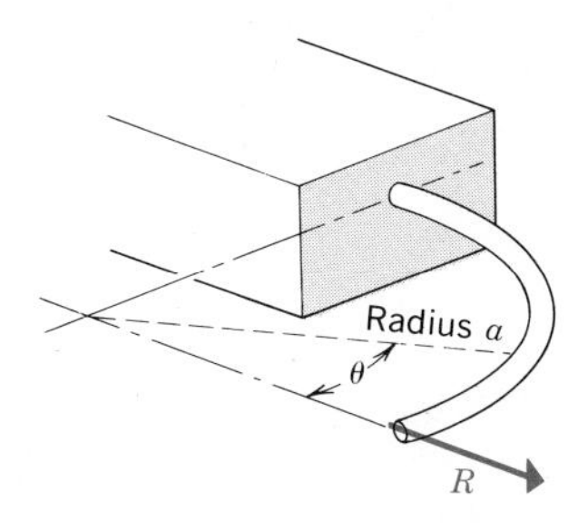

5-3 SYSTEMS WITH FRICTION

Friction forces are those that act tangential to the surface on which two objects make contact. People have gone to considerable effort to minimize them in machinery of all kinds, but depend heavily on them at contact surfaces like those between shoe soles and floors, tires and roadways, and drive belts and pulleys.

The ratio of the tangential force to the normal force between two sliding surfaces is called the *coefficient of friction*. In general, this coefficient depends on a number of variables, such as the surface materials, the surface finishes, the presence of any surface films, the velocity of sliding, the temperature, and others. The ratio of tangential force necessary to initiate sliding from a state of rest to the normal force is called the coefficient of *static* friction, whereas the force ratio as sliding continues is called the coefficient of *sliding* friction. Typically, the coefficient of static friction is somewhat greater than the coefficient of sliding friction. Furthermore, a decrease in sliding friction force with increase in speed of sliding has been observed for a number of materials, although this variation is usually small enough that it can be ignored.

A few typical values of coefficients of friction are given in Table 5-1. In using such data, we must remember that these values can be affected considerably by the factors mentioned above, and that results of predictions that depend on the amount of friction should be interpreted with this in mind.

TABLE 5-1 Some Coefficients of Friction between Dry Surfaces*

Surfaces	Coefficient of Static Friction, μ_0	Coefficient of Sliding Friction, μ_1
Hard steel on hard steel	0.78	0.42
Mild steel on mild steel	0.74	0.57
Cast iron on cast iron	1.10	0.15
Aluminum on aluminum	1.05	1.4
Hard steel on babbitt	0.42	0.33
Mild steel on lead	0.95	0.95
Aluminum on mild steel	0.61	0.47
Copper on mild steel	0.53	0.36
Brass on mild steel	0.51	0.44
Teflon on teflon or steel	0.04	0.04
Oak on oak (parallel to grain)	0.62	0.48
Oak on oak (perpendicular)	0.54	0.32

*From *American Institute of Physics Handbook*, 3rd Ed. (McGraw-Hill, 1972). Values may vary considerably.

Coulomb's Friction Law

In spite of the complexity of the mechanism of friction, for many dry surfaces, the approximation known as "Coulomb friction" has been found to lead to predictions of acceptable accuracy. The so-called Coulomb's law* of friction states that whenever sliding takes place the tangential component of force between two sliding surfaces is proportional to the normal component and acts in a direction to oppose the motion. This can be expressed analytically as

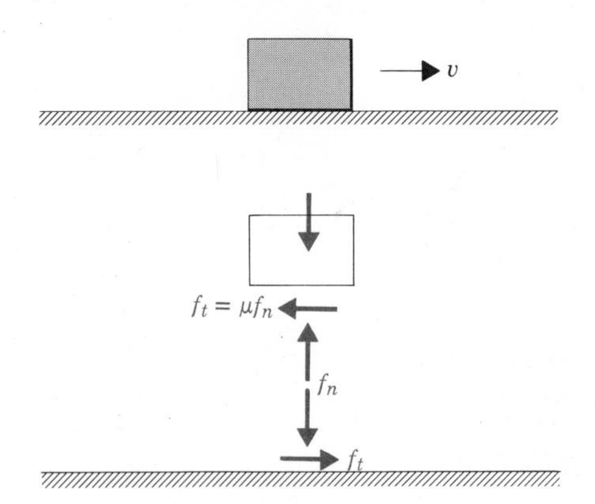

$$\mathbf{f}_t = -\mu_1 f_n \mathbf{u}_v \qquad \mathbf{v} \neq \mathbf{0} \tag{5-1a}$$

in which $\mathbf{u}_v$ is a unit vector in the direction of the relative velocity $\mathbf{v}$ of the object on which $\mathbf{f}_t$ acts, and the coefficient of sliding friction μ_1 depends on the surfaces in contact but not on the magnitude of the normal force nor on the velocity.

When the body is at rest, the friction force can have any direction required for equilibrium, but its magnitude is limited by

$$f_t \leqslant \mu_0 f_n \qquad \mathbf{v} = \mathbf{0} \tag{5-1b}$$

in which μ_0 is the coefficient of static friction. When applied forces induce a friction force that reaches this limit, motion is incipient, meaning that any further increase will cause acceleration.

*For Charles August de Coulomb, whose early work on statics and mechanics was concerned in part with friction.

EXAMPLE

What is the magnitude P of the force required to move the 80-lb block up the 15° incline shown in Figure 5-8? Also, in the absence of P, will the block remain stationary on the incline? The coefficients of static and sliding friction are equal to 0.3.

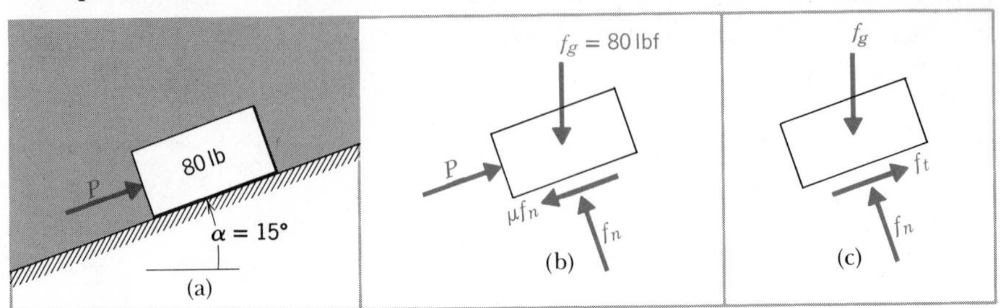

FIGURE 5-8

Referring to the free-body diagram in Figure 5-8*b*, we can write equations of equilibrium in the direction normal to the incline.

$$f_n - f_g \cos \alpha = 0$$

and in the direction along the incline:

$$P - f_g \sin \alpha - \mu f_n = 0$$

Elimination of f_n from these two equations results in

$$\begin{aligned} P &= f_g(\sin \alpha + \mu \cos \alpha) \\ &= (80 \text{ lbf})(\sin 15° + 0.30 \cos 15°) \\ &= 43.9 \text{ lbf} \end{aligned}$$

The free-body diagram in Figure 5-8*c* depicts the situation in the absence of the force P. Observe the reversal of the direction of the friction force. Now, if the block is to remain at rest,

$$f_n = f_g \cos \alpha$$

$$f_t = f_g \sin \alpha$$

But the friction force is limited by

$$f_t \leqslant \mu f_n$$

Substitution of the equilibrium equations into this inequality gives

$$f_g \sin\alpha \leqslant \mu f_g \cos\alpha$$

Or,

$$\tan\alpha \leqslant \mu$$

But $\tan\alpha = 0.27$, which is less than $\mu = 0.30$; therefore, the block will remain at rest.

EXAMPLE

In the absence of P, the angle of the incline in Figure 5-8 is slowly increased until the block begins to slide downward. The coefficient of static friction is $\mu_0 = 0.47$ and the coefficient of sliding friction is $\mu_1 = 0.44$. What will be the acceleration of the block after it breaks loose?

Prior to breakaway of the block,

$$f_t \leqslant \mu_0 f_n$$

Or, introducing the equilibrium relationships (see Figure 5-8*c*),

$$f_g \sin\alpha \leqslant \mu_0 f_g \cos\alpha$$

The equality in this relationship occurs when the critical angle α_c is reached:

$$\tan\alpha_c = \mu_0 = 0.47 \qquad \textbf{(a)}$$

from which

$$\alpha_c = 25.17°$$

After the block breaks loose,

$$f_t = \mu_1 f_n \qquad \textbf{(b)}$$

Now, the normal component of acceleration is zero;

$$f_n - f_g \cos\alpha_c = 0 \qquad \textbf{(c)}$$

but in the tangential direction, we must replace the usual static equilibrium equation with Newton's second law,

$$f_g \sin \alpha_c - f_t = ma \tag{d}$$

in which a is the downward tangential acceleration. Substituting (b) and (c) into (d),

$$f_g \sin \alpha_c - \mu_1 f_g \cos \alpha_c = ma$$

But since $f_g = mg$,

$$a = g(\sin \alpha_c - \mu_1 \cos \alpha_c)$$

$$= g \sin \alpha_c \left(1 - \frac{\mu_1}{\tan \alpha_c}\right)$$

Substituting from (a),

$$a = g \sin \alpha_c \left(1 - \frac{\mu_1}{\mu_0}\right)$$

$$= (9.807 \text{ m/s}^2)(0.4254)\left(1 - \frac{0.44}{0.47}\right)$$

$$= 0.266 \text{ m/s}^2$$

Belt Friction

Many devices incorporate a flexible line passed around a solid circular object, such as a pulley, with torque transmitted by means of the tangential (friction) forces along the contact surface. In such a mechanism, the transmission of torque results in a difference in tension between the two ends of the line, the tension varying along the portion in contact in a manner that is revealed in the following analysis.

Figure 5-9*a* shows the essential features of the device while Figure 5-9*b* shows a free-body diagram of an element of the line. We assume that Coulomb's law is valid for the surfaces and that slipping is incipient.

Equilibrium of the element in the radial direction is given by

$$N_{av} a \, \Delta\theta - T(\theta + \Delta\theta) \sin \Delta\theta = 0$$

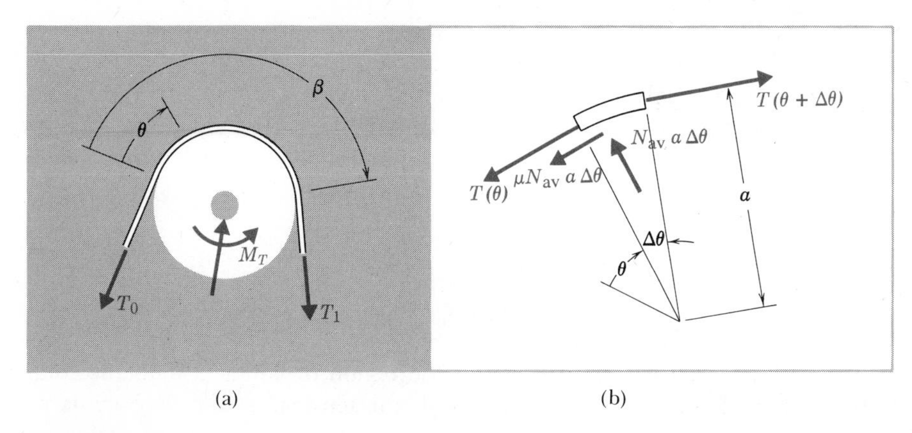

(a) (b)

FIGURE 5-9

in which N_{av} represents an average radial force per unit length of the line. Division by $\Delta\theta$ and evaluation of the limit as $\Delta\theta \to 0$ lead to

$$\mathrm{a} N(\theta) = T(\theta)$$

Equilibrium in the tangential direction is given by

$$T(\theta + \Delta\theta)\cos\Delta\theta - T(\theta) - \mu N_{av} a\,\Delta\theta = 0$$

Division by $\Delta\theta$ and evaluation of the limit as $\Delta\theta \to 0$ lead to

$$\frac{dT}{d\theta} - \mu a N = 0$$

Elimination of N from the two equilibrium relationships gives

$$\frac{dT}{d\theta} - \mu T = 0$$

This may be rearranged in the form

$$\frac{dT}{T} = \mu\, d\theta$$

and integrated to give

$$\log T = \mu\theta + C$$

The constant of integration can be related to the tension T_0 by setting $\theta = 0$:

$$\log T_0 = C$$

Then, the tension-angle relationship becomes

$$\log \frac{T}{T_0} = \mu\theta$$

Or,

$$T = T_0 e^{\mu\theta}$$

This is valid for all values of θ within the region of contact. With the total subtended angle of contact denoted by β, the tensions at the two points of tangency are related by

$$\frac{T_1}{T_0} = e^{\mu\beta} \qquad \textbf{(5-3)}$$

From Figure 5-9*a* we see that the torque transmitted is given by

$$M_T = T_1 a - T_0 a$$
$$= T_0 a(e^{\mu\beta} - 1) \qquad \textbf{(5-4)}$$

EXAMPLE

How far around a tree must a wrangler wrap a rope to hold an excited horse by pulling with a force of 20 lb? The horse can pull with a force of 650 lb and the coefficient of friction between the rope and the tree is 0.92.

If slipping is incipient, Equation 5-3 requires that

$$\frac{650 \text{ lbf}}{20 \text{ lbf}} = e^{0.92\beta}$$

Solving for the angle, we obtain

$$\beta = \frac{1}{0.92} \log 32.5$$
$$= 3.78 \text{ radians}$$
$$= 0.6 \text{ revolutions}$$

PROBLEMS

5-45 What is the value of the friction force when $\alpha = 15°$? What is the critical value of α at which the block will start sliding down the plane? *Ans.* 38.8 lb; 16.7°.

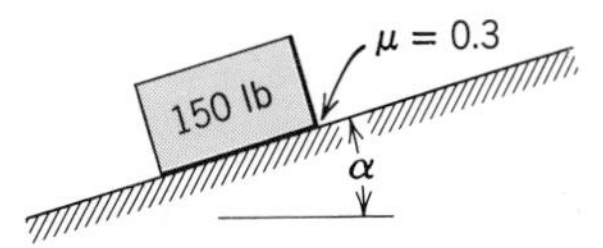

5-46 Let $\alpha = 20°$ in the system of the preceding problem. What will be the magnitude of the acceleration of the block as it slides down the plane? If it is sliding upward, having had an initial motion in this direction imparted to it, what will be the magnitude of its acceleration?

5-47 Show from the geometry of a force polygon that the relationship

$$\tan \alpha \leqslant \mu$$

limits the incline on which a friction-held body can remain at rest.

5-48 Will the crate slide down or tip over in the absence of P? The contents are such that the forces of gravity may be represented by a single force through the geometric center. What is the highest point at which P can be applied to slide the crate up without tipping it? *Ans.* Slide: 3.64 ft.

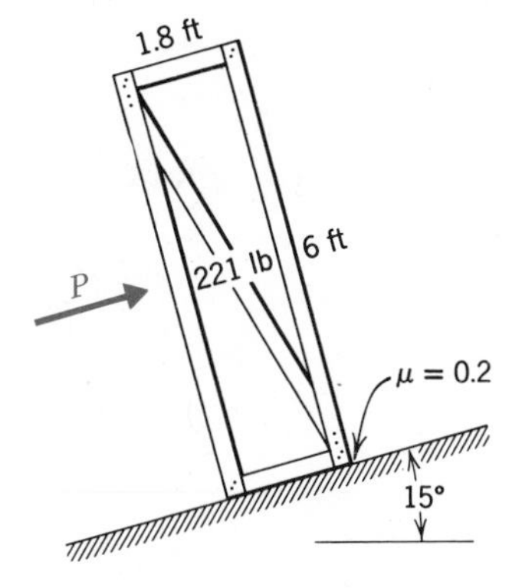

5-49 Rework Problem 5-48, with P horizontal rather than parallel to the incline.

5-50 The crate of mass m is to be moved along the floor without slipping. The contents are such that the forces of gravity may be represented by a single force through the geometric center. What is the maximum value that the angle α can have without the crate tipping?

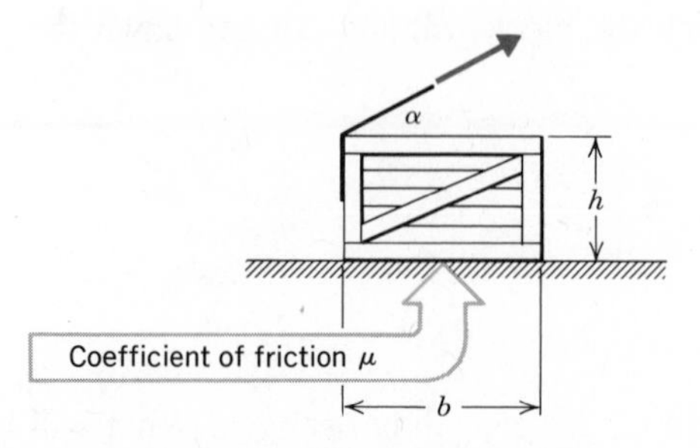

5-51 Determine the minimum value of P necessary to start motion. The coefficient of friction between all surfaces is 0.27. *Ans.* 103 N.

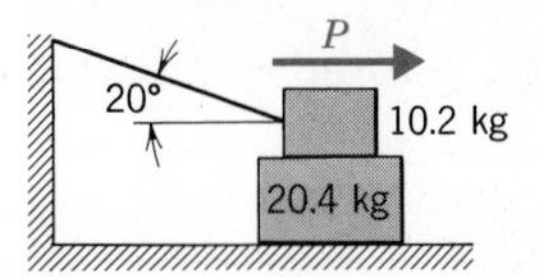

5-52 Determine the minimum value of P necessary to start motion. The coefficient of friction between all surfaces is 0.27.

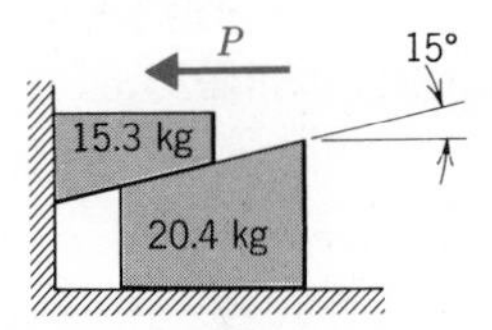

5-53 What will be the magnitude P of the horizontal force necessary to slide the ladder up the wall?

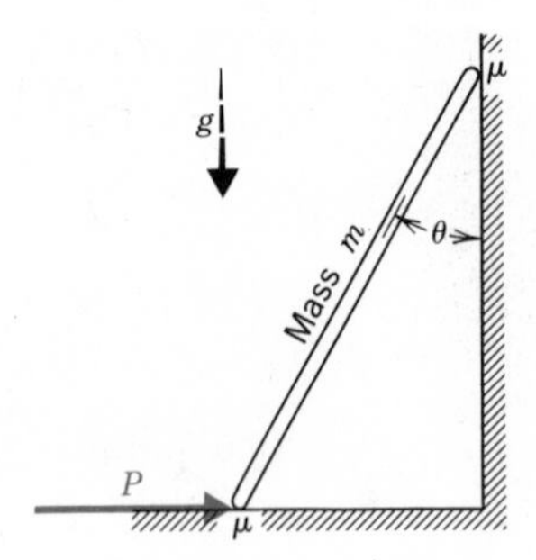

5-54 What will be the magnitude P of the horizontal force necessary to slide the ladder up the wall?

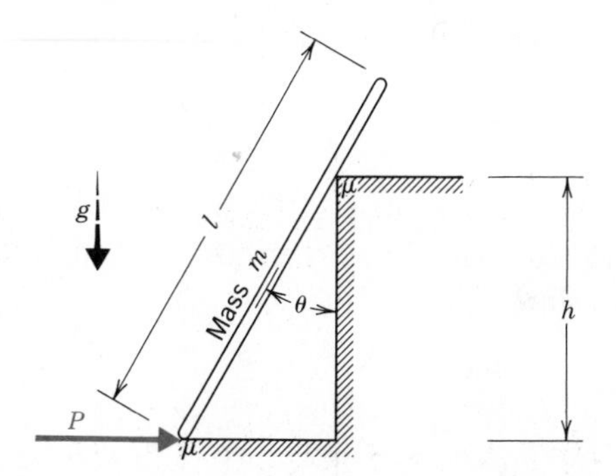

5-55 The ladder of mass m rests against the wall as shown. The coefficient of friction at both surfaces is μ. Determine how far up the ladder the man of mass M can walk before the ladder slips.

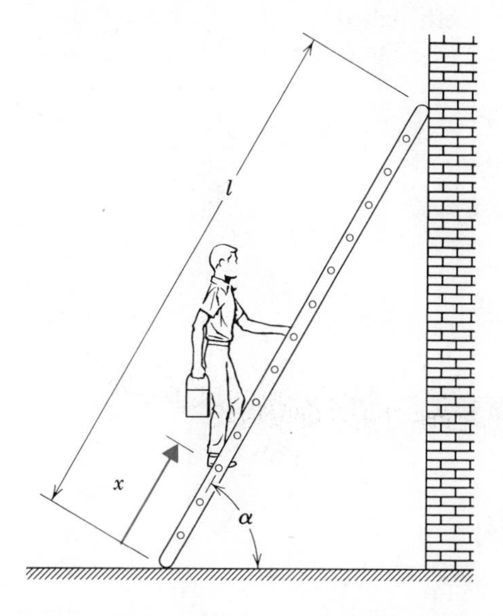

5-56 The forklift is being used to roll the 2-Mg drum up the incline, while the height of the lift is maintained as constant. The coefficient of friction between the vertical rails and the drum is 0.45, and that between the incline and the drum is 0.30. What horizontal thrust must the vehicle apply to the drum to move it?

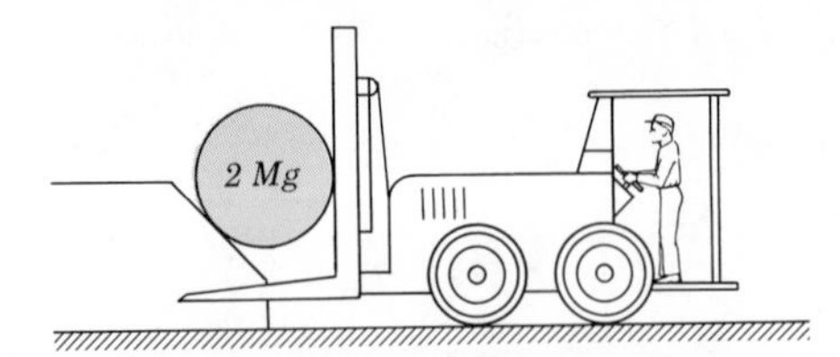

5-57 What coefficient of friction is necessary to prevent slipping between the yo-yo and the floor? Which way will the yo-yo roll, assuming there is sufficient friction to prevent slip?

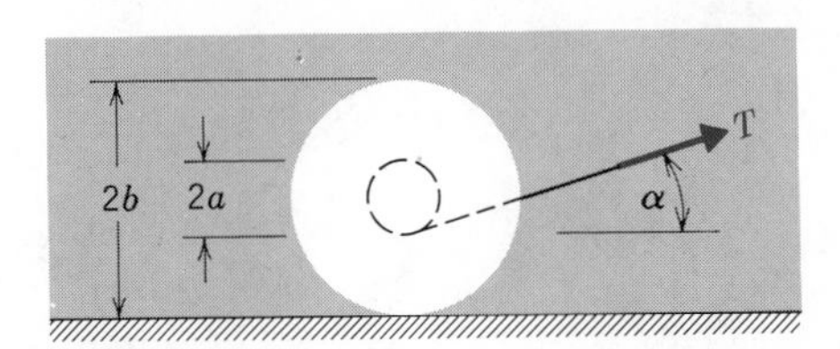

5-58 Determine the minimum coefficient of friction between the wheel and the step that will permit the wheel to roll over the step.

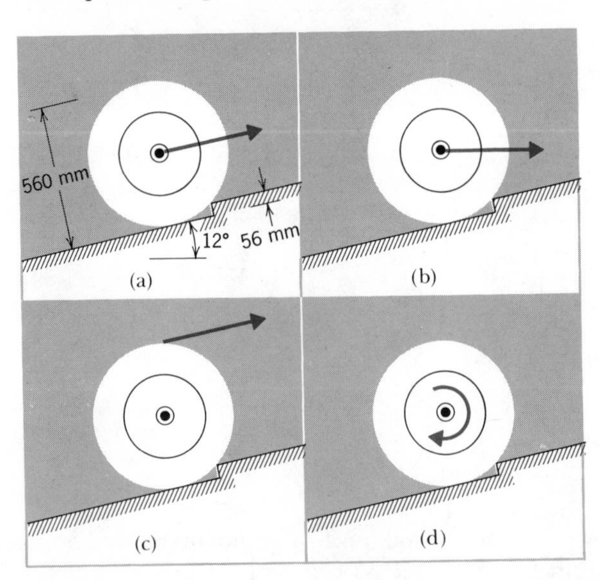

5-59 Determine the minimum coefficient of friction between the block and the bar necessary to prevent collapse. *Ans.* $\dfrac{a}{a+b}\dfrac{\sin\alpha}{\cos\beta}$.

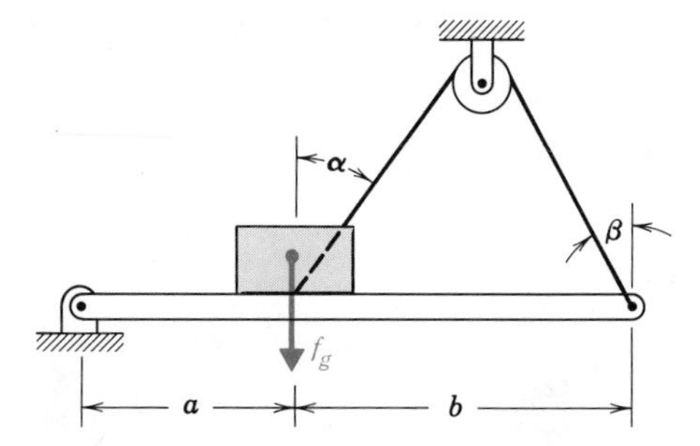

5-60 Determine the minimum coefficient of friction between the block and the bar necessary to prevent collapse.

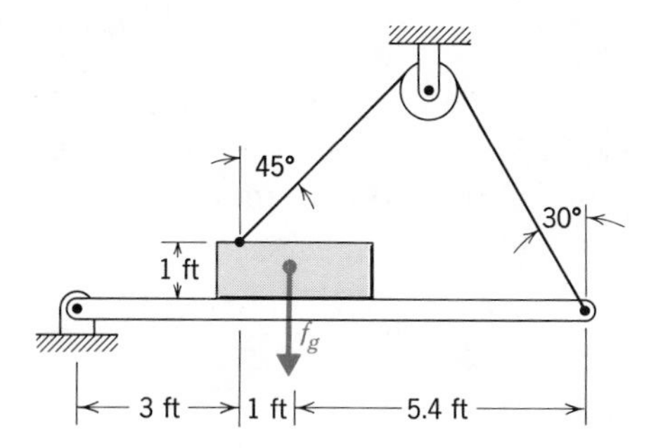

5-61 The small rollers are intended to prevent clockwise rotation of the large drum. The coefficient of friction between the rollers and drum, and between the rollers and walls, is μ. Determine the minimum distance d, so that the friction will effect a self-locking mechanism against clockwise rotation. Gravity is negligible.

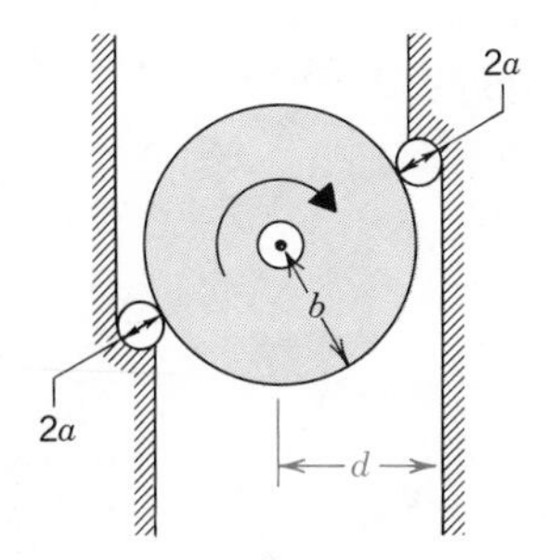

5-62 Until a clamp is tightened, the drill press table is free to slide along the column. Estimate the coefficient of friction required so that the collar will be self-locking against the column under the action of thrust from the drill. Neglect gravity.

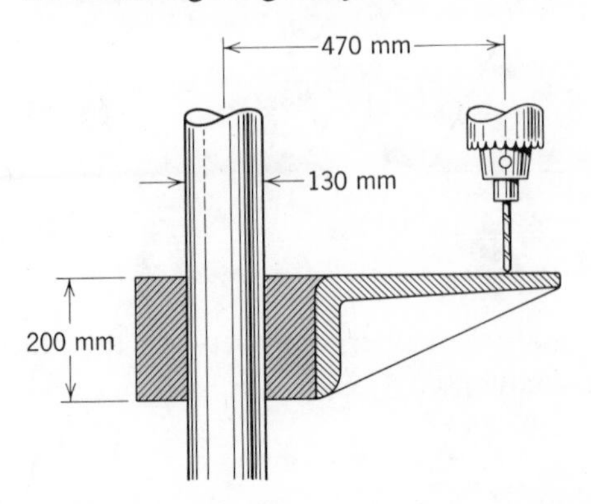

5-63 In terms of the axial force P, the coefficient of friction μ, and the inside and outside radii a and b, what is the moment M necessary to cause slipping. Assume that the normal pressure is uniform over the surface of contact.

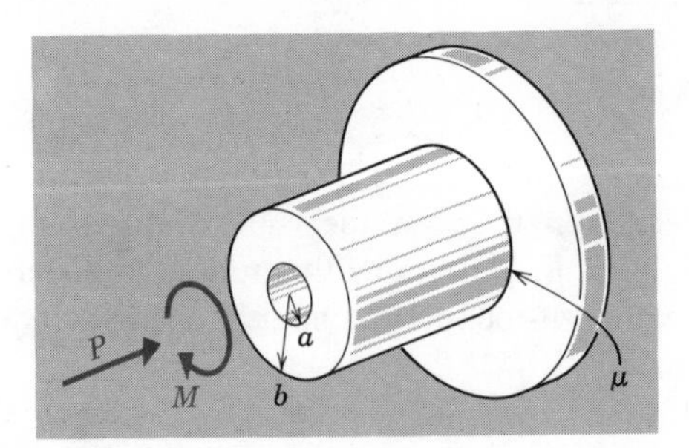

5-64 What is the magnitude P of the force required to move the cylinder along the horizontal V-groove?

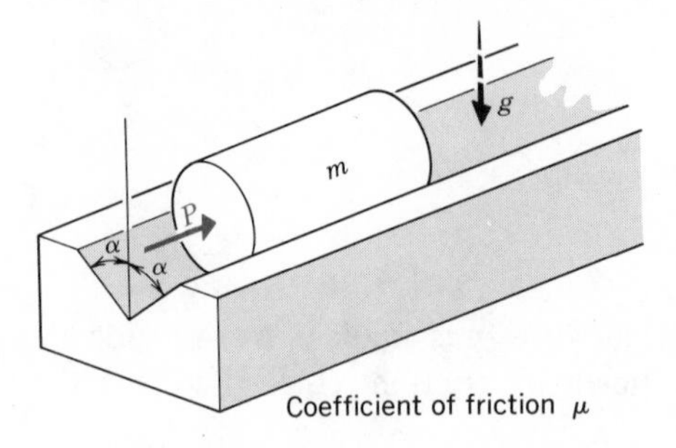

5-65 If the axis of the V-groove of the previous problem is tilted at an angle θ with the horizontal, what will be the magnitude P of the force necessary to slide the cylinder up the incline?

5-66 In terms of μ, mg, and θ, what will be the magnitude P of the horizontal force that will cause the block to slide?

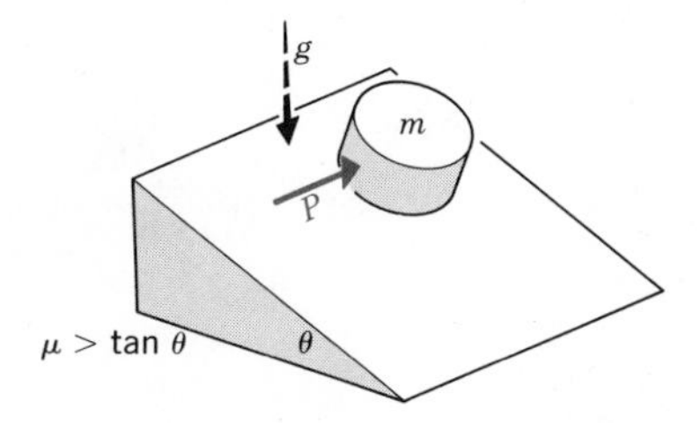

5-67 In what direction will the block of the previous problem begin to slide?

5-68 The angle θ of Problem 5-66 is half that which would allow the block to slide in the absence of P. In terms of μ and mg, what is the value of P that will cause the block to begin to slide?

5-69 A 50-lb force is required to raise the 12-lb block at a uniform rate. What is the coefficient of friction between the line and the post? What force will be required to lower the block at a uniform rate? *Ans.* 0.182; 2.88 lbf.

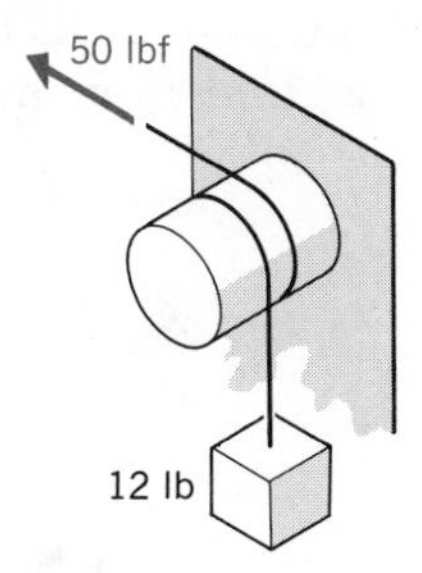

5-70 What is the ratio of the force f_2 to the force f_1 in the force amplifier?

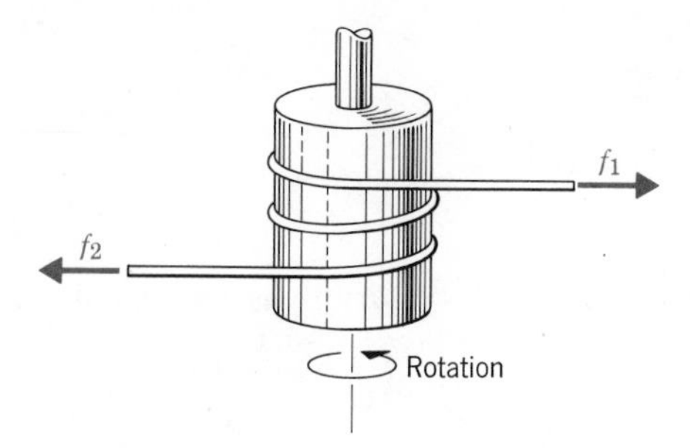

5-71 What braking force f is required to exert a friction moment of 115 N·m opposing clockwise rotation of the drum? The coefficient of friction is 0.2. What would be the required force if the rotation is counterclockwise?

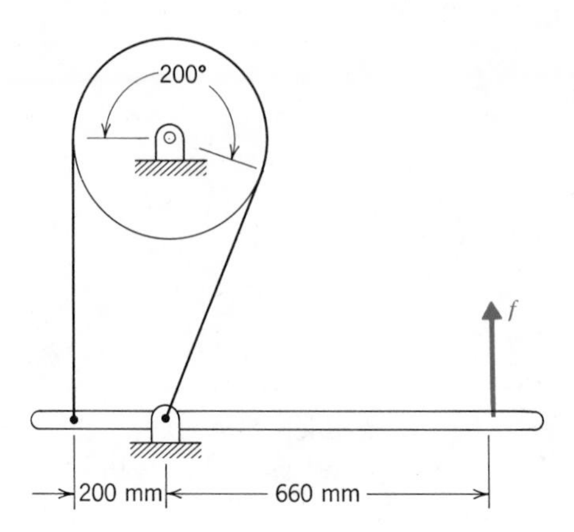

5-72 The lever arm in the differential band brake is arranged so that the tight side of the band aids the actuating force while the loose side opposes it. Show that the mechanism can be made so as to permit rotation in one direction only, while it is self-locking in the other direction. This type of device is commonly used on hoists to prevent a load from dropping.

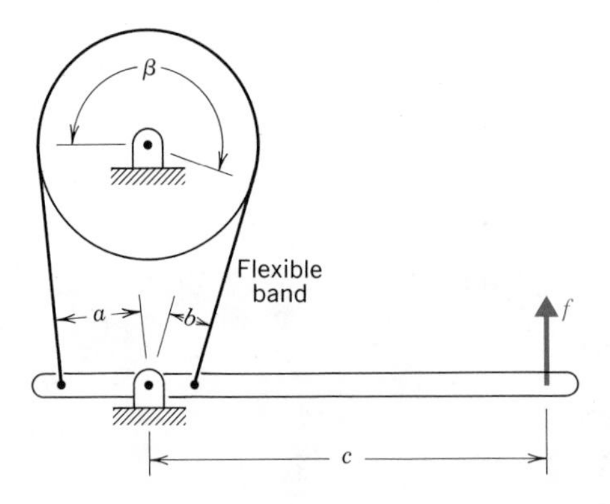

5-73 Estimate the coefficient of friction necessary in order that the oil filter wrench does not slip.

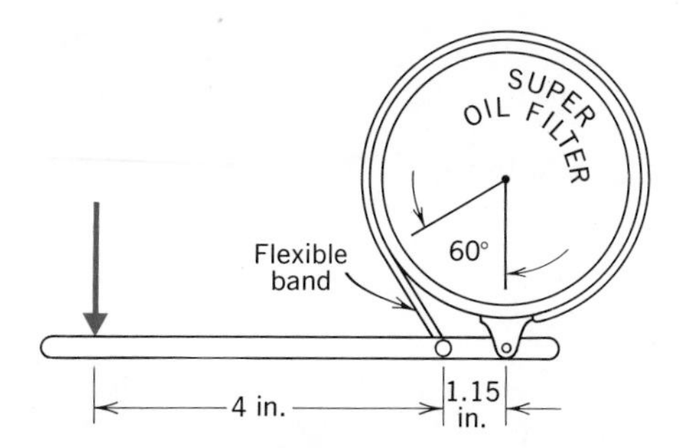

5-74 The surface over which the flexible line is strung is not circular, although the coefficient of friction μ is uniform. What will be the appropriate relationship between T_2 and T_1 for slipping in this case?

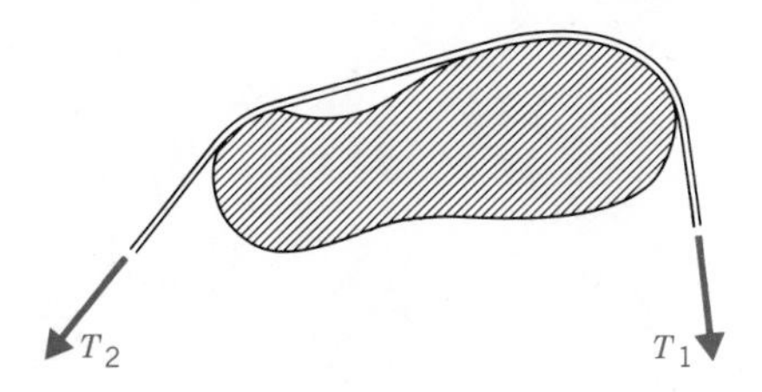

DISTRIBUTED FORCES

Each of the forces that we have considered thus far as being concentrated along a discrete line of action is actually a resultant of a set of forces distributed throughout some region. With contact forces, the distribution is over some surface area; with gravitational forces, the distribution is throughout some volume. In effect, this amounts to a composition of the components within the distributed set of forces.

If the region is sufficiently small, or the distribution has a certain regularity, the composition can be made correctly by inspection. In this chapter, we are concerned with the computation of resultants of distributed forces in cases that do not lend themselves readily to composition by inspection.

The procedures are based on the same ideas that are explained in Section 3-5 and in Chapter 4; in handling a set of distributed forces the only new feature is the computational detail of summation, which takes the form of integration.

EXAMPLE

The beam supports a load that varies in intensity along the length as indicated. The intensity (force per unit length of beam) has the values w_A and w_B at the two ends and varies linearly between these points. In terms of w_A, w_B and L, what is the resultant of the downward forces and what are the magnitudes R_A and R_B of the reactions at the supports?

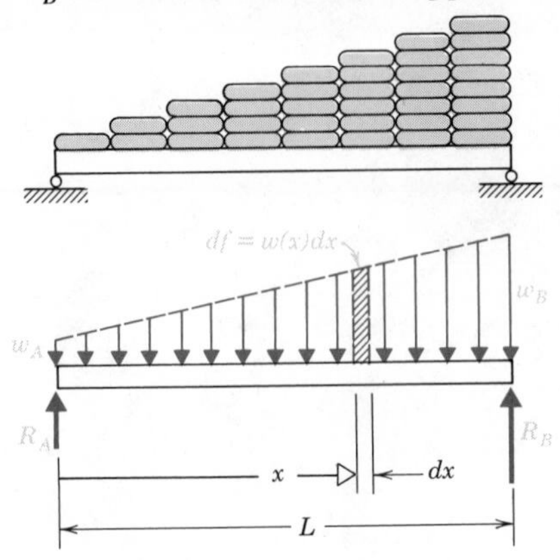

First, we write an expression for the load intensity as a function of x, the distance along the span measured from the left-hand support. This will be

$$w(x) = w_A + (w_B - w_A)\frac{x}{L}$$

Next, consider the force in the shaded portion of the load diagram, acting between the points given by x and $x + dx$. The force in this region will be the product of the intensity (force per unit length) and the length dx:

$$df = w(x)dx$$

Summing all such forces gives their resultant,

$$\begin{aligned} f &= \int w(x)\,dx \\ &= \int_0^L \left[w_A + (w_B - w_A)\frac{x}{L} \right] dx \\ &= \left[w_A x + (w_B - w_A)\frac{x^2}{2L} \right]_0^L \\ &= \tfrac{1}{2}(w_A + w_B)L \end{aligned}$$

To evaluate R_B, we equate to zero the sum of moments about A of all the forces:

$$M_A = R_B L - \int x df$$

$$= R_B L - \int_0^L x\left[w_A + (w_B - w_A)\frac{x}{L}\right] dx$$

$$= R_B L - \left[w_A \frac{x^2}{2} + (w_B - w_A)\frac{x^3}{3L}\right]_0^L$$

$$= R_B L - \left(\tfrac{1}{6}w_A + \tfrac{1}{3}w_B\right)L^2 = 0$$

This yields

$$R_B = \left(\tfrac{1}{6}w_A + \tfrac{1}{3}w_B\right)L$$

To evaluate R_A, we can use the fact that the sum of all vertical forces must be zero:

$$R_A = f - R_B$$

$$= \tfrac{1}{2}(w_A + w_B)L - \left(\tfrac{1}{6}w_A + \tfrac{1}{3}w_B\right)L$$

$$= \left(\tfrac{1}{3}w_A + \tfrac{1}{6}w_B\right)$$

Where would the line of action of a single force be equivalent to the given distributed load?

EXAMPLE

The uniform, slender, semicircular arch shown in Figure 6-1 is acted on by gravity and the reactions from the supports. Evaluate the bending moment across the section at the top of the arch.

The free-body diagram shows the desired bending moment M_b, the reaction from the other half of the arch. Because of horizontal equilibrium no axial force exists at this section. The shearing force is also zero at this section, because Newton's third law would demand an oppositely directed shearing force on the other half of the arch, and this pair of forces would be inconsistent with the symmetry of the system.

Because the arch is slender, the forces of gravity may be treated as distributed along a circular *line*. Let the cross-sectional area be denoted by A and the density (mass per unit volume) by ρ. Then the volume of the shaded

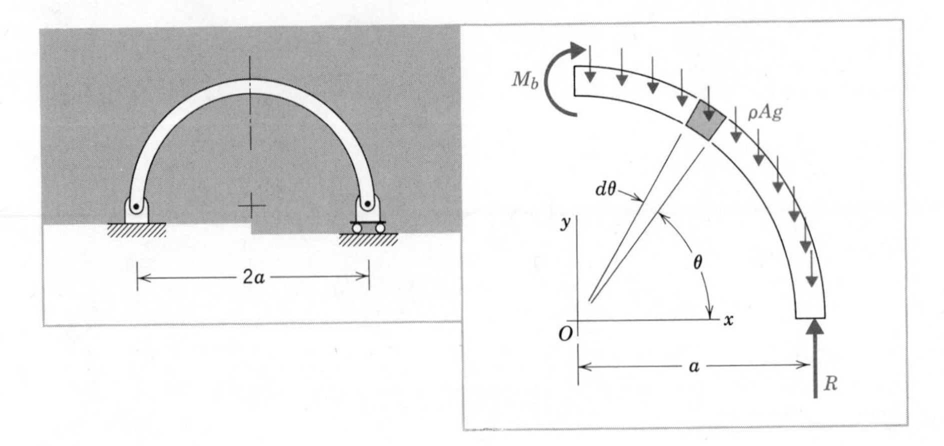

FIGURE 6-1

element of the arch will be equal to $Aa\,d\theta$, and the magnitude of the force of gravity acting on it will be given by

$$df_g = \rho(Aa\,d\theta)g$$

The resultant of the gravitational forces then has the magnitude

$$f_g = \int_0^{\pi/2} \rho Aga\,d\theta$$

$$= \tfrac{1}{2}\pi a\rho Ag$$

and vertical force equilibrium gives the support reaction as

$$R = \tfrac{1}{2}\pi a\rho Ag$$

Now, let us compute the resultant moment of the forces of gravity about the point O. The direction is into the plane of the sketch (clockwise) and the magnitude is given by

$$M_{Og} = \int a\cos\theta\,df_g$$

$$= \int_0^{\pi/2} a\cos\theta\rho Aga\,d\theta$$

$$= a^2\rho Ag$$

Finally, moment equilibrium requires that

$$Ra - M_{Og} - M_b = 0$$

which, together with the above results, gives

$$M_b = (\tfrac{1}{2}\pi - 1)a^2 \rho A g$$

In an arch with a large radius, this bending moment could easily cause failure of the structure.

6-1 SINGLE FORCE EQUIVALENTS

The setting up and evaluating of integrals such as in the examples above can be circumvented if knowledge of an equivalent discrete force is available. This information, for a variety of special cases, is available in tabulated formulas such as appear in Appendix B of this volume. These formulas are generated by such integrations.

The concepts developed in this section provide the understanding necessary for proper use of such tables and are useful in applications other than statics.

Center of Mass and Center of Gravity

Consider the forces of gravity distributed throughout an arbitrary body, as shown in Figure 6-2. If the body is small with respect to the earth, the forces will be parallel, and g will have the same value throughout the body. The force acting on the element with mass dm will be

$$d\mathbf{f}_g = dm\, g\mathbf{u}_g$$

and the resultant force will be given by

$$\begin{aligned}\mathbf{f}_g &= \int d\mathbf{f}_g \\ &= \int (dm\, g\mathbf{u}_g) \\ &= \left(\int dm\right) g\mathbf{u}_g \\ &= mg\mathbf{u}_g\end{aligned}$$

FIGURE 6-2

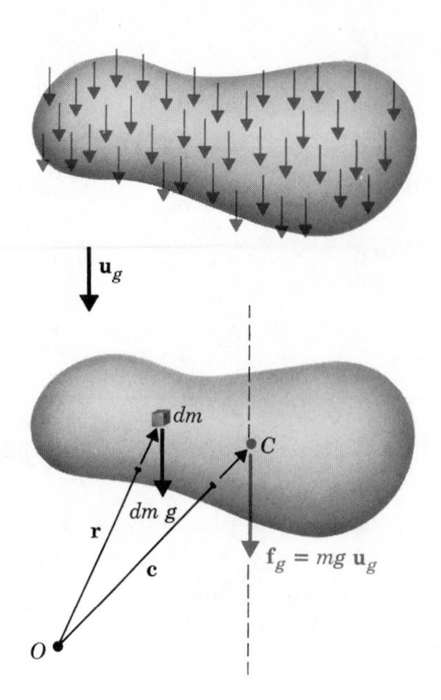

where m is the total mass of the body. The resultant moment about a point O is given by

$$\mathbf{M}_{Og} = \int \mathbf{r} \times d\mathbf{f}_g$$

$$= \int \mathbf{r} \times (g\mathbf{u}_g\, dm)$$

$$= \left(\int \mathbf{r}\, dm\right) \times g\mathbf{u}_g$$

where $\mathbf{r}$ is the position vector locating the mass element. Now, to compose the forces into a single resultant, the line of action of the equivalent force $\mathbf{f}_g$ must pass through a point located by the position vector $\mathbf{r}_f$, satisfying the moment equivalence

$$\mathbf{r}_f \times \mathbf{f}_g = \mathbf{M}_{Og}$$

Or,

$$\mathbf{r}_f \times (mg\mathbf{u}_g) = \left(\int \mathbf{r}\, dm\right) \times g\mathbf{u}_g$$

$$\mathbf{r}_f \times \mathbf{u}_g = \mathbf{c} \times \mathbf{u}_g$$

where

$$\boxed{\mathbf{c} = \frac{1}{m}\int \mathbf{r}\, dm} \qquad \textbf{(6-1)}$$

The vector **c** locates an important point C, called the *center of mass* of the body. It is clear from the equation preceding (6-1) that by choosing $\mathbf{r}_f = \mathbf{c}$, moment equivalence will be satisfied for any $\mathbf{u}_g$, which implies that the line of action of the equivalent force passes through C regardless of how the body is oriented.

The center of mass C is of fundamental importance in the study of dynamics. In statics it significance lies in the fact that the resultant moment about C, of the force of uniform gravity, is zero. That is, the body could be statically balanced by supporting it at this point only. For this reason it is also called the center of gravity.*

EXAMPLE

Find the location of the center of gravity of the portion of arch isolated as a free-body in Figure 6-1.

Using the center of the circle as a reference point, we can locate the shaded mass element with the position vector

$$\mathbf{r} = a\cos\theta\,\mathbf{u}_x + a\sin\theta\,\mathbf{u}_y$$

Then, from (6-1),

$$\begin{aligned}\mathbf{c} &= \frac{1}{m}\int (a\cos\theta\,\mathbf{u}_x + a\sin\theta\,\mathbf{u}_y)(\rho A a\, d\theta)\\ &= \frac{1}{m}\rho A a^2\left[\left(\int_0^{\pi/2}\cos\theta d\theta\right)\mathbf{u}_x + \left(\int_0^{\pi/2}\sin\theta d\theta\right)\mathbf{u}_y\right]\\ &= \frac{\rho A a^2(\mathbf{u}_x + \mathbf{u}_y)}{\rho A a\pi/2}\end{aligned}$$

*If the gravitational field varies in magnitude or direction throughout the body, the forces of gravity can produce a moment about the center of mass. This small moment can be important to the rotational motion of bodies such as satellites and planets. In this case, it is not reasonable to refer to C as the center of gravity. However, the center of mass is a property of the body itself, independent of the environment.

$$= \frac{2a}{\pi}\mathbf{u}_x + \frac{2a}{\pi}\mathbf{u}_y$$

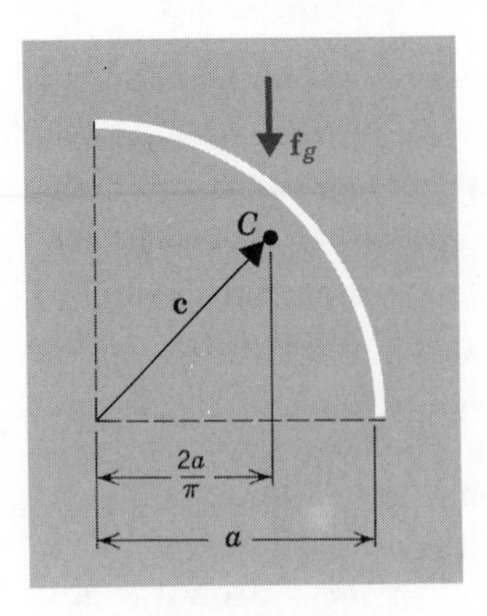

Centroids

The mass *dm* of the element used in the preceding integrals can be expressed in terms of the density ρ and the corresponding element of volume dV, as $dm = \rho\, dV$. Then Equation 6-1 can be written as

$$\mathbf{c} = \frac{\int \mathbf{r}\rho\, dV}{\int \rho\, dV}$$

Now, if the density is uniform throughout the body, ρ can be brought outside the integrals, with the result

$$\boxed{\mathbf{c} = \frac{1}{V}\int \mathbf{r}\, dV} \qquad \textbf{(6-2)}$$

The location of the point C here depends entirely on geometry, since all contribution having to do with material has been canceled. The point C, located according to (6-2), is called the *centroid of the volume V.*

Similarly, the *centroid of a surface area A* is defined as the point located by the position vector

$$\boxed{\mathbf{c} = \frac{1}{A}\int \mathbf{r}\, dA} \qquad \textbf{(6-3)}$$

and the *centroid of a line segment* of length L is defined as the point located by the position vector

$$\boxed{\mathbf{c} = \frac{1}{L}\int \mathbf{r}\, dL} \tag{6-4}$$

The calculation in the preceding example was for a uniform mass per unit length of the arch; with this density canceled out, the center of mass of the segment of arch is also the centroid of a quarter segment of a circular line.

The integrals $\int \mathbf{r} dV$, $\int \mathbf{r} dA$, and $\int \mathbf{r} dL$ are called the first moments of volume, area, and line, respectively, about the reference point O.

In carrying out the calculations, it is usually convenient to work with one rectangular Cartesian component of the position vectors at a time, that is, to evaluate separately the component equivalents of (6-2). These are, in terms of $\mathbf{r} = x\mathbf{u}_x + y\mathbf{u}_y + z\mathbf{u}_z$,

$$c_x = \frac{1}{V}\int x\, dV$$

$$c_y = \frac{1}{V}\int y\, dV$$

$$c_z = \frac{1}{V}\int z\, dV$$

EXAMPLE

Determine the location of the centroid of the triangle shown in Figure 6-3.

The area and first moment of area can be computed by evaluating the contribution to these quantities from the shaded element shown, and summing these contributions by integration with respect to y. The width of the element can be expressed in terms of y after observing that the entire triangle is similar to the triangle that lies above the shaded element:

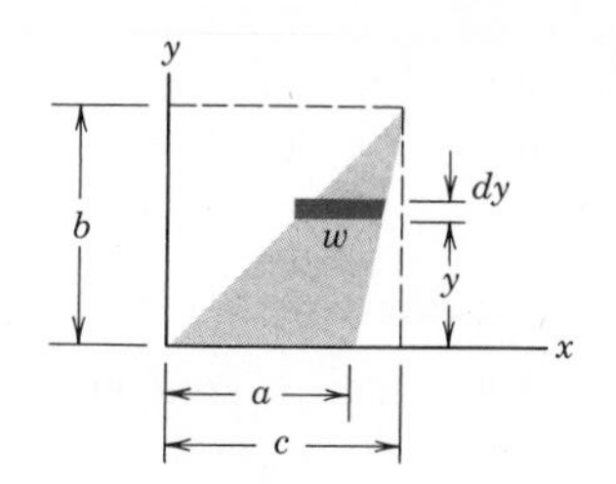

FIGURE 6-3

$$\frac{w}{b-y} = \frac{a}{b}$$

Or,

$$w = a\left(1 - \frac{y}{b}\right)$$

Thus, the area of the element is

$$dA = w\,dy = a\left(1 - \frac{y}{b}\right)dy$$

and the area of the triangle is

$$A = \int_0^b a\left(1 - \frac{y}{b}\right)dy = \frac{1}{2}ab$$

a result that you probably could have written from memory.

Now, the center of the element is located on the straight line that connects the apex of the triangle with the midpoint of the base, that is, on the line that has the equation

$$x = \frac{a}{2} + \left(c - \frac{a}{2}\right)\frac{y}{b}$$

Therefore the x-component of the first moment of the area of the triangle is

$$\int x\,dA = \int_0^b \left[\frac{a}{2} + \left(c - \frac{a}{2}\right)\frac{y}{b}\right] a\left(1 - \frac{y}{b}\right)dy$$

$$= \tfrac{1}{6}ab(a+c)$$

With this and the above value of area, the x = coordinate of the centroid is determined as

$$c_x = \frac{1}{A}\int x\,dA$$

$$= \frac{ab(a+c)/6}{ab/2}$$

$$= \tfrac{1}{3}(a+c)$$

A similar calculation yields the y-component of the first moment of area, as

$$\int y\,dA = \int_0^b ya\left(1 - \frac{y}{b}\right) dy = \frac{ab^2}{6}$$

from which we find the y-coordinate of the centroid to be

$$c_y = \frac{1}{A}\int y\,dA$$

$$= \frac{ab^2/6}{ab/2}$$

$$= \frac{b}{3}$$

EXAMPLE

Find the location of the centroid of the wedge-shaped volume shown in Figure 6-4.

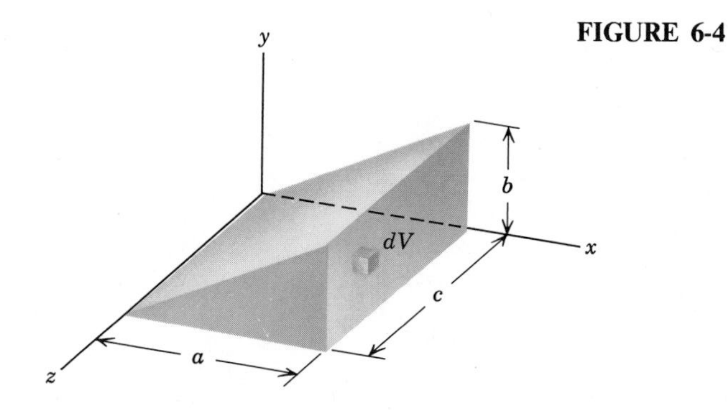

FIGURE 6-4

In terms of the rectangular Cartesian coordinates shown, the volume element can be expressed as

$$dV = dx\,dy\,dz$$

and the equation for the slanted surface is

$$x = \frac{ay}{b}$$

The integrals needed for determining c_x are evaluated next as follows:

$$
\begin{aligned}
V &= \int_0^c \int_0^b \int_{ay/b}^a dx\, dy\, dz \\
&= \int_0^c \int_0^b a\left(1 - \frac{y}{b}\right) dy\, dz \\
&= \int_0^c a\left(b - \frac{b^2}{2b}\right) dz \\
&= \frac{1}{2} abc
\end{aligned}
$$

$$
\begin{aligned}
\int x\, dV &= \int_0^c \int_0^b \int_{ay/b}^a x\, dx\, dy\, dz \\
&= \int_0^c \int_0^b \frac{1}{2}\left[a^2 - \left(\frac{ay}{b}\right)^2\right] dy\, dz \\
&= \int_0^c \frac{1}{2} a^2 \left(b - \frac{b^3}{3b^2}\right) dz \\
&= \frac{1}{3} a^2 bc
\end{aligned}
$$

Then,

$$
\begin{aligned}
c_x &= \frac{1}{V} \int x\, dV \\
&= \frac{\frac{1}{3} a^2 bc}{\frac{1}{2} abc} \\
&= \frac{2a}{3}
\end{aligned}
$$

Similar calculations lead to the other coordinates of the centroid,

$$
c_y = \frac{b}{3}
$$

$$
c_z = \frac{c}{2}
$$

Calculations like those of the last two examples have been carried out for many geometric figures and the results of a number of these summarized in Appendix B. To illustrate the use of these results, let us determine the location of the centroid of the semicircular cone of Figure 6-5.

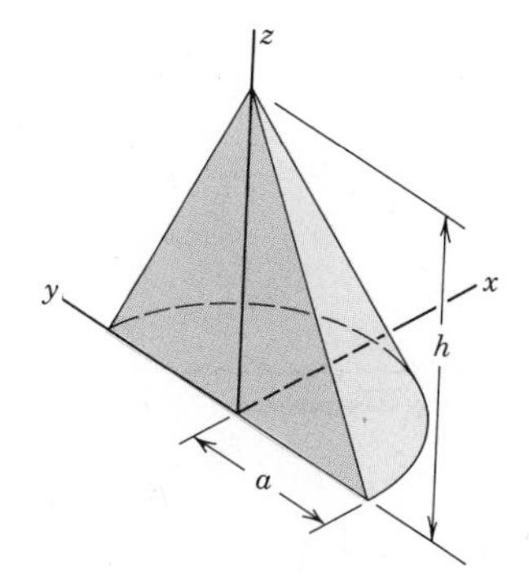

FIGURE 6-5

Referring to page 309, we see that the z-coordinate of the centroid is

$$c_z = \tfrac{1}{4}h$$

for any cone, which includes the special case of the semicircular shape we are considering here. To determine the x-coordinate of the centroid of the cone in terms of the given dimensions a and h, we need to find b_x, the x-coordinate of the semicircular base section. Referring to page 307, we see that this can be obtained from the appropriate formula for the sector of a circle by setting $\alpha = \pi/2$. Thus we have

$$b_x = \frac{2a \sin \dfrac{\pi}{2}}{3\dfrac{\pi}{2}} = \frac{4a}{3\pi}$$

Referring once again to the formulas for the cone, we find

$$\begin{aligned} c_x &= \frac{3}{4} b_x \\ &= \frac{3}{4}\left(\frac{4a}{3\pi}\right) \\ &= \frac{a}{\pi} \end{aligned}$$

A composite volume, area, or line may be made up from several parts, of which each centroid is known. In this case a summation having the same form as the integral definitions can be readily shown to be valid. In the case of an area A that is a composite formed from several areas $A_1, A_2, \ldots,$ having centroids located by $\mathbf{c}_1, \mathbf{c}_2, \ldots,$ the centroid of the composite would be located by

$$\mathbf{c} = \frac{\mathbf{c}_1 A_1 + \mathbf{c}_2 A_2 + \cdots}{A_1 + A_2 + \cdots}$$

EXAMPLE

Locate the centroid of the plane area shown in Figure 6-6.

FIGURE 6-6

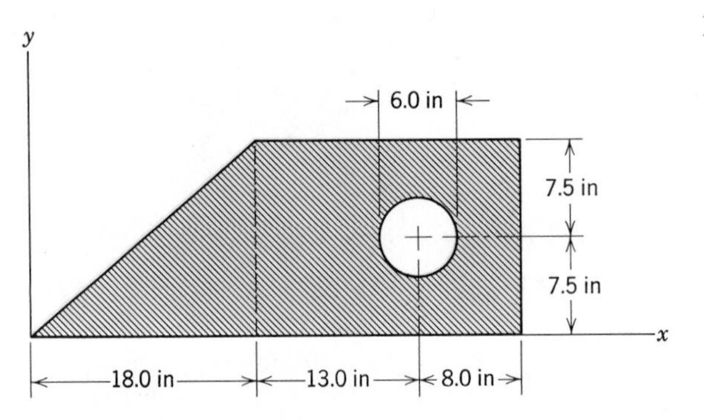

The areas and coordinates of centroids of individual parts are as follows.

	A, in^2	c_x, in	c_y, in
Triangle	135	12.0	5.0
Rectangle	315	28.5	7.5
Circle	− 28.3	31.0	7.5

The coordinates locating the centroid are then

$$c_x = \frac{(12.0)(135) + (28.5)(315) + (31.0)(-28.3)}{135 + 315 - 28.3} \text{ in}$$

$$= 23.05 \text{ in}$$

$$c_y = \frac{(5.0)(135) + (7.5)(315) + (7.5)(-28.3)}{135 + 315 - 28.3} \text{ in}$$

$$= 6.70 \text{ in}$$

PROBLEMS

6-1 For the example on page 201, calculate the reaction R_A by summing the moments of forces about B.

6-2 For the example on page 201, calculate the shear force V and the bending moment M at the section located by x.

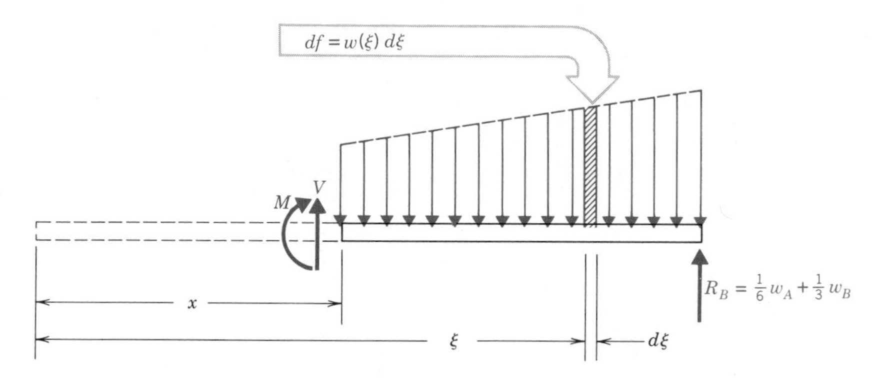

6-3 (a) Find the x-coordinate of the centroid of the trapezoid by integration. *Ans.* $c_x = \dfrac{\frac{1}{3}h_A + \frac{2}{3}h_B}{h_A + h_B} L.$

(b) Referring to the example on page 201, what would be the magnitude and where would be the line of action of a single concentrated force equivalent to the given distributed force?

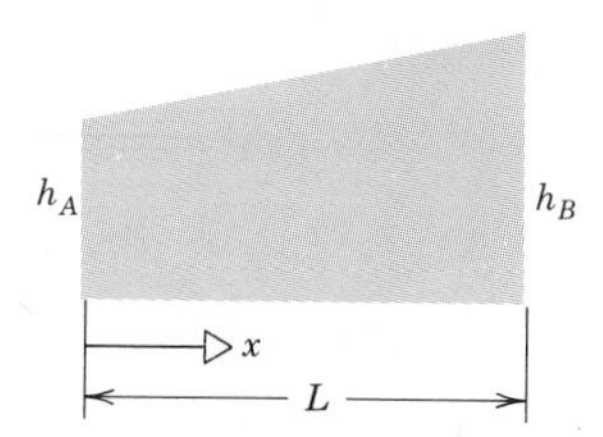

6-4 If the beam in the example on page 201 is loaded with a single concentrated force equivalent to the given distributed load, what will be the shear force V and bending moment M at the section located by x?

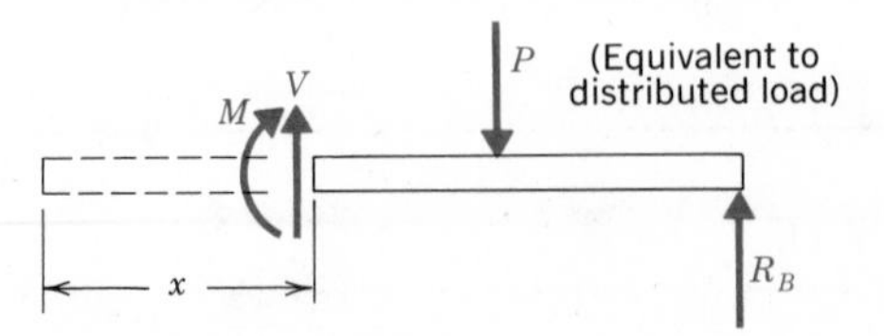

6-5 Plot V and M vs x for the loading of Problem 6-2 and for the equivalent loading found in Problem 6-4. After comparing the results, what can you say concerning the meaning of equivalent loading?

6-6 The mass of the earth is about 5.98×10^{24} kg, and that of the moon about 7.35×10^{22} kg. The center-to-center distance between the earth and the moon is 3.84×10^5 km (mean). The radius of the earth is 6370 km. With respect to the surface of the earth, where is the center of mass of the earth-moon system: *Ans.* 1710 km below the surface of the earth.

For Problems 6-7 through 6-11, find the location of the centroid of the shaded area.

6-7

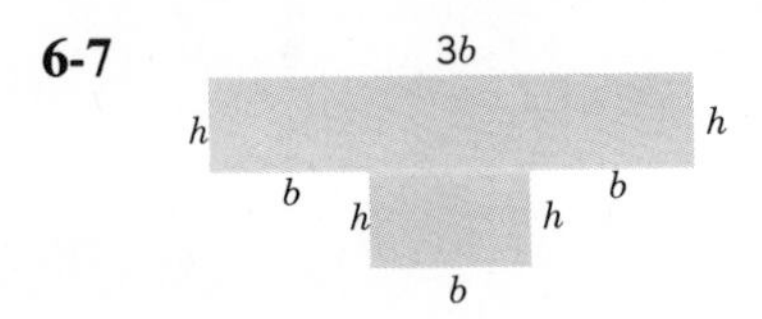

6-8

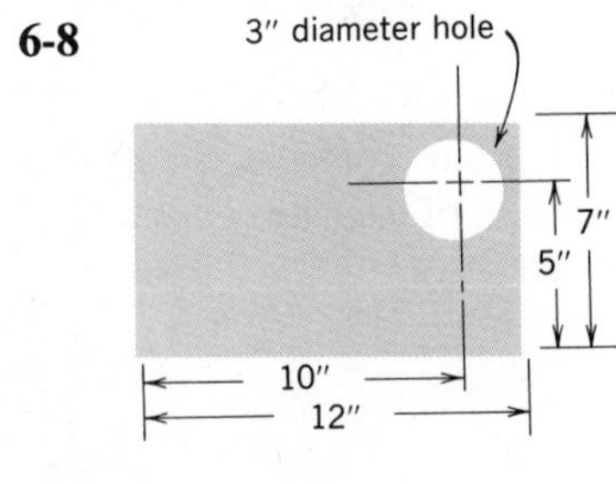

6-9

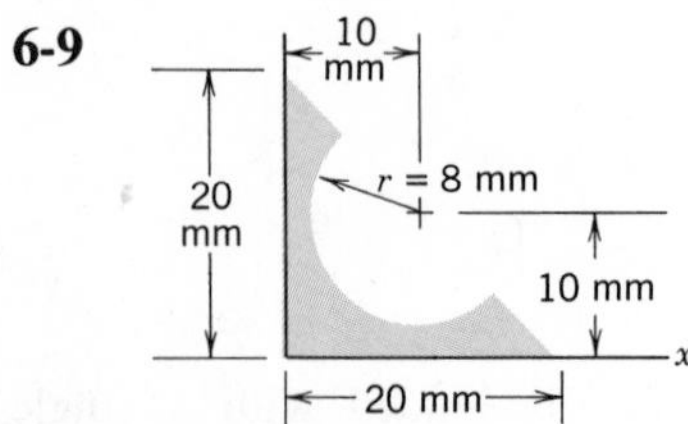

6-10

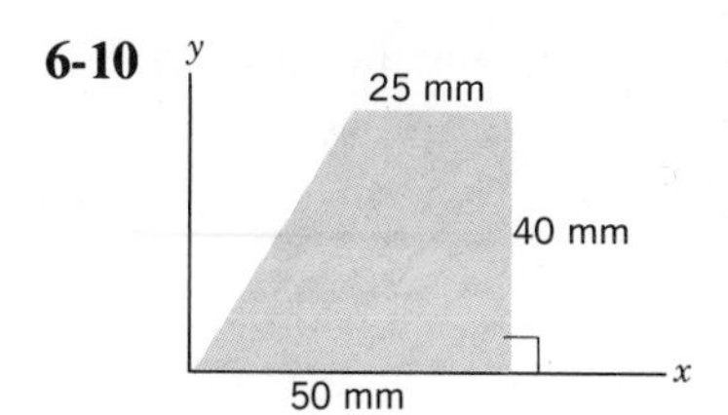

6-11

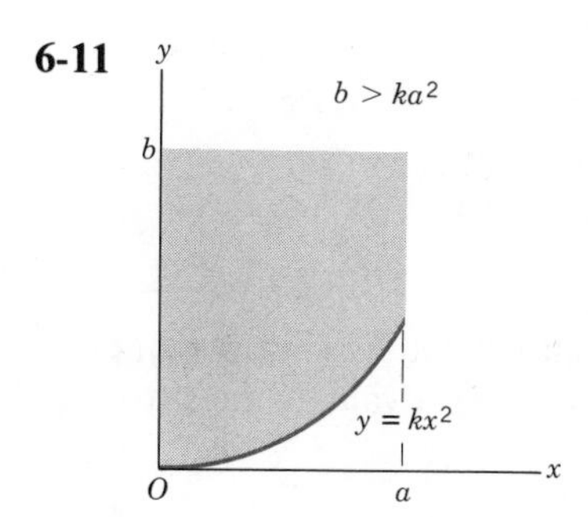

6-12 The frame shown is made of thin, homogeneous rod. Determine the location of its center of mass.

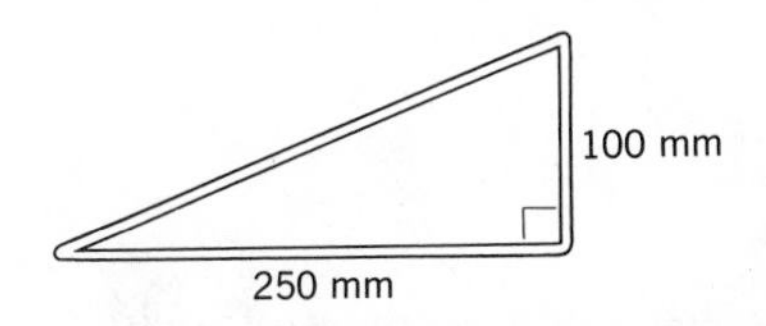

6-13 At what angle α will the plate hang, suspended at rest? *Ans.* 18.4°.

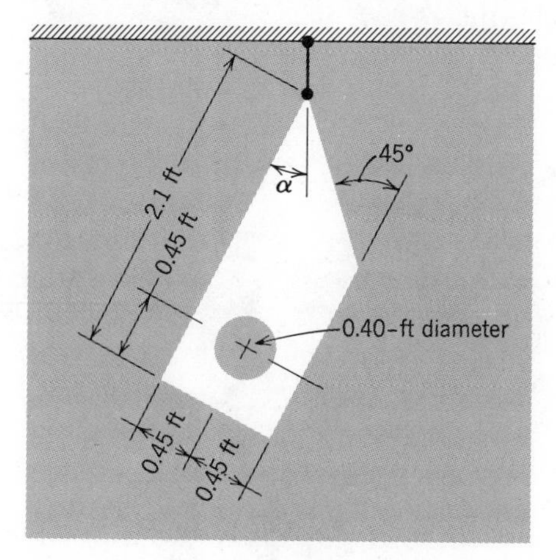

6-14 Determine the tension in each of the cables supporting the uniform slab.

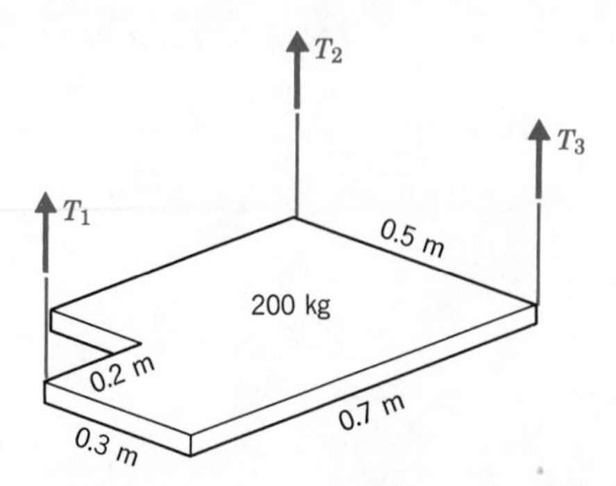

6-15 A uniform, triangular plate of mass m is suspended in a horizontal plane by three vertical wires, each attached at an apex. Evaluate the tension in each wire.

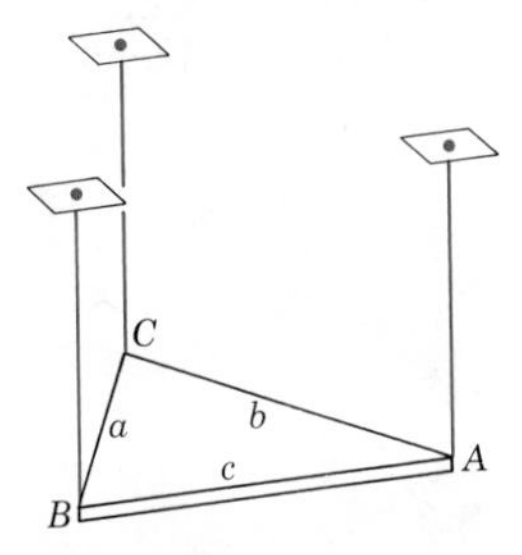

6-16 A slender rod of mass m has the shape of a segment of a circle and is suspended in a horizontal plane by three vertical wires as shown. Evaluate the tension in each wire.

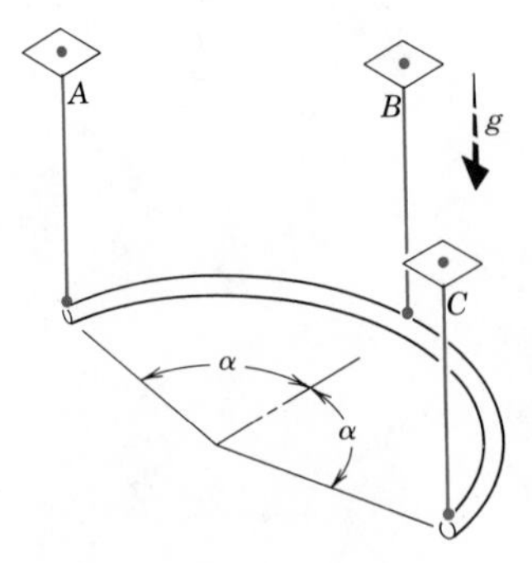

For Problems 6-17 through 6-20, verify the indicated formulas in Appendix B.

6-17 c_x and c_y for the area bounded by the x axis and the curve $y = bx^n/a^n$.

6-18 c_x and c_y for the segment of a circle

6-19 c_x for the segment of a sphere

6-20 c_x, c_y, and c_z for the cone

6-21 Locate the center of mass of the system consisting of the three homogeneous blocks. *Ans.* (2.583 ft, 3.714 ft, −2.046 ft).

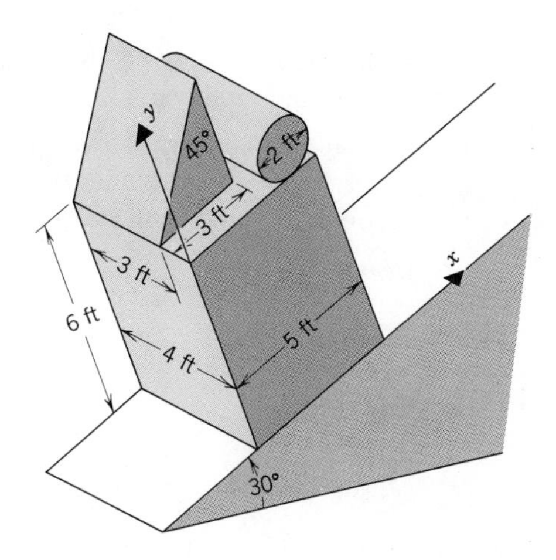

6-22 The bracket is made from uniform plate material. Determine the location of the center of mass.

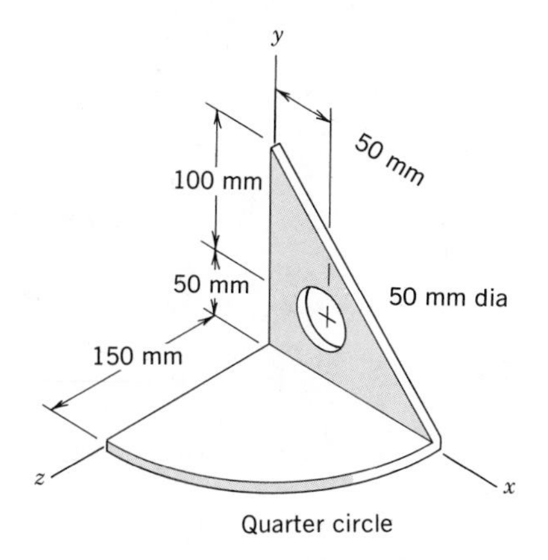

6-23 The trough is constructed of 0.060-in thick steel plate and has a semicircular cross section. What is the location of its center of mass?

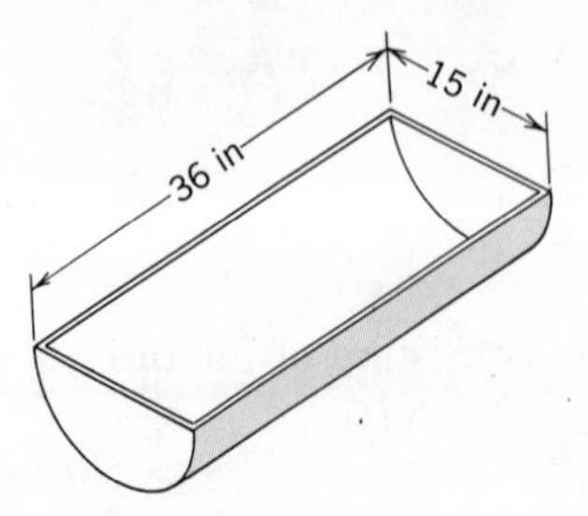

6-24 When the trough of the previous problem is half full of water, what is the location of the center of mass of the trough and water?

6-25 The uniform, thin rod is bent into a shape consisting of two quarter-circular arcs, the upper section lying in the x–y plane and the lower section in the y–z plane. Find the coordinates of the center of mass.

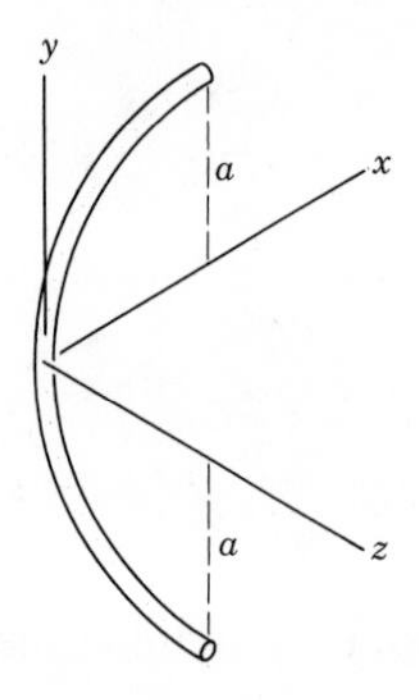

6-26 Determine the coordinates of the centroid of the pyramid. *Ans.* $(\frac{a}{4}, \frac{b}{4}, \frac{c}{4})$.

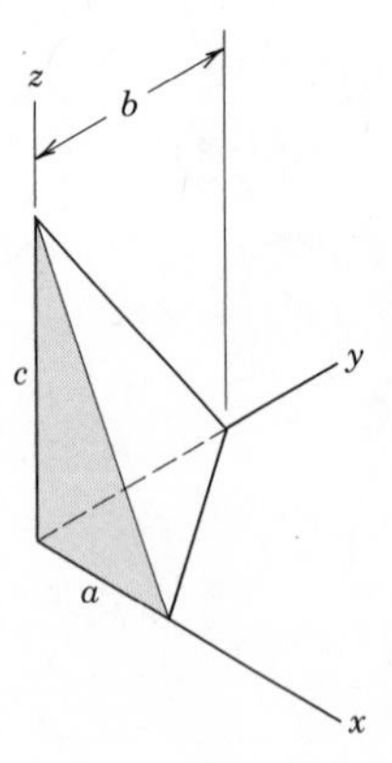

6-27 Determine the location of the centroid of the spherical wedge.

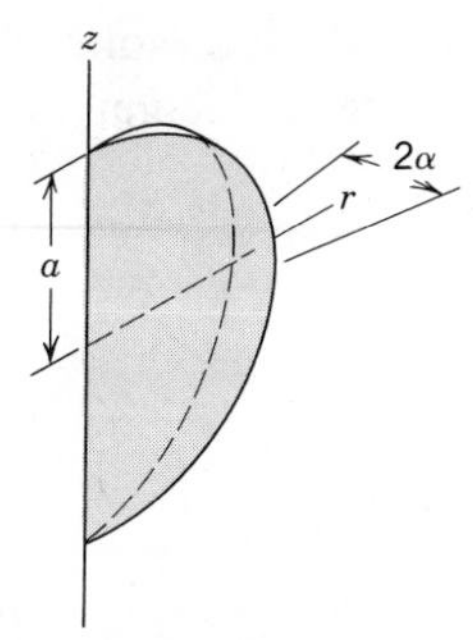

6-28 Determine the location of the centroid of the half torus.

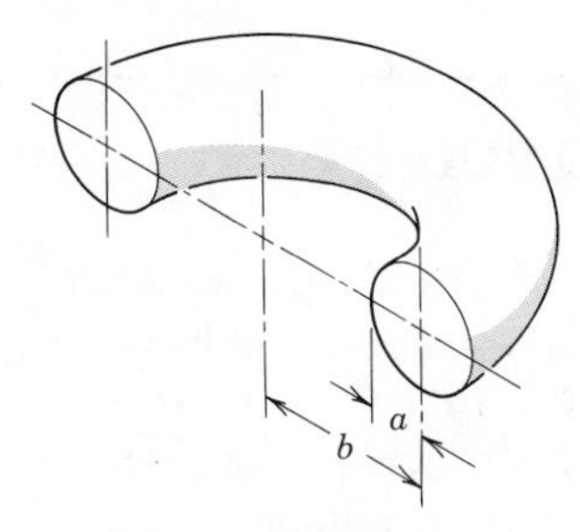

6-29 Determine the location of the center of mass of the thin-walled hemispherical shell.

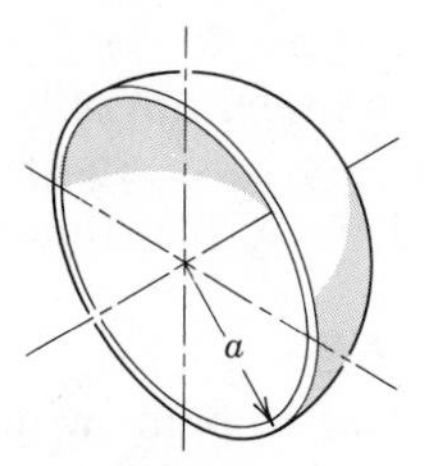

6-30 The thickness of the wall of the conical shell is uniform and much smaller than the radius a. Determine the location of the center of mass of the shell.

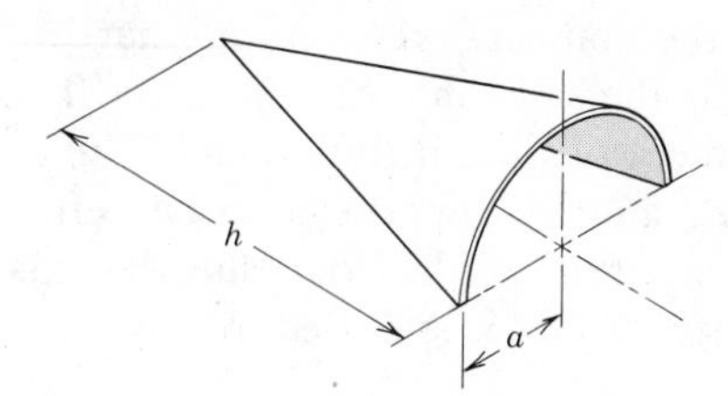

6-31 The 100-mm diameter sphere has a 40-mm hole drilled radially to its center, and the bottom of the hole is flat. Locate the center of mass. *Ans.* 3.12 mm behind bottom of hole.

6-2 SUBMERGED BODIES

Submerged structures are usually required to withstand significant forces from surrounding fluid. The simplest to predict are those that are induced by gravity when the fluid is at rest. The hydrostatic loading then acts normally to every surface and varies only with depth; from these facts we can develop some special techniques for computing force and moment resultants.

Static Fluid Pressure

When a fluid is at rest, it is capable of transmitting contact forces only in a direction perpendicular to the surface of contact. Here we refer not only to the interfaces between the fluid and some object but to any surface separating different parts of a fluid-filled region. This property distinguishes fluids from solids.

The intensity of this contact force, or force per unit of area, is called *pressure*. In general, the force intensity varies from point to point within the region, so that a definition of pressure at a point involves a limit, as the area considered approaches zero, of the ratio of force to area. In the SI system the unit of pressure is the newton per square meter, called the pascal (Pa). Atmospheric pressure at sea level is about 10^5 Pa = 100 kPa.

A consequence of the fact that the contact force acts normal to every surface is that the pressure at a point is the same in every direction. To show this, consider the free-body of the element of fluid in Figure 6-7. The area of the inclined surface is denoted as ΔA, and we suppose it to be sufficiently small that variations of pressure over each of the four surfaces may be neglected. The orientation of the inclined surface is specified by the unit vector normal to the surface:

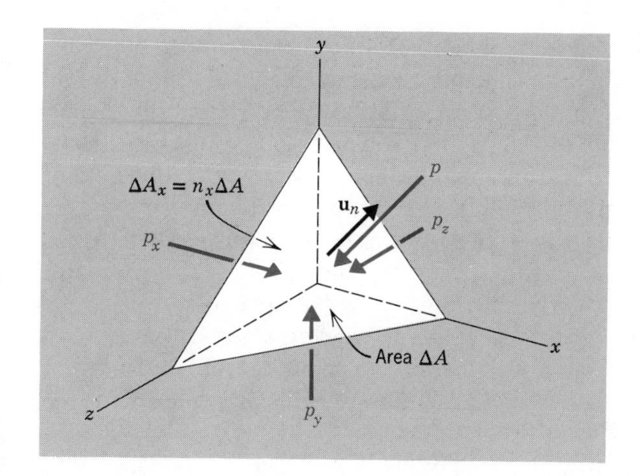

FIGURE 6-7

$$\mathbf{u}_n = n_x \mathbf{u}_x + n_y \mathbf{u}_y + n_z \mathbf{u}_z$$

That is, n_x, n_y, and n_z are the direction cosines of the normal to the surface. The areas of the three surfaces perpendicular to the axes are then (See Problem 3-35.)

$$\Delta A_x = n_x \Delta A, \quad \Delta A_y = n_y \Delta A, \quad \Delta A_z = n_z \Delta A$$

The volume of the element is $\Delta V = \frac{1}{3}\Delta A h$, where h is the perpendicular distance from the origin to the inclined surface. The force of gravity then has the magnitude $\frac{1}{3}\Delta A h \rho g$. With the x component of this force denoted as $\frac{1}{3}\Delta A h \rho g_x$, the equilibrium of forces in the x-direction gives

$$p_x(n_x \Delta A) - (p \Delta A) n_x + \tfrac{1}{3}\Delta A h \rho g_x = 0$$

Or,

$$(p_x - p) n_x + \tfrac{1}{3} h \rho g_x = 0$$

Now, as we let $h \to 0$, this reduces to

$$p_x = p$$

The same consideration in the other two directions gives

$$p_y = p_z = p$$

Thus the pressure at a point in a fluid at rest is the same in every direction.

We can make a useful interpretation of force relationships described above. The x component of the force acting on the inclined surface, $(p\Delta A)n_x$, is equal to the force $p(n_x\Delta A)$ acting on the surface perpendicular to the x-axis. But this latter area is the projection of the inclined surface onto the plane perpendicular to the x direction. Therefore, the component in any direction, of the force acting on an arbitrary element is equal to the product of the pressure and the projection of the element area onto the plane perpendicular to this direction. This *projected area interpretation* often simplifies the integration to obtain a component of the resultant of pressure over nonplanar surfaces.

An *incompressible* fluid is an idealization in which the density of the fluid is independent of pressure. Liquid water will undergo an increase in density of only 0.0023% under the influence of 10 000 atmospheres of pressure, so it is essentially incompressible for our present purposes.

In a body of incompressible fluid, gravity will induce pressure that increases linearly with depth. To verify this and determine the rate of increase, consider a free-body of a vertical cylindrical portion of fluid with cross sectional area ΔA and height $z_2 - z_1$. The horizontal forces are missing from the free-body in the sketch. However, we can deduce from the projected area interpretation that these forces are self-canceling provided that the pressure varies only with z and not in the horizontal directions. Vertical force equilibrium requires that

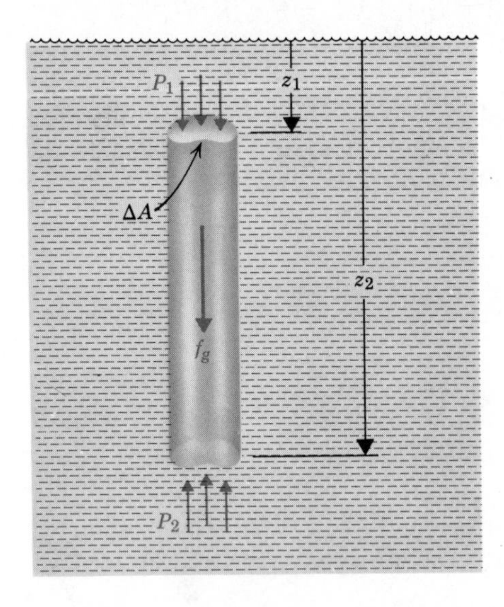

$$f_g = \rho\Delta A(z_2 - z_1)g = p_2\Delta A - p_1\Delta A$$

Or,

$$p_2 - p_1 = \rho g(z_2 - z_1) \tag{6-5}$$

For a container of arbitrary shape, the pressure change can be traced anywhere by using connected elements as above; therefore, (6-5) is valid for any two points that are connected by a region filled with the particular incompressible fluid.

For fresh water near the earth's surface,

$$\rho g = 9.8 \text{ kPa/m} = 62.4 \text{ lbf/ft}^3$$

Force Resultants

EXAMPLE

Evaluate the resultant force transmitted by the water to the dam depicted in Figure 6-8.

The force acting on the element of dam surface at depth z can be expressed as

$$d\mathbf{f} = p(z)w\,ds\left(\frac{dx}{ds}\mathbf{u}_z - \frac{dz}{ds}\mathbf{u}_x\right)$$

$$= p(z)w(dx\,\mathbf{u}_z - dz\,\mathbf{u}_x)$$

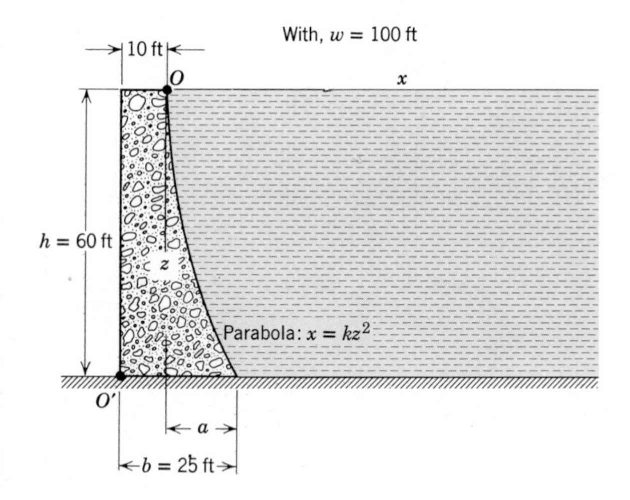

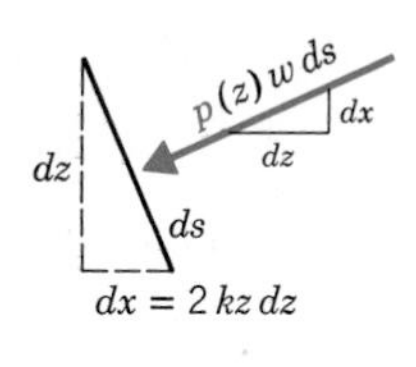

FIGURE 6-8

Notice that this last expression could be written directly by using the projected area interpretation. With atmospheric pressure denoted by p_a, the pressure at depth z is

$$p(z) = p_a + \rho g z$$

By using this and the differential relationship $dx = 2kz\,dz$ above, we can form the integral expression for the resultant force:

$$\mathbf{f} = w \int_0^h (p_a + \rho g z)(2kz\mathbf{u}_z - \mathbf{u}_x)\,dz$$

Use of the relationship $a = kh^2$ and evaluation of the integrals lead to

$$f_x = -wh\left(p_a + \tfrac{1}{2}\rho g h\right) \qquad \textbf{(a)}$$

$$f_z = wa\left(p_a + \tfrac{2}{3}\rho g h\right) \qquad \textbf{(b)}$$

Finally, substitution of the numerical values

$$p_a = 2100 \text{ lbf/ft}^2$$

$$pg = 62.4 \text{ lbf/ft}^3$$

$$w = 100 \text{ ft}$$

$$h = 60 \text{ ft}$$

$$a = 25 \text{ ft} - 10 \text{ ft} = 15 \text{ ft}$$

results in

$$f_x = -23.8 \times 10^6 \text{ lbf}$$

$$f_z = 6.9 \times 10^6 \text{ lbf}$$

Equations (a) and (b) can be obtained from a different point of view from that based on the force element shown in Figure 6-8. Consider the free-body of the portion of water shown in Figure 6-9. By Newton's third law, the force acting on the curved surface is of equal magnitude but has opposite direction from the force we wish to evaluate. Considering horizontal force equilibrium of this free-body, we can write

$$-f_x = \int_0^h p(z) w\,dz$$

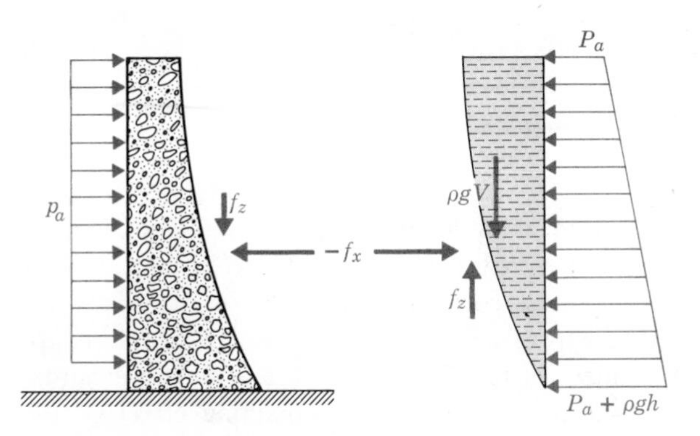

FIGURE 6-9

Now note that from the viewpoint of the geometry of the pressure distribution diagram, this is also equal to the volume under the pressure "envelope." In terms of well-known formulas for the areas of rectangles and triangles and the volume within a cylinder, we can write

$$-f_x = \left(p_a h + \tfrac{1}{2}\rho g h^2\right)w \tag{a}$$

Next, considering vertical force equilibrium, we have

$$f_z = p_a a w + \rho g V$$

where V is the volume of water in the free-body. This volume can be evaluated as

$$\begin{aligned} V &= \int_0^h (a - x)w\,dz \\ &= \int_0^h (a - kz^2)\,dz \\ &= \tfrac{2}{3}wah \end{aligned}$$

leading to the result (b).

Observe that the above calculations are evaluations of the force resultant that the *water* applies to the upstream surface of the dam. Other forces acting on the dam are from atmospheric pressure over the top and the downstream side, and the components of reaction across the bottom and ends.

Consider the effect of the air pressure. The density of air at sea level and 0° C is about 1.3 kg/m^3, so that the pressure will be approximately

$$\rho a g h = (1.3 \text{ kg/m}^3)(9.8 \text{ N/kg})(18.3 \text{ m})$$
$$= 0.23 \text{ kPa}$$
$$= 0.0023\, p_a$$

greater at the bottom of the dam than at the top. With this difference neglected, the pressure over the downstream surface yields a resultant force of magnitude $p_a wh$. When all horizontal components acting on the dam are summed, this force cancels that expressed by the first term in equation (a). Therefore, the horizontal component of reaction at the base of the dam can be calculated without regard for the air pressure. The remaining effect of the atmospheric pressure is to induce an increased vertical component of reaction at the base.

Archimedes' Principle Consider a submerged object, with its underneath surface in contact only with the fluid. Because of the pressure gradient induced by gravity, the upward acting fluid pressure on the underneath surface will be greater than the downward acting pressure on the upper surface. The net effect is an upward force called the *buoyant force* acting on the object. Now if the object were to be replaced by fluid identical to that surrounding the object, the buoyant force acting on this replacement fluid would be the same as that on the original object. But since the replacement fluid would be in equilibrium, *the magnitude of this buoyant force must be identical with the magnitude of the force of gravity on the displaced fluid*. This is known as Archimedes' Principle.

EXAMPLE

A sphere having a density ρ_s and radius a is released at the surface of a body of still water that has a density ρ. At what depth will the sphere reach equilibrium?

The volume of water displaced is equal to that of the spherical segment bounded by the plane of the water surface and the portion of the sphere below the surface. In terms of depth h, this volume is

$$V_{\text{displ}} = \pi h^2 \left(a - \frac{h}{3} \right) \qquad 0 \leqslant h \leqslant 2a$$

Therefore, the magnitude of the buoyant force is

$$f_b = \rho g \pi h^2 \left(a - \frac{h}{3} \right) \qquad 0 \leqslant h \leqslant 2a$$

and if the sphere is to be in equilibrium, this must be equal to the magnitude of the force of gravity acting on the sphere:

$$f_b = f_g$$

$$\rho g \pi h^2 \left(a - \frac{h}{3} \right) = \rho_s g \left(\frac{4\pi a^3}{3} \right) \qquad 0 \leqslant h \leqslant 2a$$

Or,

$$\left(\frac{h}{a} \right)^3 - 3\left(\frac{h}{a} \right)^2 + \frac{4\rho_s}{\rho} = 0 \qquad 0 \leqslant \frac{h}{a} \leqslant 2$$

The root of this cubic equation that lies within the region $0 \leqslant h/a \leqslant 2$ has the value

$$\frac{h}{a} = 1 + 2\cos\left(\frac{\phi + 4\pi}{3} \right)$$

where ϕ is given by

$$\cos\phi = 1 - \frac{2\rho_s}{\rho}$$

What happens when $\rho_s > \rho$?

Moment Resultants

EXAMPLE

Reconsider the dam depicted in Figure 6-8. Referring to the element shown there, we can compute the resultant moment of the hydrostatic forces about the y-axis (which is directed outward from the plane of the sketch) as follows:

$$\mathbf{M}_{Oy}(\text{water}) = -\mathbf{u}_y \int p(z) w (x\,dx + z\,dz)$$

$$= -\mathbf{u}_y w \int_0^h (p_a + \rho g z)\left[kz^2(2kz) + z\right] dz$$

$$= -\mathbf{u}_y w \int_0^h (p_a + \rho g z)\left(\frac{2a^2z^3}{h^4} + z\right) dz$$

$$= -\mathbf{u}_y w\left[\tfrac{1}{2}p_a(a^2 + h^2) + \rho g h\left(\tfrac{1}{3}h^2 + \tfrac{2}{5}a^2\right)\right]$$

Similarly, the resultant moment of the atmospheric pressure across the top surface and downstream side is

$$\mathbf{M}_{Oy}(\text{air}) = \mathbf{u}_y p_a w \frac{(b-a)^2 + h^2}{2}$$

The moment about the y-axis of all forces of contact with air and water above the base is then

$$\mathbf{M}_{Oy} = \mathbf{M}_{Oy}(\text{water}) + \mathbf{M}_{Oy}(\text{air})$$

$$= -\mathbf{u}_y w\left[p_a b\left(a - \tfrac{1}{2}b\right) + \rho g h\left(\tfrac{1}{3}h^2 + \tfrac{2}{5}a^2\right)\right]$$

The moment of these same forces about a parallel axis through the point O' can be evaluated by similar calculation or by using the above result and Equation 3-31 (p. 85). The result is

$$\mathbf{M}_{O'y} = \mathbf{u}_y w\left\{\frac{\rho g h}{30}\left[5h^2 - 4a(5b - 2a)\right] - \frac{p_a b^2}{2}\right\}$$

$$= (87.9 \times 10^6 \text{lbf}\cdot\text{ft})\mathbf{u}_y$$

This indicates the effect of the water and air pressure tending to overturn the dam. This is counteracted by the moment of the gravitational forces on the concrete, and by any capability of the ground attachment at the base and ends for providing additional moments.

EXAMPLE

The triangular gate shown in Figure 6-10 is hinged about the axis OO'. To determine the force needed to hold it closed, we calculate the resultant moment about the axis OO' of the water pressure acting on the gate.

Referring the figure, we see that the moment about this axis (the y-component of resultant moment) is given by

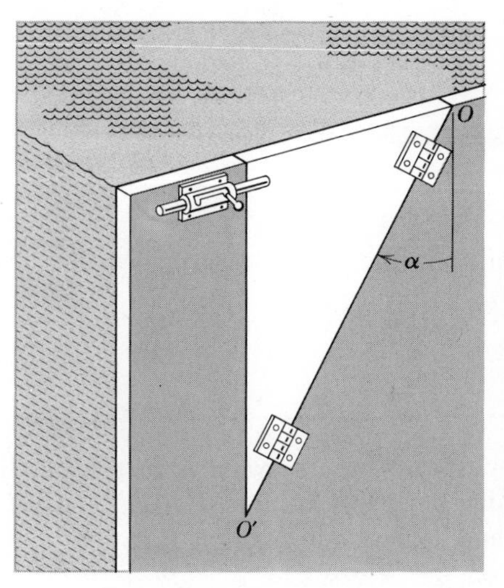

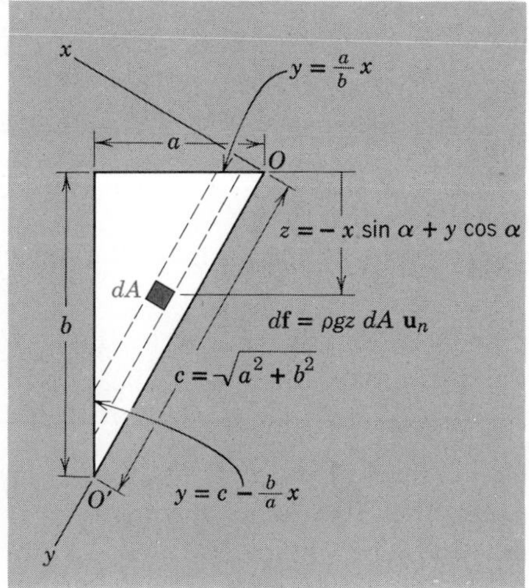

FIGURE 6-10

$$M_{Oy} = -\int x\,df$$

$$= -\int x\rho g(-x\sin\alpha + y\cos\alpha)\,dA$$

$$= \rho g\sin\alpha \mathcal{I}_{yy} + \rho g\cos\alpha \mathcal{I}_{yx}$$

where

$$\mathcal{I}_{yy} = \int x^2 dA$$

$$= \int_0^{ab/c}\int_{ax/b}^{c-(bx/a)} x^2\,dy\,dx$$

$$= \int_0^{ab/c} x^2\left(c - \frac{bx}{a} - \frac{ax}{b}\right)dx$$

$$= \frac{a^3b^3}{12c^2}$$

and

$$\mathcal{I}_{yx} = -\int yx\,dA$$

$$= -\int_0^{ab/c}\int_{ax/b}^{c-(bx/a)} xy\,dy\,dx$$

$$= -\frac{(b^2+3a^2)a^2b^2}{24c^2}$$

Substitution of the values of the integrals into the above results in

$$M_{Oy} = -\frac{\rho g a^2 b^3}{24c}$$

Values of integrals like those computed above are available in tabulated summaries like those given in Appendix B of this volume. Certain general relationships, indicating how to use and extend information found in the tables, are pointed out in the next section.

Second Moments of Area

Consider the flat surface area shown in Figure 6-11, acted on by pressure that varies linearly with distance from the axis Os. The resultant moment about O is a vector in the plane of the surface and is related to the pressure gradient through integrals such as those evaluated above. As is shown in the preceding example, the calculation of moment resultants for this situation arises in the analysis of submerged bodies; another application is in the analysis of bending moments in elastic beams, a subject outside the scope of this book.

FIGURE 6-11

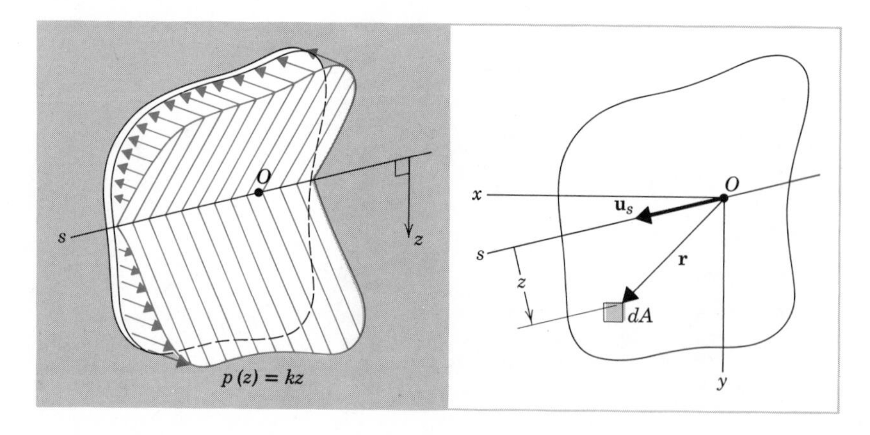

Observe from Figure 6-11 that the distance from the axis Os is given by

$$t = r \sin \measuredangle_{\mathbf{u}_s}^{\mathbf{r}} = (\mathbf{u}_s \times \mathbf{r}) \cdot \mathbf{u}_n$$

where $\mathbf{u}_n$ is directed outward and perpendicular to the plane of the plate. The force acting on the area element dA can therefore be expressed as

$$d\mathbf{f} = p\, dA \mathbf{u}_n = kt \mathbf{u}_n\, dA = k \mathbf{u}_s \times \mathbf{r}\, dA$$

and the moment about O can then be expressed as

$$\begin{aligned} \mathbf{M}_O &= \int \mathbf{r} \times d\mathbf{f} \\ &= k \int \mathbf{r} \times (\mathbf{u}_s \times \mathbf{r})\, dA \\ &= k \int [(\mathbf{r} \cdot \mathbf{r})\mathbf{u}_s - (\mathbf{u}_s \cdot \mathbf{r})\mathbf{r}]\, dA \end{aligned}$$

Now with the vectors in this equation resolved in the directions of x and y,

$$\begin{aligned} \mathbf{r} &= x\mathbf{u}_x + y\mathbf{u}_y \\ \mathbf{u}_s &= s_x \mathbf{u}_x + s_y \mathbf{u}_y \end{aligned}$$

the above can be rearranged to give the x- and y-components of the moment:

$$\begin{aligned} \mathbf{M}_O = & \left[k\mathcal{I}_{xx} s_x + k\mathcal{I}_{xy} s_y\right]\mathbf{u}_x \\ & + \left[k\mathcal{I}_{yx} s_x + k\mathcal{I}_{yy} s_y\right]\mathbf{u}_y \end{aligned} \qquad \textbf{(6-6)}$$

in which

$$\begin{aligned} \mathcal{I}_{xx} &= \int y^2\, dA \quad \mathcal{I}_{yy} = \int x^2\, dA \\ \mathcal{I}_{xy} &= \mathcal{I}_{yx} = -\int xy\, dA \end{aligned} \qquad \textbf{(6-7)}$$

These integrals are called the *second moments of area*,* their values depending on the size and shape of the area, and the locations of the axes.

Calculations of second moments of area by integration is carried out in the same manner as are those for area and first moments of area (pp. 209–12). The procedure is illustrated on page 231 for a triangular shape. Careful attention is paid to the figure as each detail of the integral is written: The area

*$\mathcal{I}_{ii}$ is also commonly called the moment of inertia of the area about the axis i, and $\mathcal{I}_{xy}$ the *product of intertia* of the area, about the axes x and y. These terms derive from mathematically analogous integrals that describe inertial resistance to moments in the study of rigid body dynamics. A definition of product of inertia is sometimes made without the minus sign.

element is expressed in terms of the variables to be used for integration, $dA = dx\,dy$. The limits for the first integration (with respect to y) are written in terms of x, the variable for the subsequent integration, from the equations of the lines that bound the triangular area. The more mechanical steps of carrying out the integrations are usually less troublesome, but would still be instructive to verify.

Values of second moments of area for many figures have been computed and a number of these results tabulated in Appendix B-2. In the following we give some relationships that are useful for extending the information in these tables.

Translation of Axes Let the axes X and Y emanate from the centroid C of a plane area, and let C be located with respect to a parallel pair of axes x and y by the coordinates (c_x, c_y). Then the second moments of the area with respect to the two parallel sets of axes are related as follows:

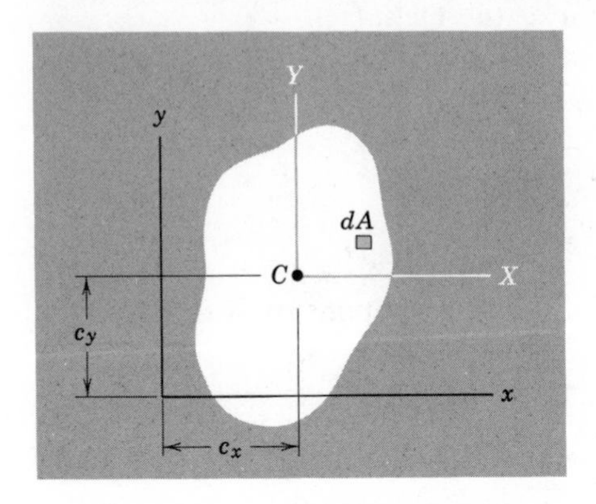

$$\mathcal{I}_{xx} = \int y^2\, dA$$

$$= \int (Y + c_y)^2\, dA$$

$$= \int Y^2 dA + 2c_y \int Y\, dA + c_y^{\,2} \int dA$$

Because C is the centroid, $\int Y\,dA = 0$. With the second moment $\int Y^2\,dA$ denoted by $\mathcal{I}_{XX}$, the above may be written as

$$\mathcal{I}_{xx} = \mathcal{I}_{XX} + c_y^{\,2}A \qquad \textbf{(6-8a)}$$

Similar manipulations lead to

$$\mathcal{I}_{yy} = \mathcal{I}_{YY} + c_x^2 A \tag{6-8b}$$

$$\mathcal{I}_{xy} = \mathcal{I}_{XY} - c_x c_y A \tag{6-8c}$$

Thus each of the second moments may be computed by adding, to the corresponding moment with respect to parallel centroidal axes, a term representing the second moment the area would have if it were concentrated at the centroid.

EXAMPLE

Compute the second moments with respect to x and y of the rectangle in Figure 6-12.

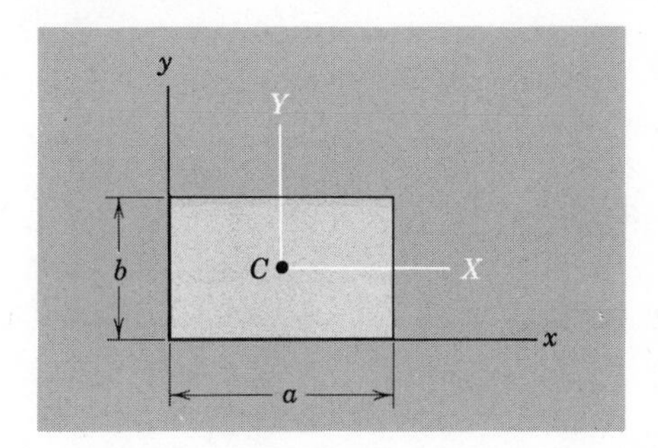

FIGURE 6-12

From Appendix B we find that

$$\mathcal{I}_{xx} = \frac{ab^3}{3}$$

$$\mathcal{I}_{YY} = \frac{ba^3}{12}$$

and from symmetry,

$$\mathcal{I}_{XY} = 0$$

Then, from (6-8),

$$\mathcal{I}_{yy} = \frac{ba^3}{12} + \left(\frac{a}{2}\right)^2 (ab)$$

$$= \frac{ba^3}{3}$$

$$\mathcal{I}_{xy} = 0 - \left(\frac{a}{2}\right)\left(\frac{b}{2}\right)$$

$$= -\frac{a^2b^2}{4}$$

Rotation of Axes Let the coordinate axes $\bar{x}$ and $\bar{y}$ be rotated counterclockwise through the angle θ, with respect to the axes x and y. Then the coordinates of the element of area dA with respect to the two sets of axes are related as follows:

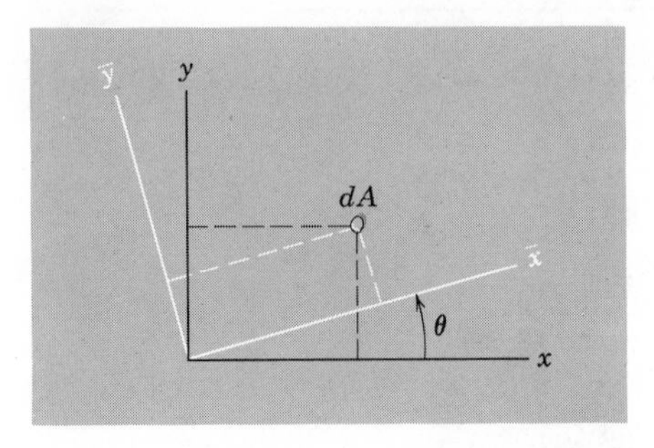

$$\bar{x} = x\cos\theta + y\sin\theta$$

$$\bar{y} = -x\sin\theta + y\cos\theta$$

These formulas then can be used to obtain the relationships among the second moments of area with respect to the two sets of axes:

$$\bar{\mathcal{I}}_{xx} = \int \bar{y}^2\, dA$$

$$= \int (-x\sin\theta + y\cos\theta)^2\, dA$$

$$= \cos^2\theta \int y^2\, dA - 2\sin\theta\cos\theta \int xy\, dA + \sin^2\theta \int x^2\, dA$$

$$= \cos^2\theta\, \mathcal{I}_{xx} + 2\cos\theta\sin\theta\, \mathcal{I}_{xy} + \sin^2\theta\, \mathcal{I}_{yy}$$

Similar manipulations lead to

$$\bar{\mathcal{I}}_{yy} = \sin^2\theta\, \mathcal{I}_{xx} - 2\cos\theta\sin\theta\, \mathcal{I}_{xy} + \cos^2\theta\, \mathcal{I}_{yy}$$

$$\bar{\mathcal{I}}_{xy} = \cos\theta\sin\theta\left(\mathcal{I}_{yy} - \mathcal{I}_{xx}\right) + (\cos^2\theta - \sin^2\theta)\mathcal{I}_{xy}$$

Introducing the trigonometric identities

$$\cos^2\theta = \tfrac{1}{2}(1 + \cos 2\theta)$$

$$\sin^2\theta = \tfrac{1}{2}(1 - \cos 2\theta)$$

$$\cos\theta \sin\theta = \tfrac{1}{2}\sin 2\theta$$

into the above results in

$$\bar{\mathcal{I}}_{xx} = \tfrac{1}{2}(\mathcal{I}_{xx} + \mathcal{I}_{yy}) + \tfrac{1}{2}(\mathcal{I}_{xx} - \mathcal{I}_{yy})\cos 2\theta + \mathcal{I}_{xy}\sin 2\theta \qquad \textbf{(6-9a)}$$

$$\bar{\mathcal{I}}_{yy} = \tfrac{1}{2}(\mathcal{I}_{xx} + \mathcal{I}_{yy}) - \tfrac{1}{2}(\mathcal{I}_{xx} - \mathcal{I}_{yy})\cos 2\theta - \mathcal{I}_{xy}\sin 2\theta \qquad \textbf{(6-9b)}$$

$$\bar{\mathcal{I}}_{xy} = \tfrac{1}{2}(\mathcal{I}_{yy} - \mathcal{I}_{xx})\sin 2\theta + \mathcal{I}_{xy}\cos 2\theta \qquad \textbf{(6-9c)}$$

In the paragraph that follows we describe a useful geometric intepretation of these formulas.

Mohr's Circle Otto Mohr (1835–1918), a German structural engineer, pointed out that Equations 6–9 can be represented by the diagram shown in Figure 6-13. This *Mohr's circle* has the advantages that some of the basic properties hidden in the equivalent Equations 6-9 become more readily apparent, and most people find the diagram easier to remember than the equations.

Each point (M, P) on the circle represents a pair of values of second moments for a particular orientation of axes. The abscissa M is equal to the moment of inertia about one axis, and the ordinate P is equal to the product of inertia with respect to that same axis and the axis oriented 90° clockwise from it.

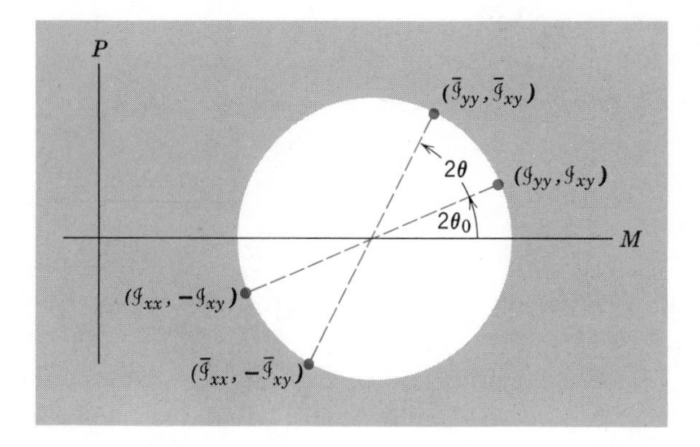

FIGURE 6-13

Given values of the second moments with respect to a particular set of axes, the circle may be constructed as follows. A point is located with abscissa equal to $\mathcal{I}_{yy}$ and ordinate equal to $\mathcal{I}_{xy}$. Next, another point is located with abscissa equal to $\mathcal{I}_{xx}$ and ordinate equal to $-\mathcal{I}_{xy}$. (The product of inertia with respect to x and the axis 90° clockwise from x is equal to $-\mathcal{I}_{xy}$). These two points are diametrically opposite on the circle.

Now, any other point on the circle, reached by a rotation of 2θ from any established point corresponds to the second moments with respect to axes rotated θ from those corresponding to the established point on the circle. The rotation in the plane of the Mohr's circle is in the same direction as the rotation of axes in the plane of A.

EXAMPLE

Construct the Mohr's circle for second moments of area with respect to various sets of axes passing through the corner O, of the rectangle shown in Figure 6-14.

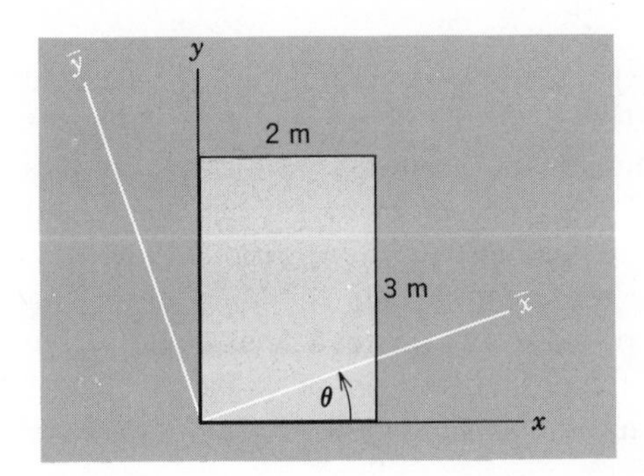

FIGURE 6-14

From the preceding example, we have

$$\mathcal{I}_{xx} = \tfrac{1}{3}(2\text{ m})(3\text{ m})^3 = 18\text{ m}^4$$

$$\mathcal{I}_{yy} = \tfrac{1}{3}(3\text{ m})(2\text{ m})^3 = 8\text{ m}^4$$

$$\mathcal{I}_{xy} = -\tfrac{1}{4}(3\text{ m})^2(2\text{ m})^2 = -9\text{ m}^4$$

These are used to establish the points (8, −9) and (18, 9) as shown in Figure 6-15. The center of the circle is then at the intersection of the M axis and the line connecting these two points. The second moments $\bar{\mathcal{I}}_{xx}$ and $\bar{\mathcal{I}}_{xy}$, corresponding to the $(\bar{x}, \bar{y})$ axes shown in Figure 6-14, are the coordinates of the point so

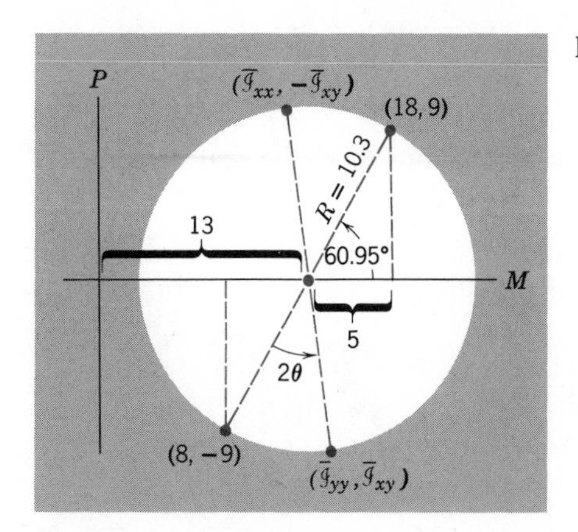

FIGURE 6-15

labeled in Figure 6-15. It is evident from the circle that the maximum moment of inertia is equal to 23.3 m^4 about an axis rotated $60.95°/2 = 30.47°$ clockwise from x. The minimum moment of inertia is equal to 2.7 m^4, about an axis oriented 90° from that of maximum moment of inertia. Furthermore, the product of inertia with respect to these two axes is zero.

To verify the correspondence between Figure 6-13 and Equations 6-9, let the radius of the circle be denoted by R, and the angle from the M axis to $(\mathcal{I}_{yy}, \mathcal{I}_{xy})$ by $2\theta_0$, as shown in the figure. From the construction of the circle, the center is established at $M = \frac{1}{2}(\mathcal{I}_{xx} + \mathcal{I}_{yy})$. Also, from the figure, we have

$$\begin{aligned}\bar{\mathcal{I}}_{yy} &= \tfrac{1}{2}(\mathcal{I}_{xx} + \mathcal{I}_{yy}) + R\cos(2\theta_0 + 2\theta)\\ &= \tfrac{1}{2}(\mathcal{I}_{xx} + \mathcal{I}_{yy}) + (R\cos 2\theta_0)\cos 2\theta - (R\sin 2\theta_0)\sin 2\theta\end{aligned}$$

and

$$R\cos 2\theta_0 = \tfrac{1}{2}(\mathcal{I}_{yy} - \mathcal{I}_{xx})$$

$$R\sin 2\theta_0 = \mathcal{I}_{xy}$$

Taken together, these relationships from the figure imply that

$$\bar{\mathcal{I}}_{yy} = \tfrac{1}{2}(\mathcal{I}_{xx} + \mathcal{I}_{yy}) + \tfrac{1}{2}(\mathcal{I}_{yy} - \mathcal{I}_{xx})\cos 2\theta - \mathcal{I}_{xy}\sin 2\theta$$

which agrees with Equation 6-9b. In a similar manner, Equations 6-9a and 6-9c can be established from the figure.

PROBLEMS

6-32 Determine c, the correction that must be applied to the scale reading to obtain the water surface elevation. *Ans.* 2 ft.

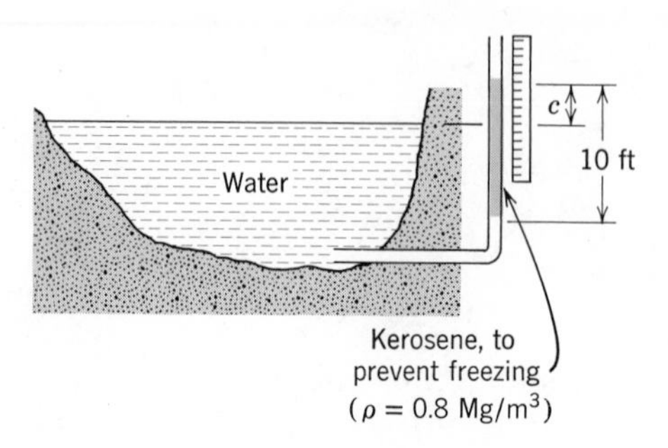

6-33 The gate G is hinged along the top. What force must be applied at the bottom of the gate to prevent its opening?

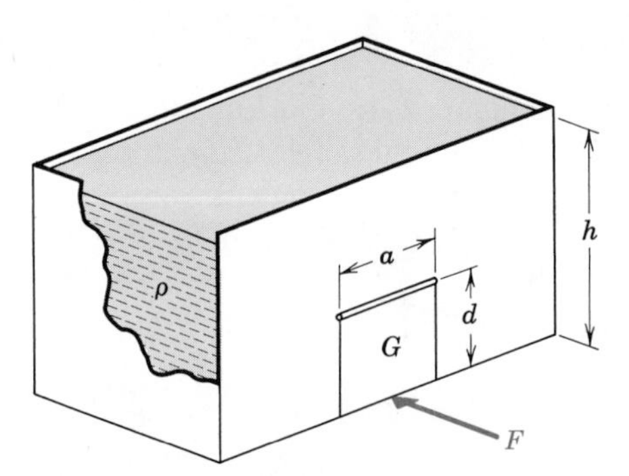

6-34 Locate the line of action of the horizontal component of force on the dam face due to water pressure.

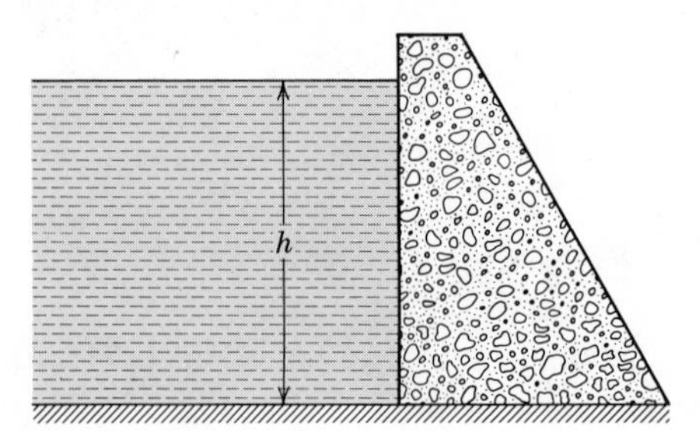

6-35 What is the density of fluid x?

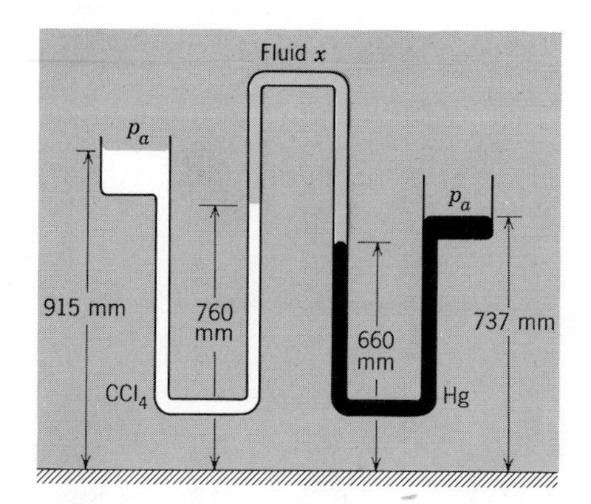

6-36 Determine the air pressure inside the chamber. *Ans.* $p_{atm} + 0.68$ lbf/in^2.

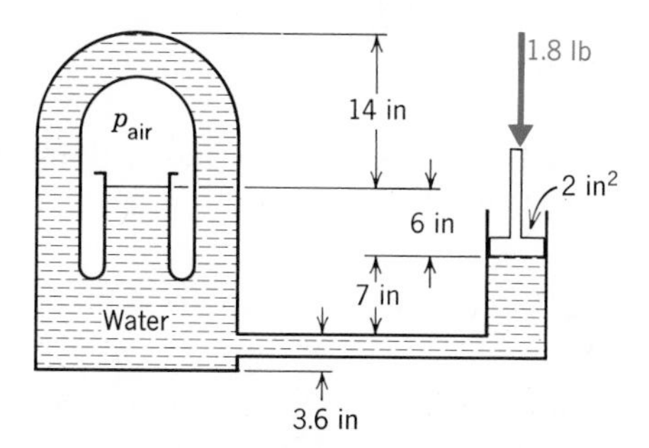

6-37 A hollow sphere, made of steel with a density of 7.8 Mg/m^3, has an outside diameter of 600 mm and a wall thickness of 50 mm. How much water will it displace as it floats on the surface?

6-38 The width of the tank (in the direction perpendicular to the plane of the sketch) is w. Determine the tension in the line AB.

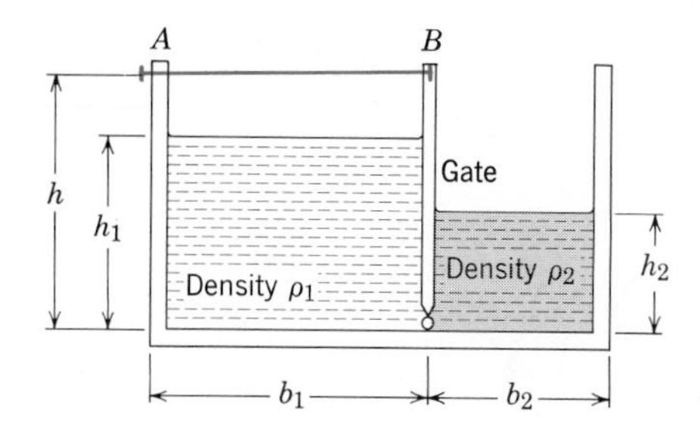

6-39 At what angle θ will the heavy gate be in equilibrium without any support at the top? If it were placed in that position and support removed, would you be willing to stand on the downstream side? Explain carefully.

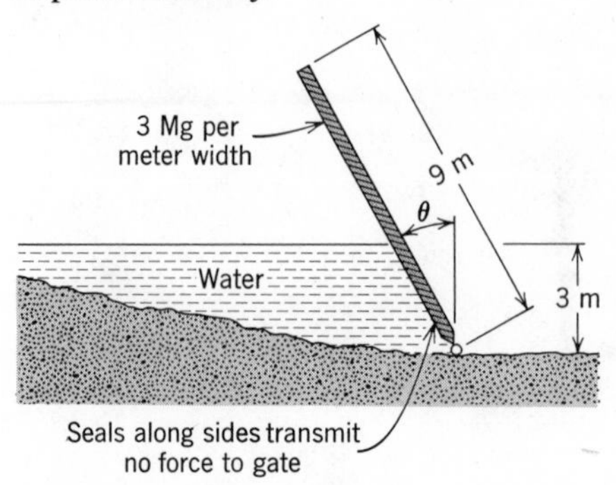

6-40 Consider a free-body consisting of the dam in Figure 6-8 together with the portion of water shown in Figure 6-9. Use equations of equilibrium for this composite body to compute the reaction components at the base of the dam.

6-41 Locate the line of action of the vertical reaction at the bottom of the dam in Figure 6-8, assuming there is no support along the sides of the dam.

6-42 If the trough of Problem 6-23 is filled with water, what will be the force from the water tending to push one end off the tank?

6-43 The form for the concrete retaining wall is essentially hinged at O and held in place by horizontal braces AB located every three feet. When poured, the wet concrete has a density of 2.4 Mg/m^3. Estimate the force in each brace AB, due to the pressure from the concrete.

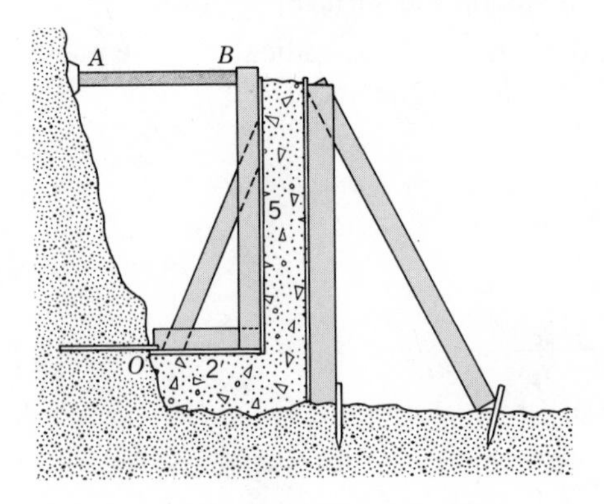

For Problems 6-44 through 6-48, verify the indicated formulas for $\mathcal{I}_{xx}$, $\mathcal{I}_{xy}$, and $\mathcal{I}_{yy}$ in Appendix B-2.

6-44 The triangular area on page 306.

6-45 The area bounded by the x-axis, the line $x = a$, and the line $y = \frac{bx^n}{a^n}$ shown on page 306.

6-46 The quarter ellipse shown on page 307.

6-47 The sector of a circle shown on page 307.

6-48 The segment of a circle shown on page 307.

6-49 Show that if a plane area has an axis of symmetry, the second moment $\mathcal{I}_{xy}$ with respect to that axis and a perpendicular axis is zero.

6-50 Use the geometry of Figure 6-13 to establish consistency with Equations 6-9a and 6-9c.

6-51 Determine the orientation of the axes passing through the corner of the triangle, such that $\mathcal{I}_{x_1x_2} = 0$.

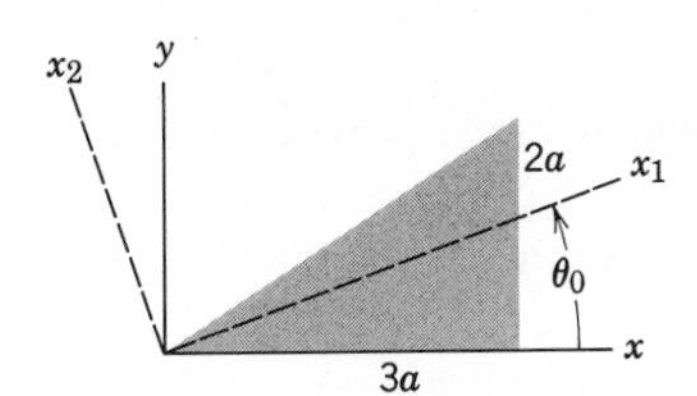

6-52 Determine the orientation of the axes passing through the centroid of the triangle, such that $\mathcal{I}_{x_1x_2} = 0$.

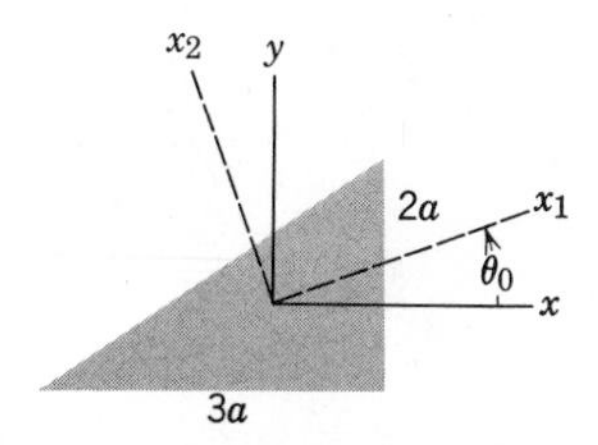

6-53 Find the orientation of the axes passing through the corner of the quarter ellipse, such that $\mathcal{I}_{x_1x_2} = 0$.

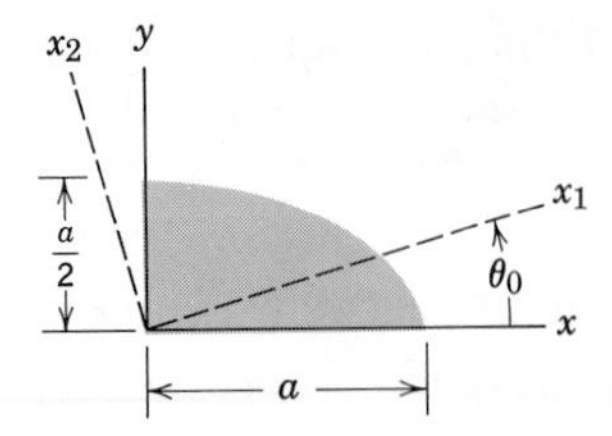

6-54 Determine the second moments of the sector of the circle with respect to the axes shown.

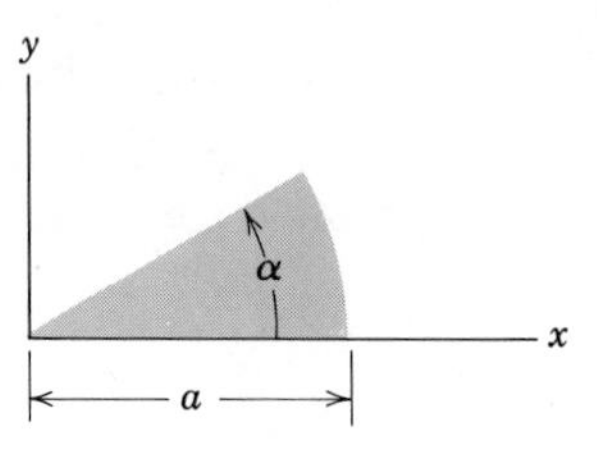

6-55 The quarter circular plate is held against the tank with bolts at the three corners. What is the minimum force in the bolt at A, necessary to hold the plate against the pressure from the liquid in the tank?

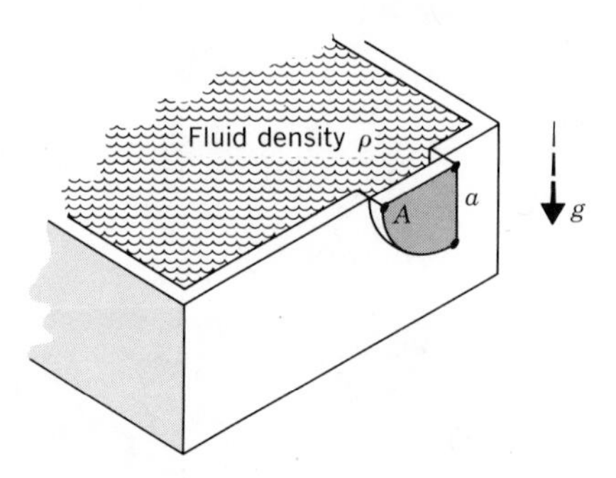

6-56 A section of a bent beam has normal pressure acting over the surface that varies linearly with the distance from the axis Ox. Evaluate the moment about O of this pressure.

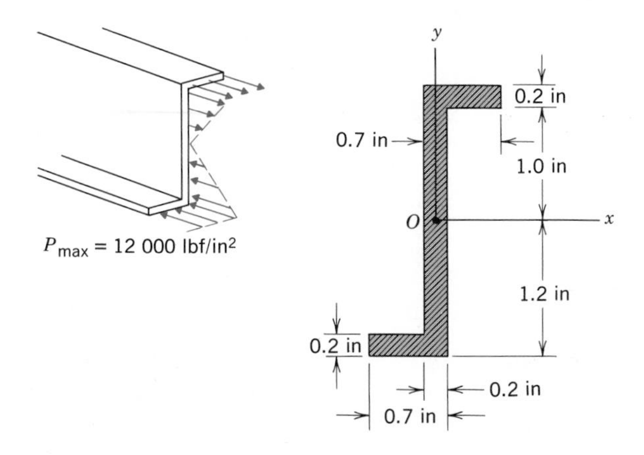

6-57 For the cross section of the previous problem, determine the orientation of axes x_1, x_2 such that $\mathcal{I}_{x_1 x_2} = 0$.

6-3 BENDING MEMBERS

We learned in Section 5-2 that loads lateral to the axis of a stiff rod will induce internal shear forces and bending moments at various sections of the rod. We also learned that to evaluate them at a particular section, a free-body of the portion of the rod on either side of the section can be subjected to equilibrium analysis.

Here, we are concerned with the manner in which the shear force and bending moment vary from section to section, the usual goal of this analysis being the determination of the maximum bending moment. We describe an approach slightly different from that of Section 5-2, as it sometimes produces the result more easily.

Differential Equations of Equilibrium

Consider a beam loaded as is shown in Figure 6-16. The loading w is specified as force per unit length of the beam (SI units: newtons per meter). We use the distance x to specify a particular section along the beam and recognize that in general the loading w, the shear force V, and the bending moment M are all functions of x.

Analysis will incorporate the sign convention indicated in the figure: x increases toward the right; positive values of w indicate a downward acting

FIGURE 6-16

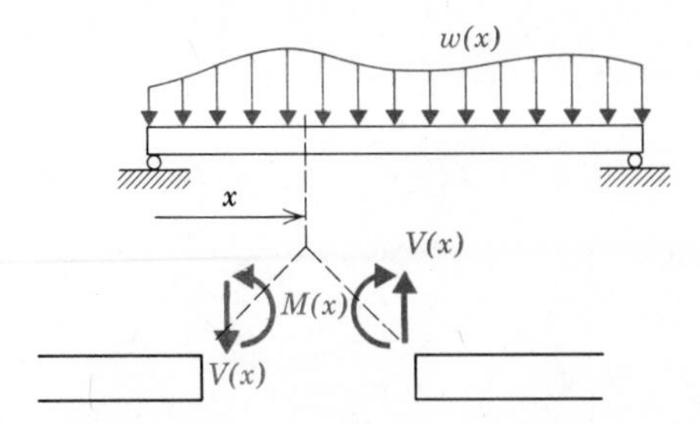

loading; positive values of V indicate a downward acting force on the portion of the beam to the left of the section and an upward acting force on the portion of the beam to the right of the section; positive values of M indicate a counterclockwise acting couple on the portion of the beam to the left of the section and a clockwise acting couple on the portion of the beam to the right of the section.

A short element of the beam is shown in Figure 6-17. Vertical force equilibrium requires that

$$V(x + \Delta x) - V(x) = -w_{av}\Delta x$$

where w_{av} is the average load intensity over the element. After division by Δx this equation becomes, in the limit at $\Delta x \to 0$,

$$\boxed{\frac{dV}{dx} = -w(x)} \tag{6-10}$$

Moment equilibrium of the element requires that

$$M(x + \Delta x) - M(x) = V(x)\Delta x + (w_{av}\Delta x)\xi\,\Delta x$$

where $\xi\,\Delta x$ is the distance from the single force equivalent of the distributed load, to the right-hand end of the element; we note that $0 < \xi < 1$. After division by Δx this equation becomes, in the limit as $\Delta x \to 0$,

$$\boxed{\frac{dM}{dx} = V(x)} \tag{6-11}$$

From the equations of equilibrium (6-10) and (6-11), we see that the shear force $V(x)$ can be obtained by integration of the load intensity $w(x)$, and the bending moment $M(x)$ obtained in turn by integration of the shear force $V(x)$.

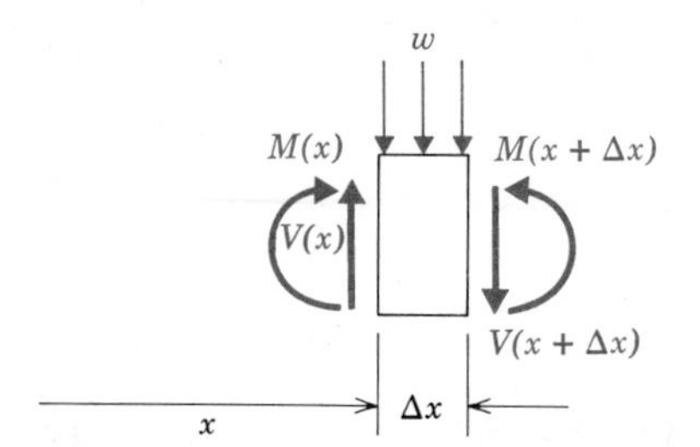

FIGURE 6-17

EXAMPLE

Determine the distribution of shear force and bending moment in the cantilever with uniformly distributed load, as depicted in Figure 6-18a.

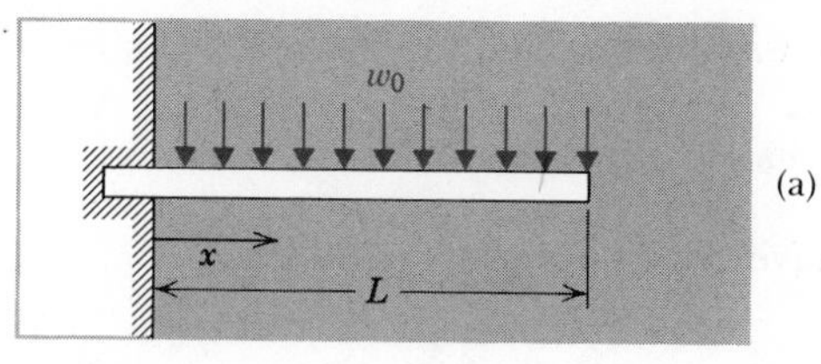

(a)

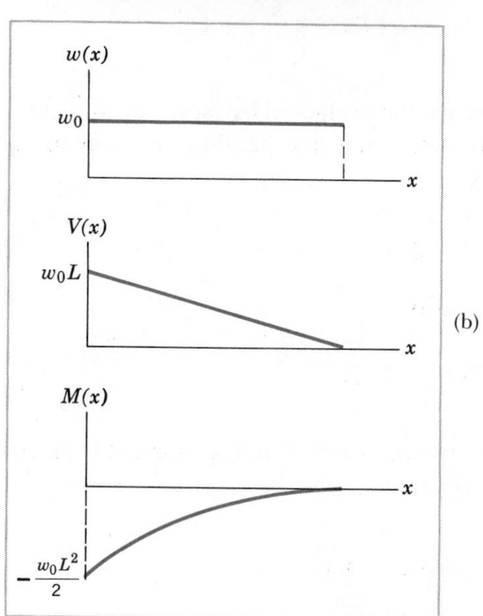

(b)

FIGURE 6-18

The load intensity is constant with respect to x:

$$w(x) = w_0$$

Then from (6-10),

$$V(x) = -\int w_0 \, dx = -w_0 x + C_1$$

The constant of integration can be evaluated by means of the boundary condition at the right-hand end; we note that since this end is free of forces

$$V(L) = 0$$

$$-w_0 L + C_1 = 0$$

$$C_1 = w_0 L$$

With this value substituted above, the distribution of shear force is given by

$$V(x) = w_0(L - x)$$

Next, using this result in (6-11), we have

$$M(x) = \int w_0(L - x)\, dx = w_0\left(Lx - \tfrac{1}{2}x^2\right) + C_2$$

The constant of integration arising here can be evaluated by means of the other boundary condition at the right-hand end, where the bending moment must vanish:

$$M(L) = 0$$

$$\tfrac{1}{2} w_0 L^2 + C_2 = 0$$

$$C_2 = -\tfrac{1}{2} w_0 L^2$$

With this value substituted above, the distribution of bending moment is given by

$$M(x) = -\tfrac{1}{2} w_0 (L - x)^2$$

Curves showing the variations of bending moment and shear are shown in Figure 6-18*b*.

Graphical Interpretations

It is often easier to construct the shear and moment diagrams directly, by means of graphical interpretations of Equations 6-10 and 6-11, than through detailed analytical integration as is shown above.

Recall that on integration of a function that is represented by a given curve, the integrated function exhibits a difference in its values between x_1 and x_2 equal to the area under the given curve between x_1 and x_2. Applying this idea to the first integration in the above example, we see that, between $x = 0$ and some value greater than zero, the value of the shearing force must change by an amount equal to the corresponding area under the loading curve. Because of the minus sign in Equation 6-10 this change must be negative, representing a *decrease* in V. This, together with the observation that V must vanish at $x = L$, permits the drawing of the shear diagram and evaluation of the maximum shear force without resorting to equations. Application of this idea to the next integration gives the rule for constructing the bending moment diagram: The change in M is equal to the area under the shearing force curve.

Another interpretation is often helpful for determining the forms of the shear and bending moment diagrams, and in checking them for errors. The slope of the bending moment curve is at every point equal to the value of the shearing force at that point. Likewise, the slope of the shearing force curve is at every point equal to the negative of the value of the load intensity at that point. These interpretations follow from the differential forms (6-10) and (6-11).

Concentrated Forces and Couples

A concentrated force is our idealization for a very high intensity of loading over a very short interval of the beam. The magnitude of the force is equal to the area under the short portion of the load diagram over which the high-intensity load acts. In integrating to construct the shear diagram, we must show a decrease in the value of V equal to this area as the integration passes through this region. Thus, consistent with the idealization of a concentrated force, the shear diagram exhibits a discontinuity equal to the magnitude of the force, as is illustrated in Figure 6-19. To verify this fact from another point of view, you will find it helpful to consider the equilibrium of a short element of the beam on which the force acts.

An externally applied, concentrated couple is also shown in Figure 6-19. This results in a discontinuity in the bending moment curve, as can be verified by considering the equilibrium of a short element of beam to which the couple is applied.

FIGURE 6-19

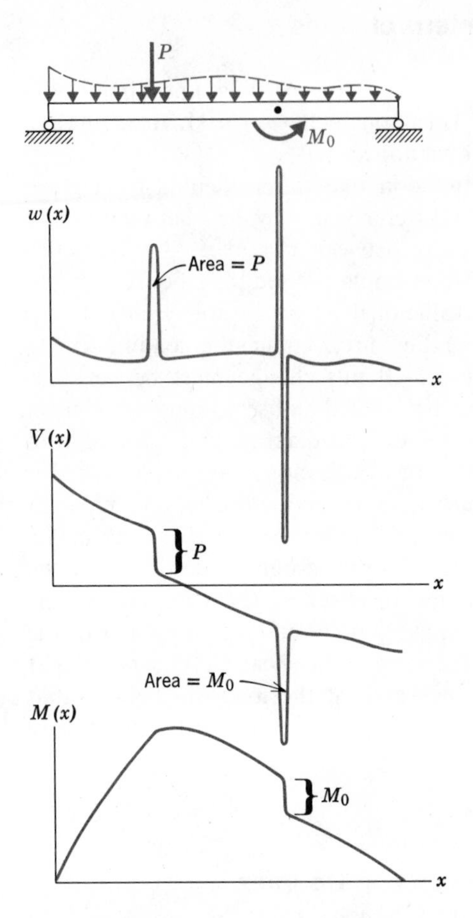

EXAMPLE

Determine the distribution of shearing force and bending moment along the beam shown in Figure 6-20.

The values of the support reactions may be determined by considering equilibrium of the entire beam and are shown on the free-body diagram in the figure.

The reaction at the left gives the value of the shearing force at $x = 0$ as

$$V(0) = 560 \text{ lb}$$

The shape of the shearing force diagram in the region $0 < x < 17$ ft can be deduced from the fact that the slope of this curve must be equal to the negative

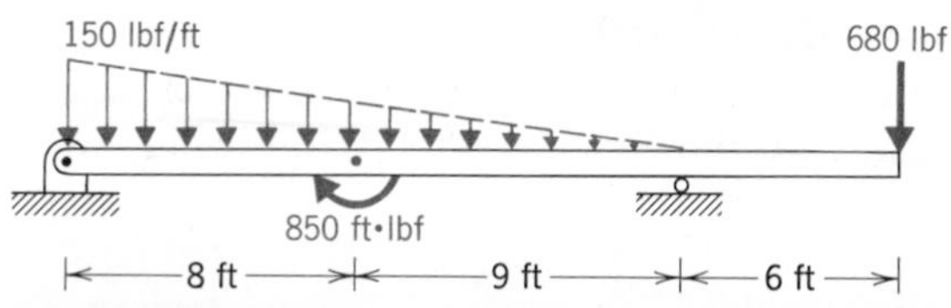

FIGURE 6-20

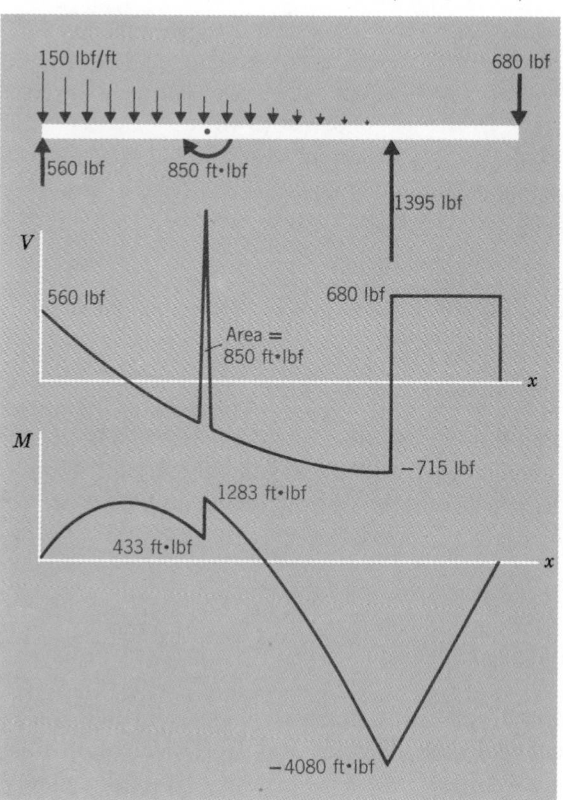

of the value of $w(x)$ at each point. The equation for this curve may be obtained by integrating Equation 6-10:

$$V = -\int (150 \text{ lb/ft})\left(1 - \frac{x}{17 \text{ ft}}\right) dx \qquad 0 < x < 17 \text{ ft}$$

$$= 560 \text{ lb} - (150 \text{ lb/ft})\left(x - \frac{x^2}{34 \text{ ft}}\right) \qquad 0 < x < 17 \text{ ft}$$

This expression does not include an evaluation of the local, high-intensity shearing force that arises from the 850-ft·lbf couple. By considering the equilibrium of a short element on which this couple acts, we find that through this local region the bending moment must increase 850 ft·lbf, while the net change in the shearing force is negligible.

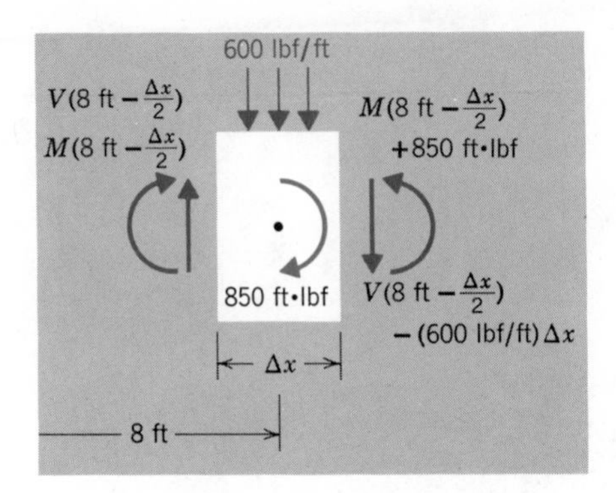

The value of the shearing force just to the left of the right-hand support may be determined by substitution of $x = 17$ ft into the above expression, or by subtracting the area under the loading curve from the value $V(0) = 560$ lbf. The result is

$$V(17\text{ ft} -) = -715\text{ lbf}$$

The notation (17 ft $-$) is to indicate the value of $V(x)$ at a point just to the left of the support.

At this support the 1395-lb reaction induces an increase in the shearing force, as may be seen from a free-body diagram of a short element on which this force acts.

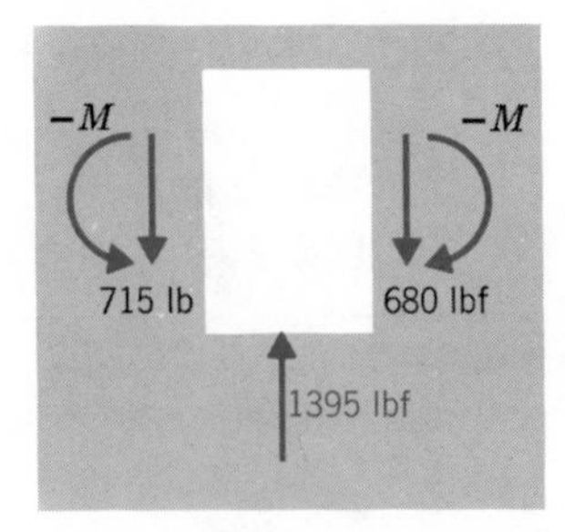

In the region 17 ft $< x <$ 23 ft, the load, and hence the slope of the shearing force curve, is zero. At $x = 23$ ft, the concentrated force induces a change in the shearing force of -680 lbf. Observe that this gives the correct value of shearing force for the region $x > 23$ ft.

The pin support at $x = 0$ implies that the bending moment must be zero at that point. Starting with this initial value, integration of $V(x)$ will give the bending moment at each section. In the region $0 < x < 8$ ft, the result is

$$M(x) = \int \left[560 \text{ lbf} - (150 \text{ lbf/ft}) \left(x - \frac{x^2}{34 \text{ ft}} \right) \right] dx \qquad 0 < x < 8 \text{ ft}$$

$$= (560 \text{ lbf})x - (150 \text{ lbf/ft}) \left(\frac{x^2}{2} - \frac{x^3}{102 \text{ ft}} \right) \qquad 0 < x < 8 \text{ ft}$$

This function reaches a maximum at $x = 4.27$ ft, where the value is $M(4.27) =$ 1138 ft·lbf. At $x = 8$ ft, the expression yields

$$M(8 \text{ ft} -) = 433 \text{ ft} \cdot \text{lbf}$$

The concentrated couple applied at $x = 8$ ft induces an increase in the bending moment, as we pointed out above, so that the bending moment just to the right of that point has the value

$$M(8 \text{ ft} +) = 433 \text{ ft} \cdot \text{lbf} + 850 \text{ ft} \cdot \text{lbf}$$

$$= 1283 \text{ ft} \cdot \text{lbf}$$

In the region 8 ft $< x <$ 17 ft, the bending moment is then given by

$$M(x) = 433 \text{ ft} \cdot \text{lbf} + (560 \text{ lbf})x - (150 \text{ lbf/ft}) \left(\frac{x^2}{2} - \frac{x^3}{102 \text{ ft}} \right)$$

Substitution of $x = 17$ ft into this expression yields the value of the bending moment at the right-hand support:

$$M(17 \text{ ft}) = -4080 \text{ ft} \cdot \text{lbf}$$

In the region 17 ft $< x <$ 23 ft, the integration of V yields the straight line with slope of 680 lbf. Observe that the value implied by this curve at $x = 23$ ft agrees with the value we can determine from a free-body of the portion of the beam to the right of $x = 23$ ft.

PROBLEMS

6-58 Sketch the shear and bending moment diagrams.

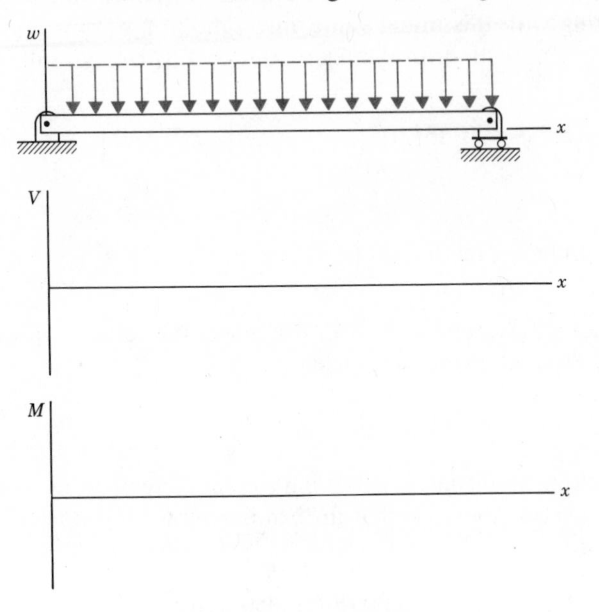

6-59 Sketch the shear and loading diagrams.

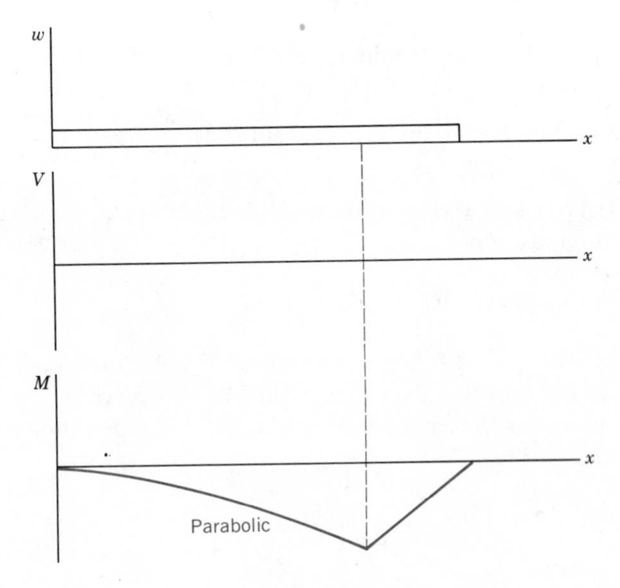

6-60 Sketch the loading and moment diagrams.

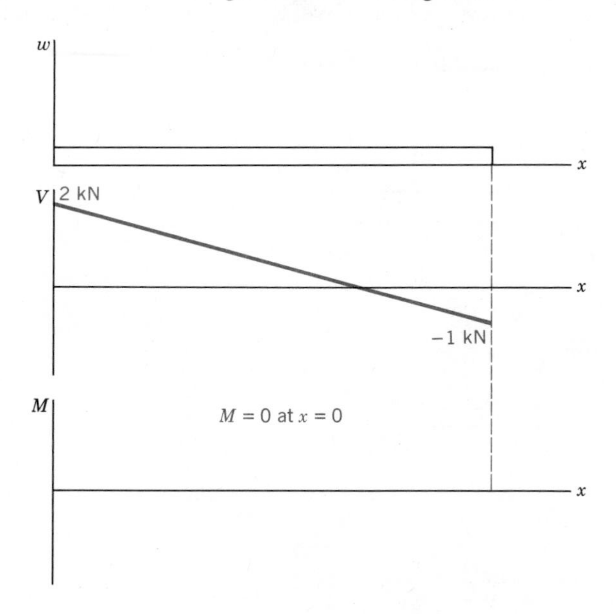

6-61 Evaluate $V(x)$ and $M(x)$ and compare with the result of Problem 6-2.

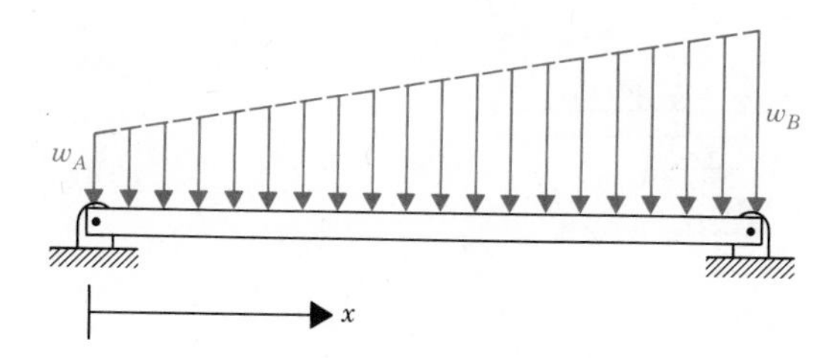

6-62 Show a force system different from that acting on the beam, but that is equivalent in the sense discussed in Sections 2-5 and 3-5. Draw shear and bending moment diagrams for the given force system and the equivalent force system.

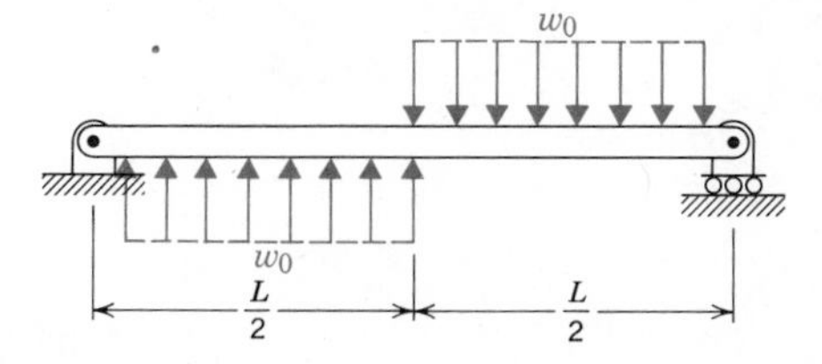

For Problems 6-63 through 6-78, draw shear and bending moment diagrams, labeling all values at important points. Give the value of the maximum bending moment.

6-63

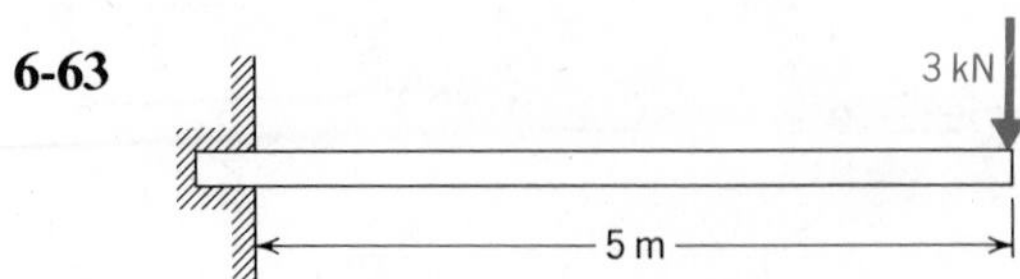

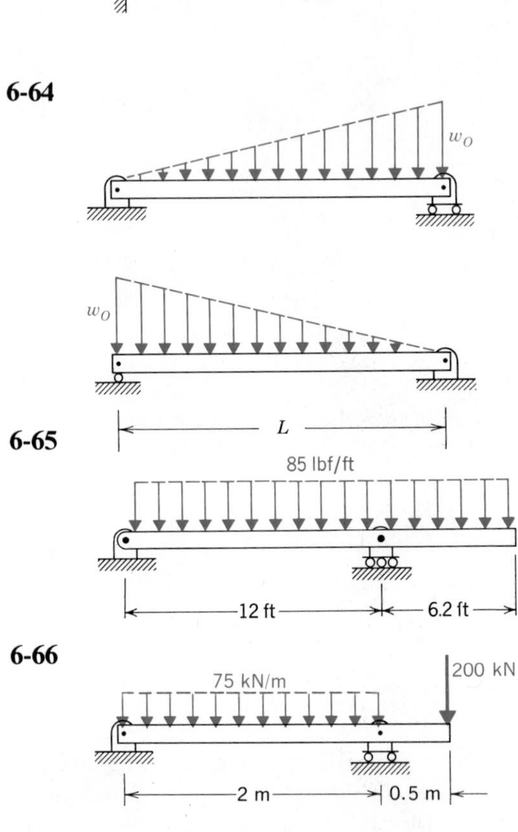

6-67

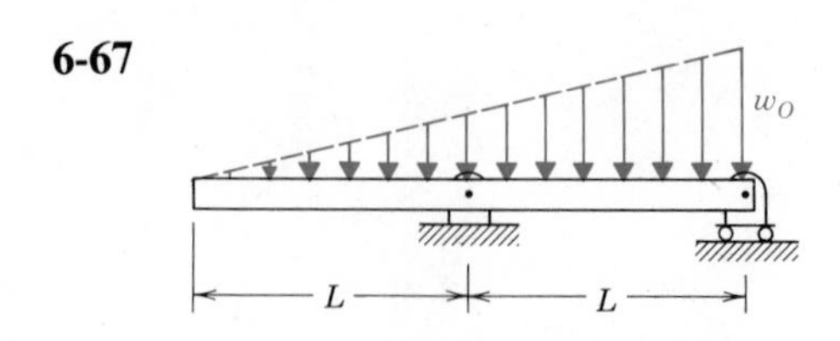

6-68

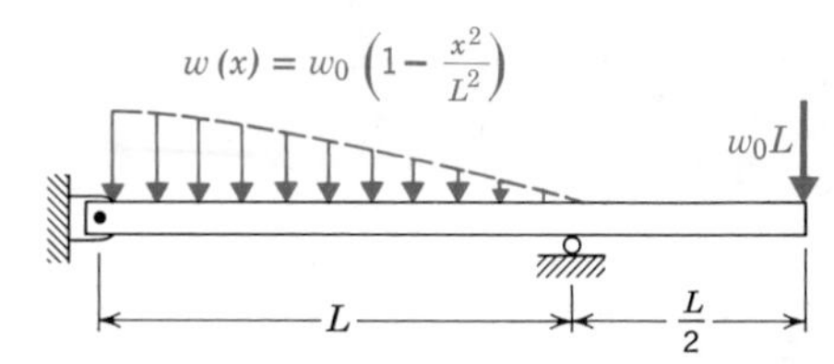

6-69

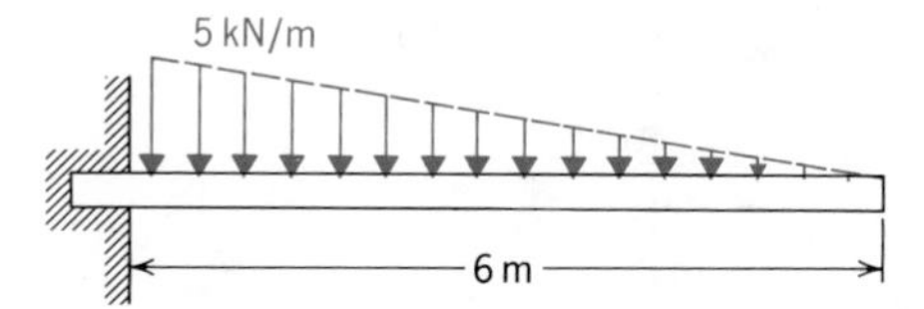

6-70

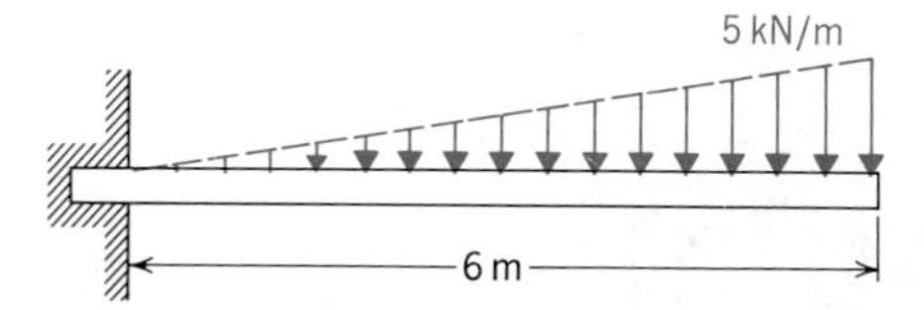

6-71 Same total load as that of the previous two problems.

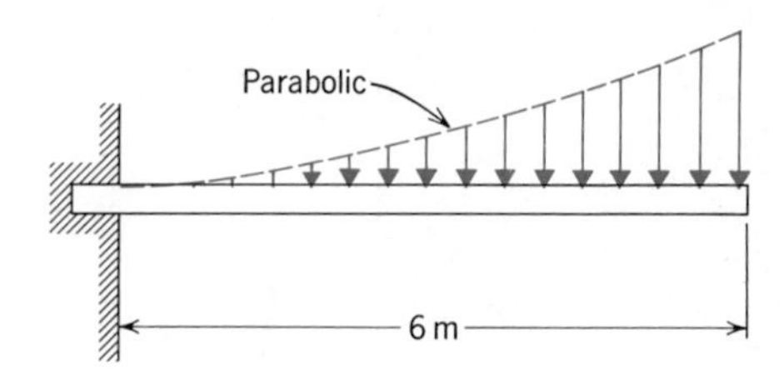

6-72 (a)

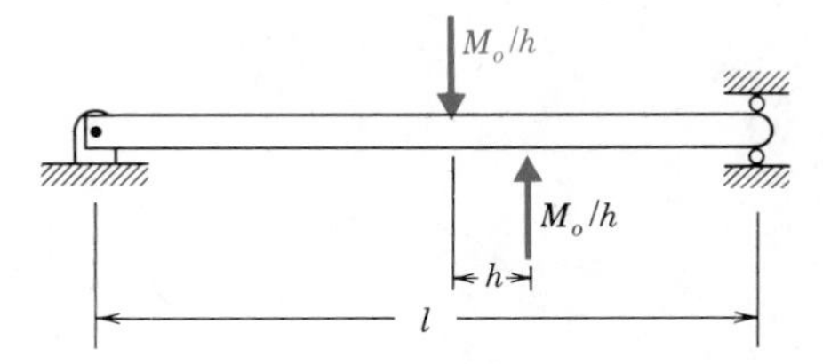

(b)

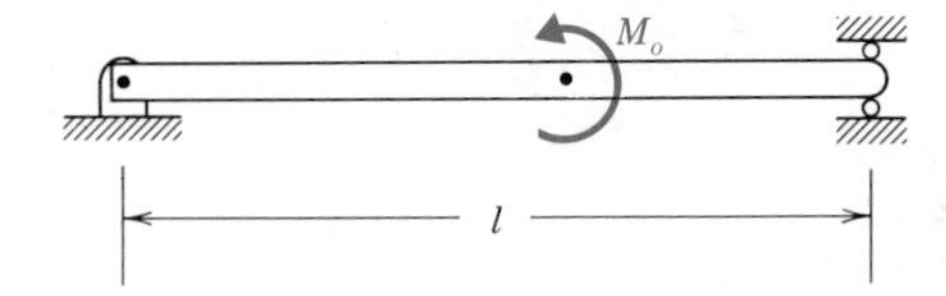

6-73

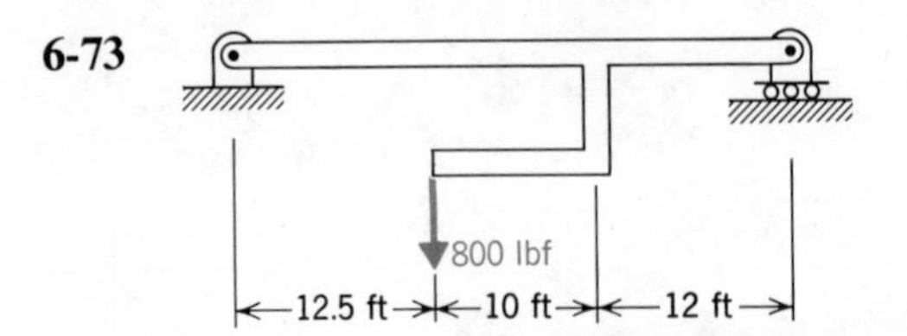

6-74

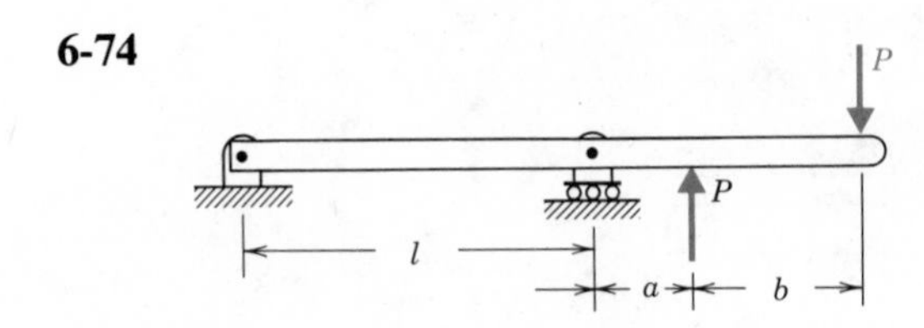

6-75

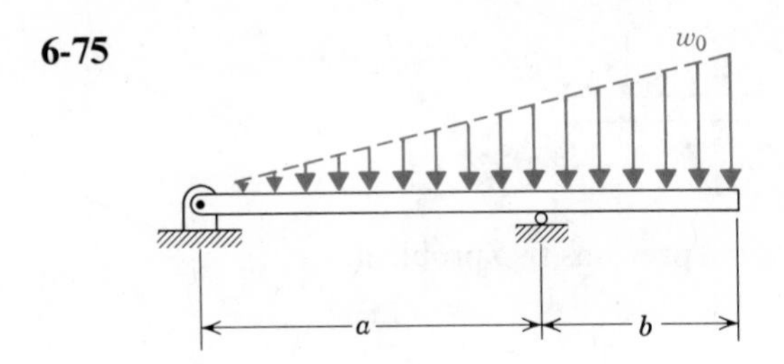

6-76

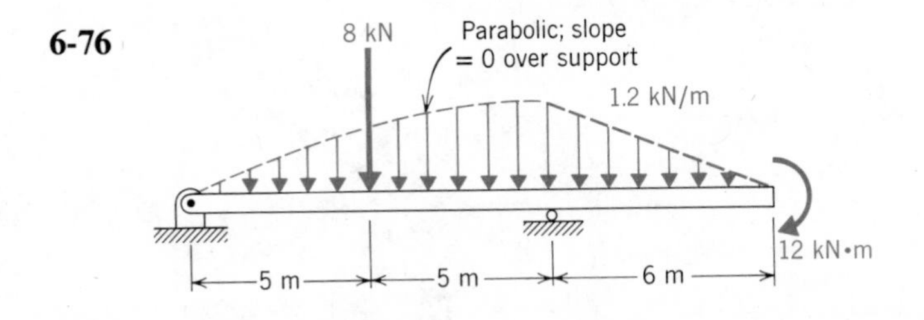

6-77

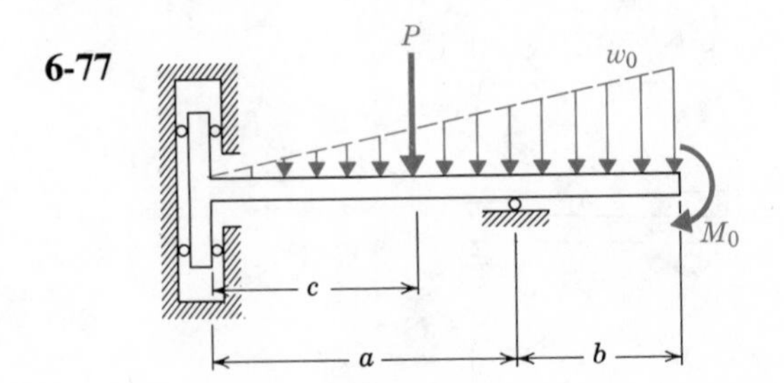

6-78

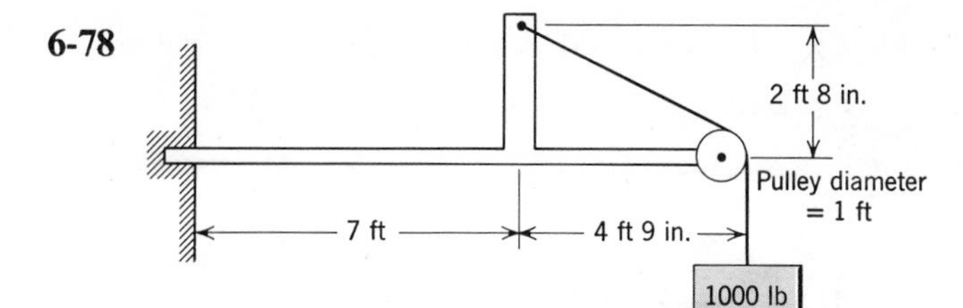

6-79 Radial loads $w(\theta)$ will induce internal forces V (shear), T (hoop tension), and M (bending moment) in the circular ring, as is indicated on the element. Derive the differential relationships connecting $w(\theta)$, $V(\theta)$, $T(\theta)$, and $M(\theta)$.

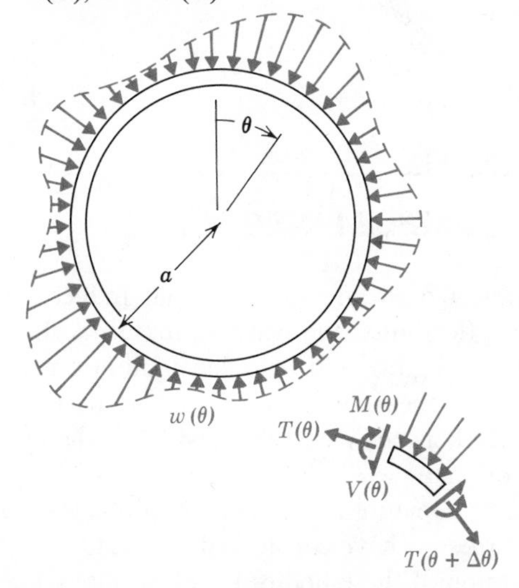

6-80 Loads $w(\theta)$ perpendicular to the plane of the circular rod will induce internal forces V (shear), M_b (bending moment), and M_t (twisting moment). Derive the differential relationships connecting $w(\theta)$, $V(\theta)$, $M_b(\theta)$, and $M_t(\theta)$.

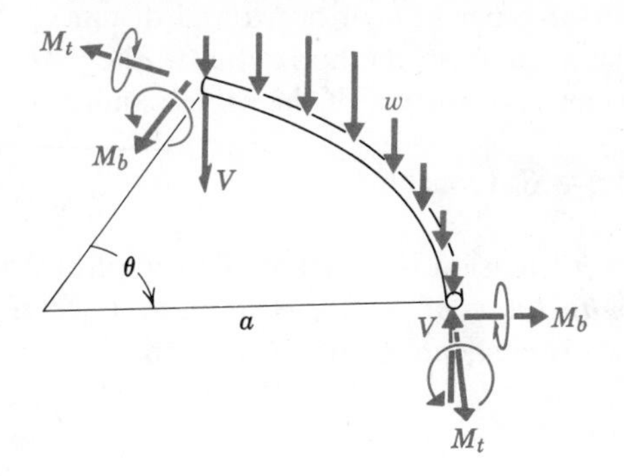

6-81 Evaluate $V(\theta)$, $M_b(\theta)$, and $M_t(\theta)$, as defined in the previous problem.

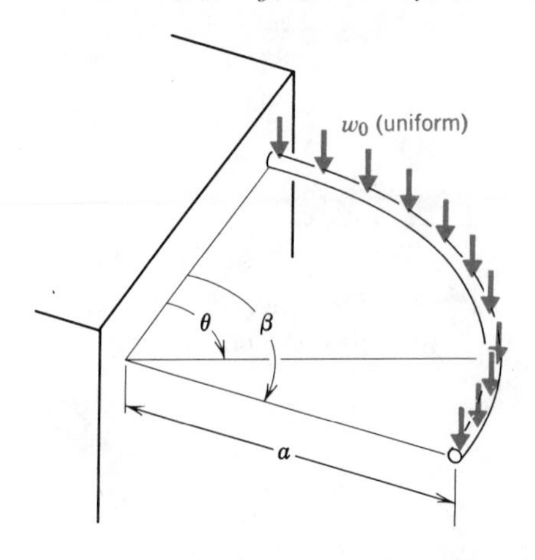

6-4 FLEXIBLE LINES

Cables and chains are often subjected to lateral forces that induce internal tension. Because of their inherent flexibility, the bending moments that these members can transmit are usually negligible, and from the equilibrium analysis of beams we know that this implies shearing forces must also be negligible. That is, the resultant internal force between elements must be in the direction tangent to the axis of the cable or chain.

The flexibility of the line adds a feature to equilibrium problems that has been absent from the other examples we have considered thus far. In addition to unknown forces, the configuration of these bodies is not normally known a priori; therefore, an additional unknown quantity enters such problems. In this section we examine the configuration and tension forces induced in flexible lines by distributed forces that are all parallel.

Consider the suspended line shown in Figure 6-21, subjected to a distributed vertical load of intensity w (force per unit of horizontal distance). With the tension denoted by S, and the angle from the horizontal to the tangent to the line by α, horizontal equilibrium of a portion of the cable requires that

$$S \cos \alpha = S_0 \text{ (constant)} \qquad \textbf{(6-12)}$$

That is, the horizontal component of tension is constant. This implies that *the tension is maximum where the configuration of the line is steepest.*

Vertical equilibrium of the short element of line requires that

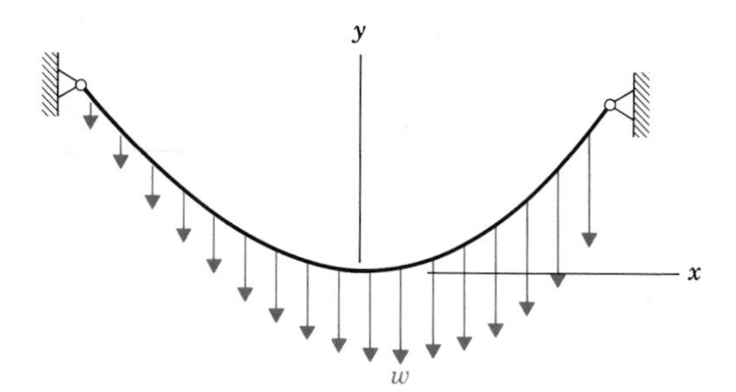

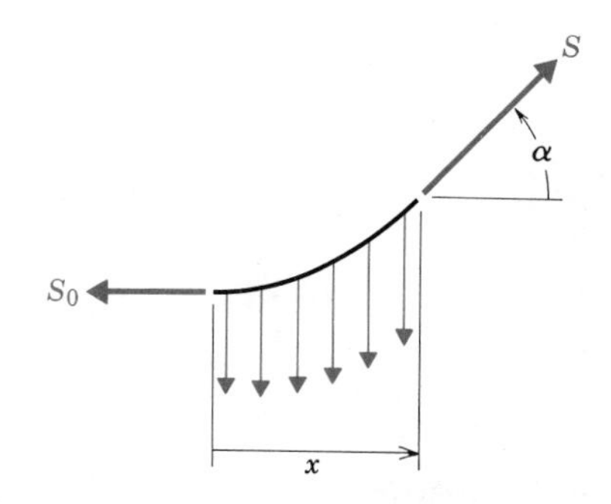

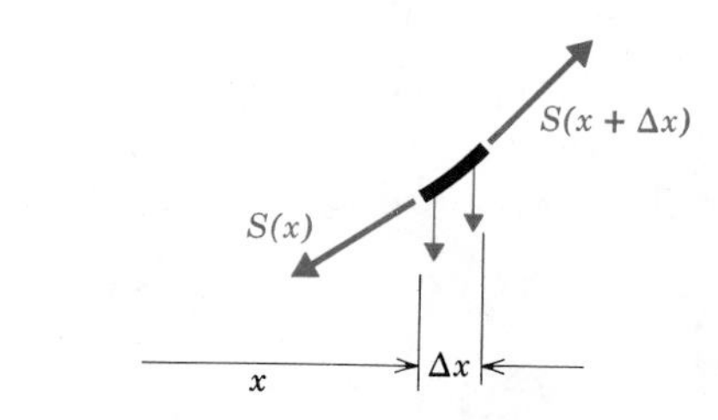

FIGURE 6-21

$$(S \sin \alpha)_{x+\Delta x} - (S \sin \alpha)_x = w_{av} \Delta x$$

where w_{av} is the average load intensity within the interval Δx. After division by Δx, this equation becomes, in the limit as $\Delta x \to 0$,

$$\frac{d}{dx}(S \sin \alpha) = w(x)$$

Or, after using (6-12) to eliminate the tension resultant,

$$\frac{d}{dx}(S_0 \tan \alpha) = w(x)$$

But $\tan\alpha = dy/dx$, so that

$$S_0 \frac{d^2y}{dx^2} = w(x) \tag{6-13}$$

Given the load distribution and support points, this equation determines the configuration by specifying $y(x)$. Once this is determined, the tension can be evaluated by means of Equation 6-12:

$$\begin{aligned} S &= S_0 \sec\alpha \\ &= S_0\sqrt{1 + \tan^2\alpha} \\ &= S_0\sqrt{1 + (dy/dx)^2} \end{aligned} \tag{6-14}$$

We now examine two commonly encountered types of load distribution. Under more complicated load distributions, the differential equation 6-13 may require numerical integration for its solution.

Horizontally Uniform Loading

Suspension bridge decks are normally suspended from the main cables by closely spaced vertical lines. These lines transmit to each main cable a load that is of nearly uniform intensity in the horizontal direction. In this case, we have

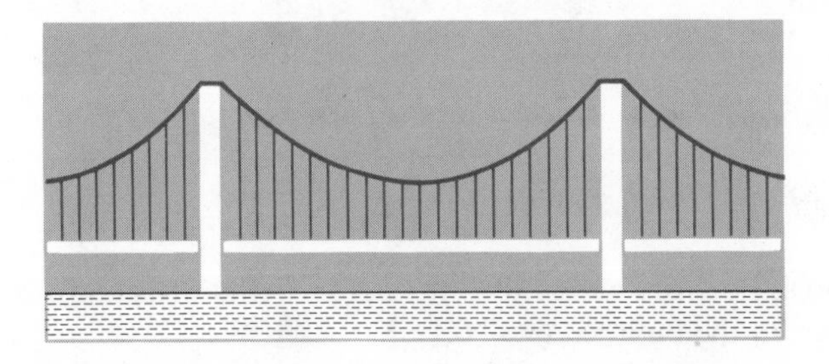

$$w(x) = w_0 \qquad \text{(constant)}$$

and Equation 6-13 may be readily integrated to give the parabolic configuration

$$S_0 y(x) = \tfrac{1}{2} w_0 x^2 + C_1 x + C_2$$

If we select the origin of the x-y coordinates at the apex of the parabola, the constants of integration C_1 and C_2 are both zero, and

$$y = \frac{w_0 x^2}{2S_0} \tag{6-15}$$

EXAMPLE

The deck of a suspension bridge has a lineal density of 42 tons per foot of length. The two main cables are suspended between towers 2800 ft apart. The top of each tower is 320 ft above the lowest point of a main cable. What is the maximum tension in each main cable?

The maximum tension will occur at the attachment with the tower, where $x = 1400$ ft. The load intensity on each of the two cables is

$$\begin{aligned} w_0 &= \frac{1}{2}(42 \text{ tons/ft}) \\ &= 21 \text{ tons/ft} \end{aligned}$$

Using the coordinates of the attachment point in Equation 6-15, the horizontal component of tension is computed as

$$\begin{aligned} S_0 &= \frac{(21 \text{ tons/ft})(1400 \text{ ft})^2}{2(320 \text{ ft})} \\ &= 64\,300 \text{ tons} \end{aligned}$$

The tension at the tower may now be determined from Equation 6-14:

$$\begin{aligned} S &= S_0\sqrt{1 + (dy/dx)^2} \\ &= \sqrt{S_0^2 + (w_0 x)^2} \\ &= \sqrt{(64\,300 \text{ tons})^2 + [(21 \text{ tons/ft})(1400 \text{ ft})]^2} \\ &= 70\,700 \text{ tons} \end{aligned}$$

Uniform Line Under Gravitational Forces

The gravitational forces exerted on the suspended line itself are of varying intensity with respect to horizontal distance, because of the varying slope of the line. For a uniform cable with density ρ and cross sectional area A, this load intensity is given by

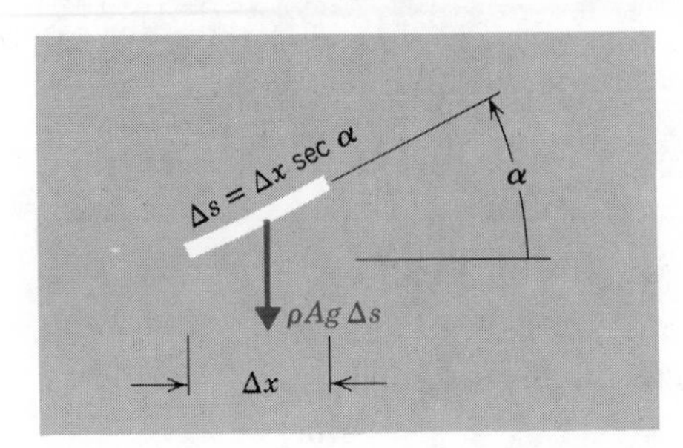

$$w_g(x) = \rho A g \sec \alpha$$

We see from the equation preceding (6-13) that when these are the only significant forces, the equilibrium curve must satisfy

$$S_0 \frac{d}{dx} \tan \alpha = \rho A g \sec \alpha$$

With the abbreviations

$$u = \tan \alpha = \frac{dy}{dx}$$

$$a = \frac{S_0}{\rho A g} \tag{6-16}$$

this equation can be rewritten as

$$\frac{du}{\sqrt{1 + u^2}} = \frac{dx}{a}$$

Integration results in

$$\sinh^{-1} u = \frac{x}{a} + C_1$$

Or,

$$u = \sinh\left(\frac{x}{a} + C_1\right)$$

Another integration gives

$$y = \int u\,dx$$

$$= a\cosh\left(\frac{x}{a} + C_1\right) + C_2$$

Now, if we select the origin of the x-y coordinates at the apex of the curve, where $y = \frac{dy}{dx} = 0$

$$C_1 = 0, \qquad C_2 = -a$$

and the equilibrium curve is then given by

$$y = a\left(\cosh\frac{x}{a} - 1\right) \tag{6-17}$$

The curve given by this equation is called the *common catenary*. Several of its properties, important in the analysis of suspended cables, are developed in the following.

Useful Relationships of Catenaries The angle α between the tangent to the catenary and the horizontal is given by

$$\sec\alpha = \frac{ds}{dx}$$

$$= \sqrt{1 + \left(\frac{dy}{dx}\right)^2}$$

$$= \sqrt{1 + \sinh^2\left(\frac{x}{a}\right)}$$

$$= \cosh\left(\frac{x}{a}\right) \tag{6-18}$$

Integration then gives the arc length, measured from the apex, as

$$s = \int\left(\frac{ds}{dx}\right)dx$$

$$= \int \cosh\left(\frac{x}{a}\right) dx$$

$$= a \sinh\left(\frac{x}{a}\right) \tag{6-19}$$

But since $\tan\alpha = dy/dx = \sinh x/a$, we also have

$$\tan\alpha = \frac{s}{a} \tag{6-20}$$

which tells us that *the parameter a is equal to the arc length from the apex to the point where* $\alpha = \pi/4$.

The tension at any point is related to the horizontal component S_0 by

$$S = S_0 \sec\alpha$$

and, in view of Equation 6-18, is given by

$$S = S_0 \cosh\left(\frac{x}{a}\right) \tag{6-21}$$

We next study the determination of the catenary size parameter a in terms of the length of line and the positions of the supports. Let the coordinates (x_1, y_1) and (x_2, y_2) locate the end points, as shown in Figure 6-22. The length of line l and support elevation difference h may be expressed by

$$l = s_2 - s_1 = a \sinh\left(\frac{x_2}{a}\right) - a \sinh\left(\frac{x_1}{a}\right)$$

$$h = y_2 - y_1 = a \cosh\left(\frac{x_2}{a}\right) - a \cosh\left(\frac{x_1}{a}\right)$$

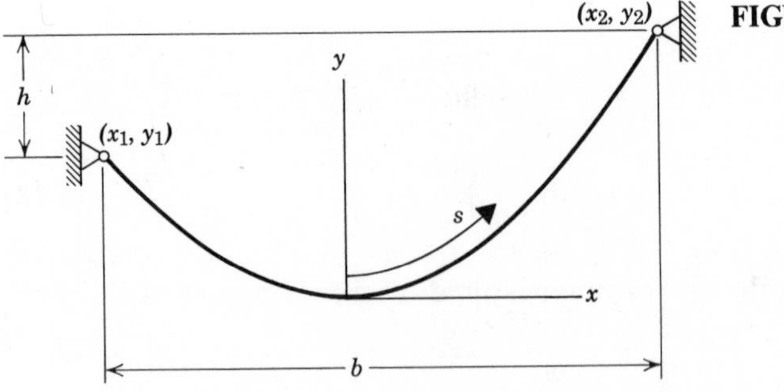

FIGURE 6-22

Addition and subtraction of these two equations lead to

$$e^{x_2/a} - e^{x_1/a} = \frac{(1+h)}{a}$$

$$-e^{-x_2/a} + e^{-x_1/a} = \frac{(l-h)}{a}$$

These may then be solved simultaneously, with the result

$$\frac{x_2}{a} = \log\left[(p+1)\frac{(l+h)}{2a}\right] \tag{6-22a}$$

$$\frac{x_1}{a} = \log\left[(p-1)\frac{(l+h)}{2a}\right] \tag{6-22b}$$

in which

$$p = \sqrt{1 + \frac{4a^2}{l^2 - h^2}} \tag{6-23a}$$

Now, the horizontal distance between the end points may be expressed by

$$\frac{b}{a} = \frac{x_2}{a} - \frac{x_1}{a} = \log\frac{p+1}{p-1}$$

from which

$$p = \frac{e^{b/a} + 1}{e^{b/a} - 1} = \operatorname{ctnh}\left(\frac{b}{2a}\right) \tag{6-23b}$$

Finally, elimination of p from Equations 6-23a and 6-23b leads to

$$\frac{b/2a}{\sinh(b/2a)} = \frac{b}{\sqrt{l^2 - h^2}} \tag{6-24}$$

This equation defines the ratio a/b for each value of the parameter $b/\sqrt{l^2 - h^2}$. Viewed in this manner, Equation 6-24 is transcendental; that is, determination of a/b requires a trial-and-error procedure. Table 6-1 and the corresponding curve of Figure 6-23 give values satisfying Equation 6-24.

TABLE 6-1

$b/\sqrt{l^2 - h^2}$	a/b	$b/\sqrt{l^2 - h^2}$	a/b
0.00	0.000 00	0.50	0.229 64
0.01	0.068 64	0.55	0.249 39
0.02	0.077 25	0.60	0.271 95
0.03	0.083 48	0.65	0.298 34
0.04	0.088 62	0.70	0.330 15
0.05	0.093 12	0.75	0.370 03
0.10	0.111 11	0.80	0.422 75
0.15	0.125 97	0.85	0.498 28
0.20	0.139 73	0.90	0.622 33
0.25	0.153 20	0.95	0.896 70
0.30	0.166 81	0.96	1.006 19
0.35	0.180 93	0.97	1.166 05
0.40	0.195 87	0.98	1.433 22
0.45	0.211 97	0.99	2.034 08

FIGURE 6-23

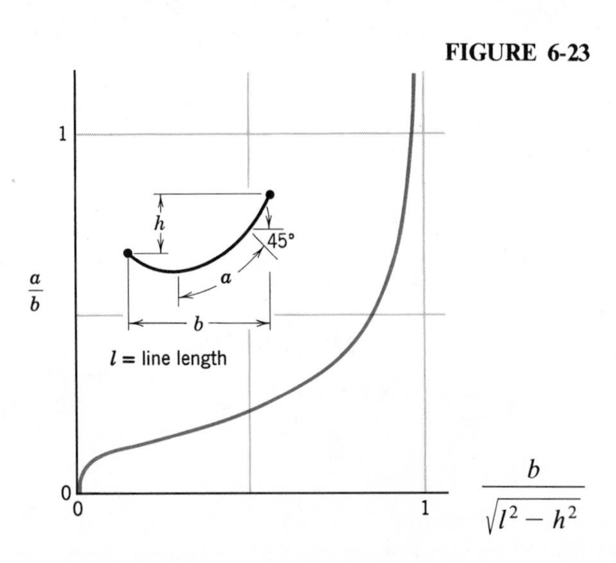

By using the first three terms in the power series expansion

$$\sinh\theta = \theta + \frac{\theta^3}{3!} + \frac{\theta^5}{5!} + \cdots$$

the following approximation can be developed:

$$\frac{a}{b} \approx \left(40\left\{\left[1 + \frac{6}{5}\left(\frac{1-u}{u}\right)\right]^{1/2} - 1\right\}\right)^{-1/2} \tag{6-24a}$$

in which

$$u = \frac{b}{\sqrt{l^2 - h^2}}$$

This approximation is 0.38% low at $u = 0.65$ and becomes increasingly accurate for larger u.

EXAMPLE

A 1300-ft, 6.5-ton cable is to be suspended between two supports that are 780 ft apart horizontally and are at an elevation difference of 500 ft. Determine the maximum tension induced by gravity.

The sag parameter in Equation 6-24 has the value

$$\frac{b}{\sqrt{l^2 - h^2}} = \frac{(780)}{\sqrt{(1300)^2 - (500)^2}}$$
$$= 0.650$$

Then from Table 6-1, we obtain

$$\frac{a}{b} = 0.2983$$

from which

$$a = (0.2983)(780 \text{ ft}) = 232.7 \text{ ft}$$

The largest tension occurs at $x = x_2$, which may be determined as follows. From Equation 6-23b,

$$p = \frac{e^{1/0.2983} + 1}{e^{1/0.2983} - 1} = 1.0726$$

Then, from Equation 6-22a

$$\frac{x_2}{a} = \log\left[\frac{(2.0726)(1300 + 500)}{2(232.7)}\right]$$

$$= 2.081$$

The tension at this point may now be computed by using Equations 6-16 and 6-21.

$$S_2 = (\rho A g)a \cosh\left(\frac{x_2}{a}\right)$$

$$= \frac{6.5 \text{ tons}}{1300 \text{ ft}}(232.7 \text{ ft}) \cosh 2.081$$

$$= 4.74 \text{ tons}$$

$$= 42.1 \text{ kN}$$

EXAMPLE

A uniform cable is suspended between two points that are at the same level and 120 m apart horizontally. The midpoint of the cable sags 20 m below the attachment points. What is the length of the cable?

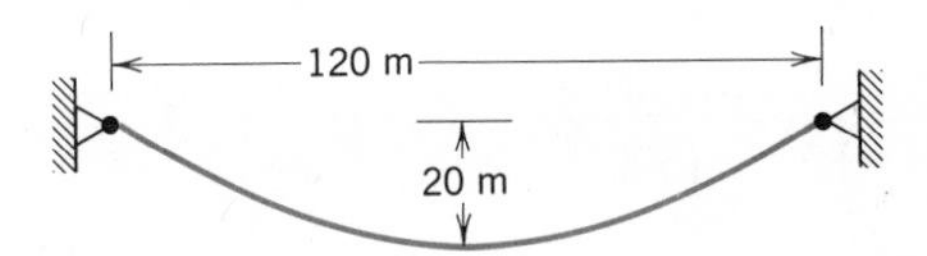

Equation 6-17, written for the support point, is

$$20 \text{ m} = a\left(\cosh\frac{60 \text{ m}}{a} - 1\right)$$

By trial and error, we find the value of the size parameter satisfying this equation to be

$$a = 93.15 \text{ m}$$

The length of the cable is now the only unknown in Equation 6-24. Setting $h = 0$ and solving this equation for the length, we obtain

$$l = 2a \sinh\frac{b}{2a}$$

$$= 2(93.15\text{ m})\sinh\frac{120}{2(93.15)}$$

$$= 128.5\text{ m}$$

A Problem in Cable Laying Ocean moorings and other flexible line installations have given rise to the following problem: when one of the attachment points is on a solid horizontal floor, an unknown portion of the cable may lie along the floor, and the remainder will form a catenary (in the absence of all but gravitational loading, of course). Given the total length of cable L and the relative positions of the attachment points, in terms of B and h as shown in Figure 6-24, we wish to determine the geometry of the catenary portion so that force analysis can be carried out.

The length of the portion of cable in contact with the floor is equal to the difference between the total length L and the length l of the catenary portion:

$$L - l = B - b$$

Or,

$$l - b = L - B$$

Equilibrium of a line element at the point where the cable parts from the floor shows that the catenary is tangent to the floor at that point. Therefore, we can consider the apex of this particular catenary as one of its attachment points; that is, we can use the equations for the general catenary with $x_1 = 0$

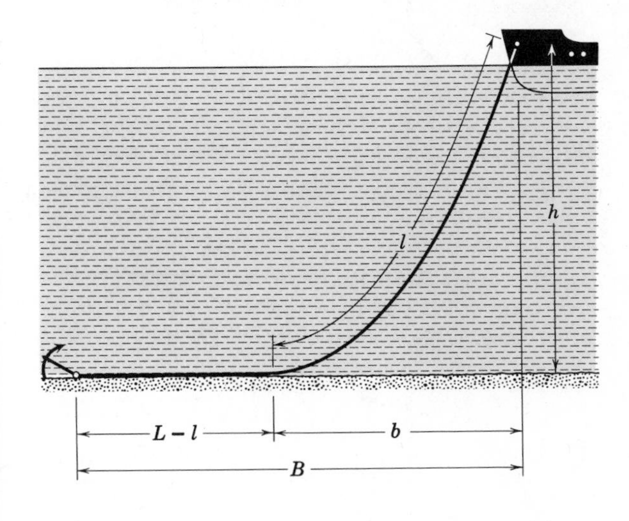

FIGURE 6-24

and $x_2 = b$. With these values, the equations for l and h, at the bottom of p. 266, then yield

$$\frac{l}{h} = \frac{\sinh(b/a)}{\cosh(b/a) - 1} \tag{6-25}$$

$$\frac{b}{h} = \frac{b/a}{\cosh(b/a) - 1} \tag{6-26}$$

Substitution into the above length relationship then gives

$$\frac{\sinh(b/a) - b/a}{\cosh(b/a) - 1} = \frac{L - B}{h} \tag{6-27}$$

Given L, B, and h, this equation determines the ratio a/b. This done, the values of l/h and b/h can be computed from Equations 6-25 and 6-26. To facilitate solution of the transcendental equation (6/27) for a/b, the curves in Figure 6-25 have been plotted. With a value of $(L - B)/h$ entered on the vertical scale on the right, the value of a/b may be read on the abscissa. Corresponding values of l/h and b/h may then be read by using the other curves.

Approximations can be constructed by using the lead terms in the power series developments of the functions of b/a in Equations 6-25, 6-26, and 6-27. This procedure results in

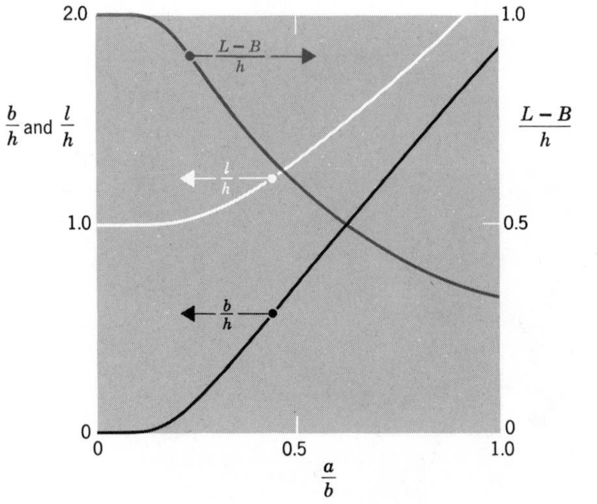

FIGURE 6-25

$$\frac{l}{h} \approx \frac{2h}{3(L-B)}\left[1+\frac{9}{20}\left(\frac{L-B}{h}\right)^2\right] \qquad \textbf{(6-25a)}$$

$$\frac{b}{h} \approx \frac{2h}{3(L-B)}\left[1-\frac{21}{20}\left(\frac{L-B}{h}\right)^2\right] \qquad \textbf{(6-26a)}$$

$$\frac{a}{b} \approx \frac{h}{3(L-B)}\left[1-\frac{3}{10}\left(\frac{L-B}{h}\right)^2\right] \qquad \textbf{(6-27a)}$$

At a value of $(L - B)/h = 0.4197$ (where $a/b = 0.75$), the above approximation for l/h is 0.09% low, the approximation for b/h is 0.12% low, and the approximation for a/b is 0.30% high. The accuracies improve with decreasing values of $(L - B)/h$.

EXAMPLE

One end of an 1870-m cable is anchored on the horizontal sea floor 750 m below the surface. The upper end is attached to a ship at a point 1600 m south of the anchor. The difference between the gravitational and buoyant forces acting on the submerged cable is 131 N/m of cable length. What is the force exerted by the cable on the ship?

The net force intensity is

$$\rho_n A g = 131 \text{ N/m}$$

and the lengths shown in Figure 6-24 are

$$L = 1870 \text{ m}$$
$$B = 1600 \text{ m}$$
$$h = 750 \text{ m}$$

The ratio appearing in the right-hand side of (6-27) has the value

$$\frac{(L-B)}{h} = \frac{1870 - 1600}{750} = 0.360$$

Using this value in Figure 6-25, we obtain

$$\frac{l}{h} = 1.96$$

$$\frac{b}{h} = 1.60$$

$$\frac{a}{b} = 0.89$$

Then,

$$l = (1.96)(750 \text{ m}) = 1470 \text{ m}$$

$$b = (1.60)(750 \text{ m}) = 1200 \text{ m}$$

$$a = (0.89)(1200 \text{ m}) = 1068 \text{ m}$$

Now, from (6-16), the horizontal component of tension is

$$S_0 = (131 \text{ N/m})(1068 \text{ m}) = 140 \text{ kN}$$

and from (6-21) the resultant tension at the attachment with the ship is

$$S_2 = (140 \text{ kN}) \cosh \frac{1200}{1068} = 238 \text{ kN}$$

This force acts at an angle α_2 with the horizontal given by

$$\cos \alpha_2 = \frac{140}{238}$$

Or,

$$\alpha_2 = 54°$$

PROBLEMS

6-82 The lineal density of the cable in the example on page 263 is 9 Mg/m. What is the additional tension at the tower caused by the forces of gravity on the cable itself? *Ans.* 10 530 tons.

6-83 Evaluate the compressive force in the tower of the suspension bridge discussed in the example on page 263.

6-84 The towers of a suspension bridge are 430 m apart, and the points of cable attachment are 50 m above the lowest points on the cable. The maximum tension in each cable is 150 MN. What is the minimum tension in each cable, and what is the deck loading?

6-85 The cable supports a load with a lineal density (with respect to horizontal distance) of 200 kg/m. The lineal density of the cable is negligible. Evaluate the minimum and maximum tensions in the cable.

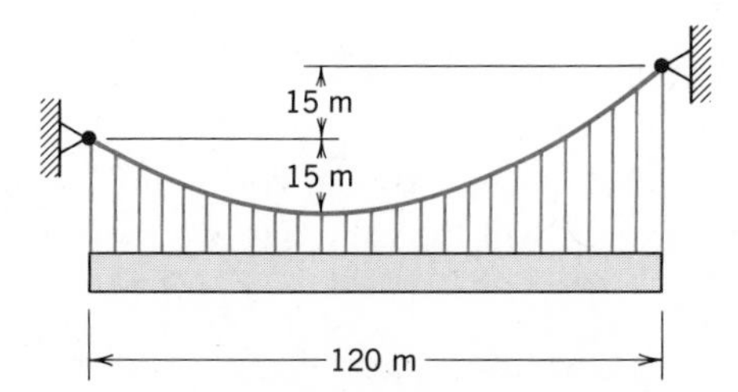

6-86 What is the length of the cable of the previous problem?

6-87 The intensity of the load acting on the cable varies linearly with horizontal distance as indicated. Gravity forces on the cable are negligible. The tangent to the cable at the left support is horizontal. What is the maximum tension in the cable?

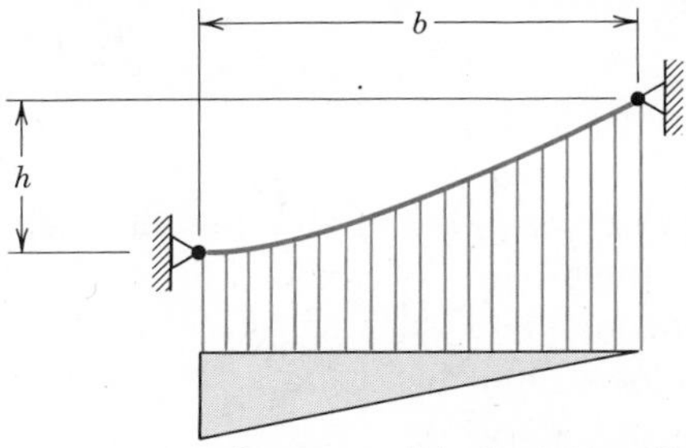

6-88 A uniform cable is to be strung between the two points as is shown below. The cable is to be tangent with the horizontal at the lower attachment point. What length cable is required, and what is the maximum tension if the lineal density of the cable is 10 lbm/ft? *Ans.* 183.8 ft; 2188 lb.

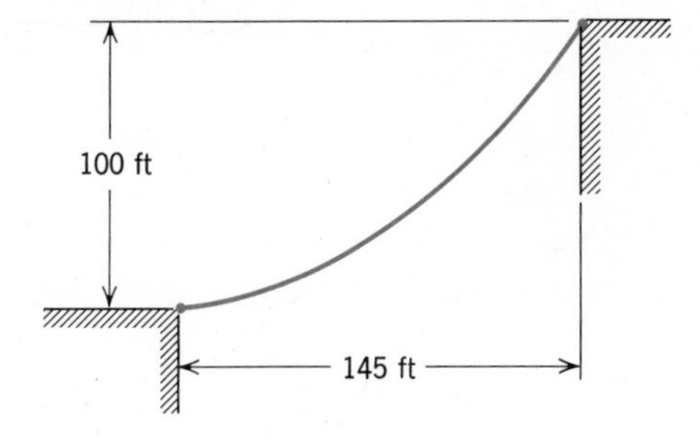

6-89 If the cable of the example on page 270 has a lineal density of 0.88 kg/m, what is the maximum tension in the cable?

6-90 What angle does the cable of the example on page 270 make with the horizontal, at the attachment points?

6-91 How far below the lower support will the 1300-ft cable of the example on page 269 sag?

6-92 A 125-m long cable is suspended between the two points as shown. How far below the lower support will the cable sag?

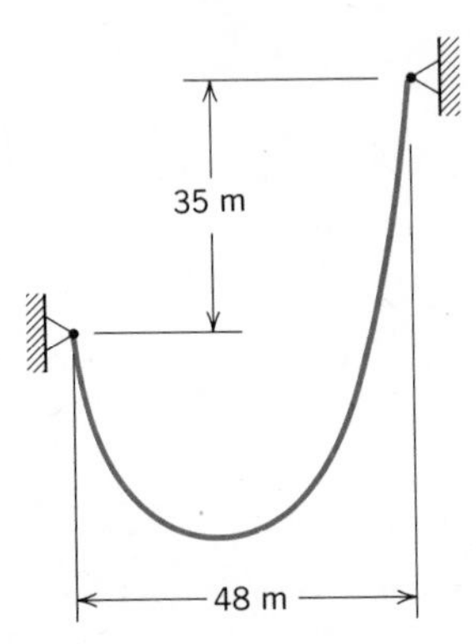

6-93 A logging balloon is tethered with a 90-m long line that has a lineal density of 0.78 kg/m. At the ground attachment, the cable makes an angle of 57° with the horizontal and has a tension of 2 kN. Calculate the height h of the balloon, and the tension in the line at the attachment with the balloon.

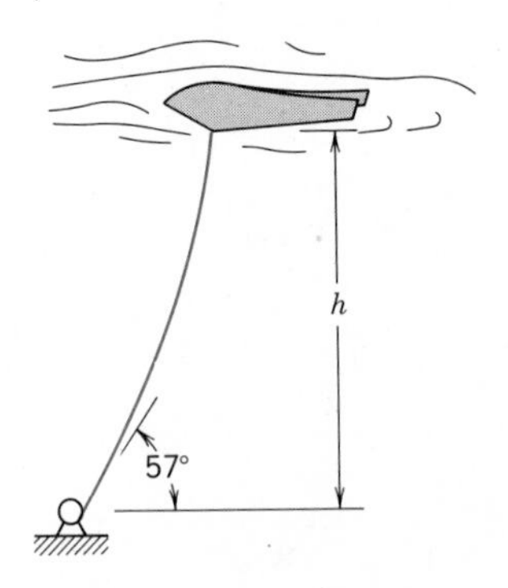

6-94 The 735-kg cable supports the 400-kg load at mid-span. Before the concentrated load was added, the sag in the cable was 30 m. Determine the final sag h_1, and the maximum tension in the cable. *Ans.* 33.05 m; 3.328 kN.

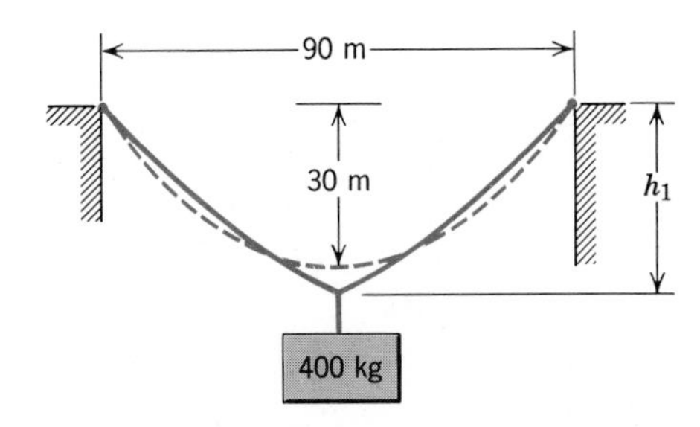

6-95 Show that for a given cable length and material, and for given support positions, the stress (tension divided by cross-sectional area) in the cable loaded only by gravitational forces on the cable itself, is independent of the cable diameter.

6-96 What is the largest value the parameter $b/\sqrt{l^2 - h^2}$, appearing in Equation 6-24, can have?

6-97 For given values of B and h, shown in Figure 6-24, what is the shortest length L for which some of the cable will lie along the ocean floor?

6-98 Show that the apex of a common catenary will lie between the support points (x_1, y_1) and (x_2, y_2), provided that

$$\frac{ah}{l^2 - h^2} \leqslant \frac{1}{2}$$

Taking account of (6-24), this criterion has the form

$$\frac{b}{h} \geqslant f\left(\frac{b}{\sqrt{l^2 - h^2}}\right)$$

Plot this criterion and Equation 6-24, using a common abscissa $b/\sqrt{l^2 - h^2}$, and explain how this can be used to solve Problem 6-88 and others with different support positions.

6-99 A 140-ft line having a density of 0.9 Mg/m^3 runs from the dock to the anchor as shown. The density of seawater is about 1.03 Mg/m^3. Estimate the line tension at the anchor.

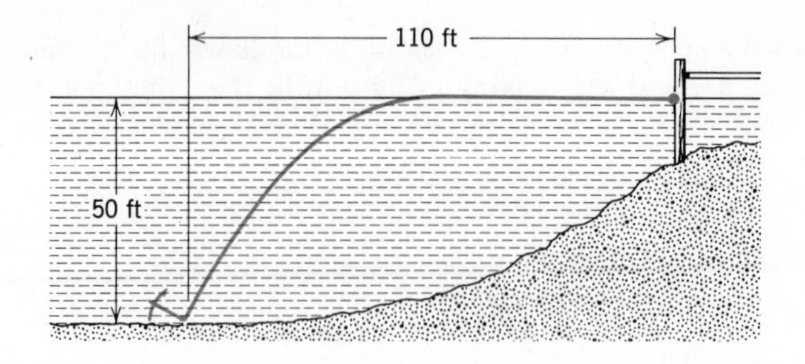

6-100 Derive the approximation (6-24a), and plot the error versus u through the range of values given in Table 6-1.

6-101 Show that the catenary size parameter a is related to the dimensions shown by

$$\frac{b}{a} = \log \frac{1 + \frac{h_2}{a} + \sqrt{\frac{h_2}{a}\left(2 + \frac{h_2}{a}\right)}}{1 + \frac{h_1}{a} - \sqrt{\frac{h_1}{a}\left(2 + \frac{h_1}{a}\right)}}$$

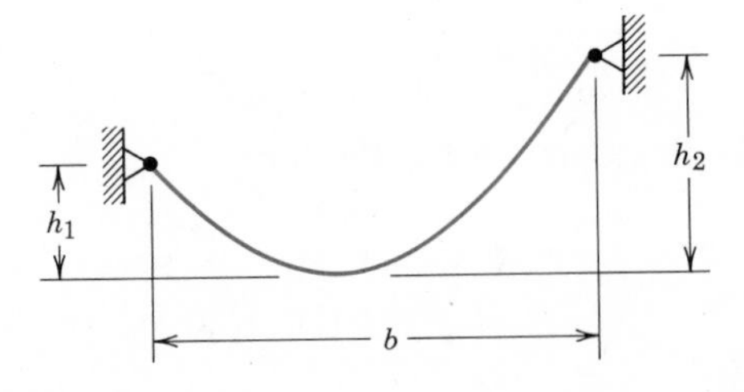

6-102 The cable is to sag 90 m below the tower at A. How long must the cable be, and what will be the tensions at A and B if the cable has a lineal density of 16 kg/m?

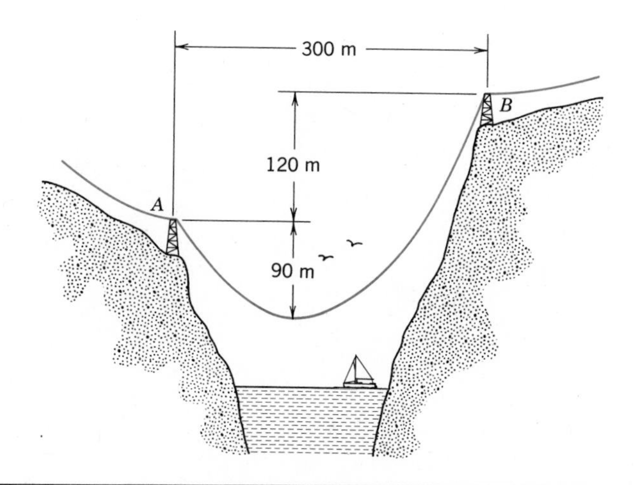

7

VIRTUAL WORK

Some mechanical systems, particularly complex mechanisms, are better suited to an approach to equilibrium analysis which is somewhat different from that taken in Chapters 4 through 6. The main advantage of this alternative approach is that it often gives directly an equation that would otherwise require a number of intermediate steps for the elimination of unknown forces from a system of equations. The idea is based on the notion of work, which we will pursue in some detail in the study of dynamics.

7-1 WORK DONE BY A FORCE

The work done by a force **f**, as the particle on which it acts undergoes a change in its position, is defined as

$$\boxed{W_{1-2} = \int_{\mathbf{r}_1}^{\mathbf{r}_2} \mathbf{f} \cdot d\mathbf{r}} \tag{7-1}$$

where **r** is a varying position vector that locates the particle relative to some stationary reference point. The work increment,

$$dW = \mathbf{f} \cdot d\mathbf{r}$$

is a scalar quantity that can be positive, negative, or zero, depending on the angle between the force and displacement increment vectors. Recalling the geometric definition of the dot product, we see that the work increment is the product of the displacement increment and the component $\mathbf{f}_{dr}$ of the force in the direction of the displacement increment. Thus, the component of force perpendicular to the displacement increment does not contribute to the work increment. Alternatively, the work increment is the product of the magnitude of the force and the component of the displacement increment in the direction of the force.

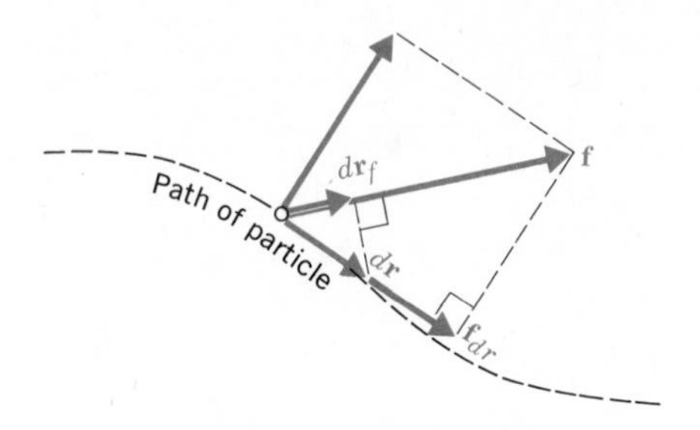

As an example, let us evaluate the work done by the force of gravity on a 7.26-kg shotput as it travels from a height of 2.0 m to a height of 4.5 m. The work increment in this case is

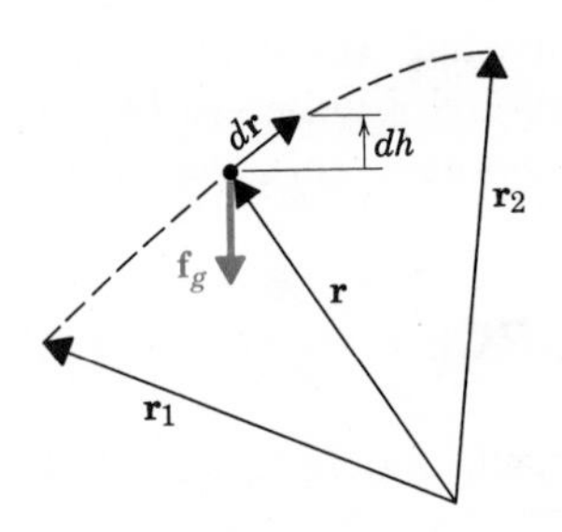

$$\mathbf{f}_g \cdot d\mathbf{r} = f_g dr \cos \measuredangle_{\mathbf{f}_g}^{d\mathbf{r}}$$
$$= -f_g dh$$

so that the work will be

$$W_{1-2} = \int_{\mathbf{r}_1}^{\mathbf{r}_2} \mathbf{f}_g \cdot d\mathbf{r}$$

$$= \int_{2.0m}^{4.5m} -(7.26\ \text{kg})(9.81\ \text{N/kg})\,dh$$

$$= -178\ \text{N}\cdot\text{m}$$

The SI unit for work is the newton meter, called the *joule* ($\text{J} = \text{N} \cdot \text{m}$).

As another example, let us evaluate the work done by a force f as it stretches a spring from its relaxed position to an extended configuration, in terms of the extension x and the stiffness k (force per unit displacement). The diagram shows how the force varies as the spring is stretched. At an intermediate extension ξ, the magnitude of the corresponding force is $k\xi$. Noting that the force and displacement increments are parallel throughout the process, we see that the work increment will be

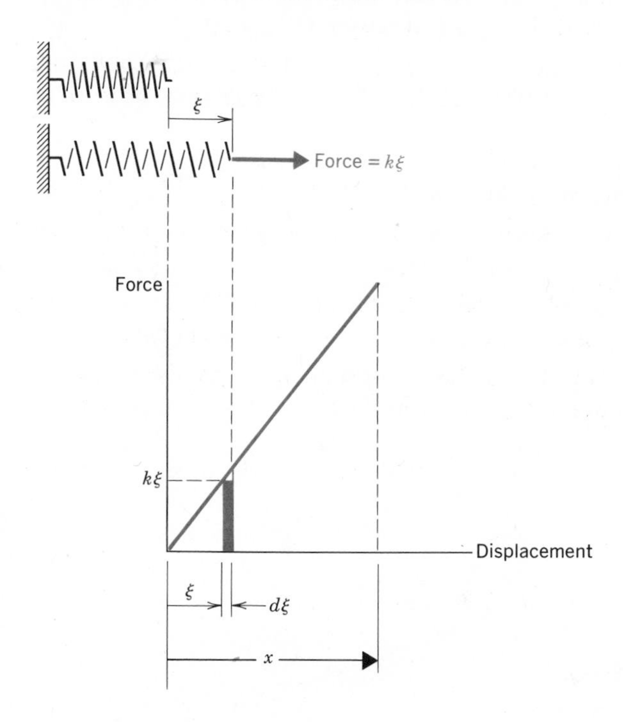

$$dW = \mathbf{f} \cdot d\mathbf{r} = f d\xi$$

Thus, the work will be

$$W = \int_0^x f(\xi)\,d\xi$$

$$= \int_0^x k\xi d\xi$$

$$= \tfrac{1}{2}kx^2$$

Observe that the work is equal to the area under the curve of force versus displacement.

PROBLEMS

7-1 What is the work done by a force in stretching a spring of stiffness k from its relaxed position to an extended position, in terms of k and the magnitude of the force at the extended position? *Ans.* $f^2/2k$.

7-2 Two springs have different stiffnesses, $k_2 > k_1$. Both are stretched from their relaxed positions. Which force does the greater amount of work in stretching the spring
(a) if the springs are stretched by the same amount?
(b) if they are stretched by the same force?

7-3 As a force *compresses* a spring from its relaxed position to a shorter length, is the work done positive or negative?

7-4 Show that the work increment is equal to the product of the force and the component of the displacement increment in the direction of the force.

7-5 The block is pushed up the incline at constant speed by the horizontal force P. What is the work of this force as the block moves to a height h?

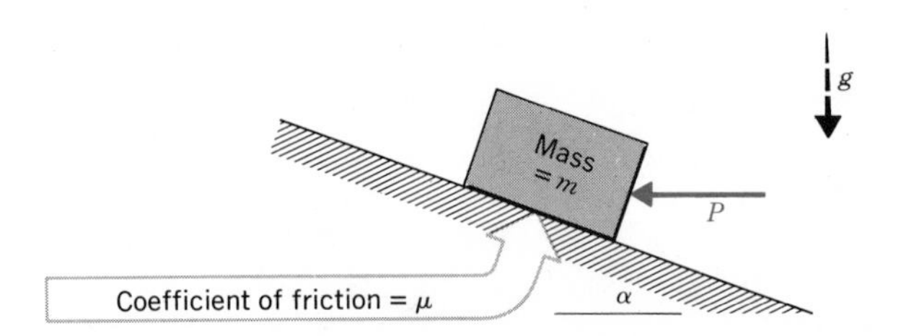

7-6 The block is pulled at constant speed over the horizontal surface by the force F. The angle α is maintained constant. What work is done by F as the block moves a distance d?

Ans. $W = \dfrac{\mu}{1 + \mu \tan\alpha} mgd.$

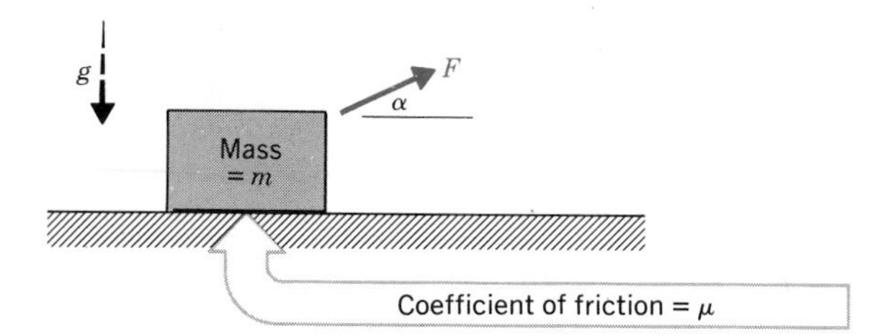

7-7 If the shotput in the example on page 282 is slowly *lifted* from 2 m to 4.5 m, what is the work done by the upward-directed force necessary to overcome the gravitational force? What is the sum of the works done by the lifting and gravitational forces?

7-8 What is the work done by a lifting force as it slowly raises an object of mass m from the surface of the earth to an altitude of 1000 km? Compare this with an estimate made under the assumption that the magnitude of the force is constant, equal to mg.

7-9 What is the work done by the force $P(x)$ in compressing the gas within the cylinder? The gas pressure is given by

$$p = p_0\left(\frac{V_0}{V}\right)^{\gamma}$$

$V =$ volume; p_0, V_0, γ all constant

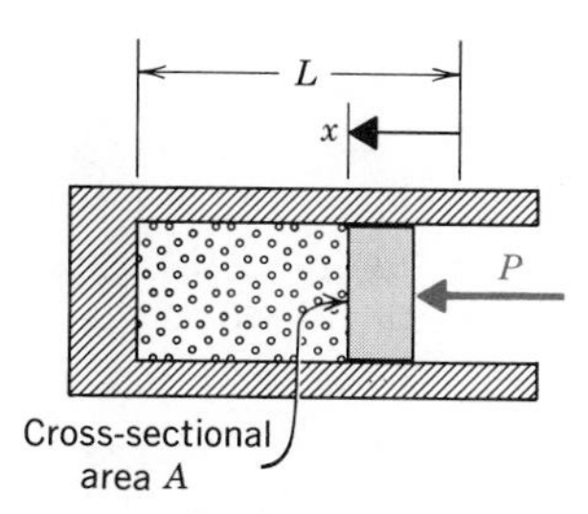

7-2 WORK DONE BY FORCES ON A RIGID BODY

As a rigid body undergoes an incremental change in position, the increment of work done by the forces acting on the body is defined as the sum of the work increments done by the individual forces. Referring to Figure 7-1a, we write this as

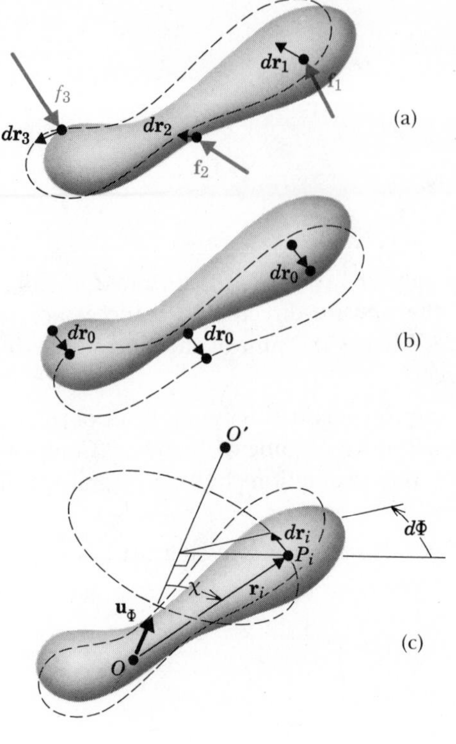

FIGURE 7-1

$$dW = \sum_i (\mathbf{f}_i \cdot d\mathbf{r}_i)$$

Consider first a simple displacement increment in which every point moves parallel to all other points as depicted in Figure 7-1b; that is, no rotation takes place. The displacement of all points will then be equal,

$$d\mathbf{r}_i = d\mathbf{r}_O$$

and the increment of work becomes

$$\begin{aligned} dW &= \sum_i (\mathbf{f}_i \cdot d\mathbf{r}_O) \\ &= \left(\sum_i \mathbf{f}_i \right) \cdot d\mathbf{r}_O \\ &= \mathbf{f} \cdot d\mathbf{r}_O \end{aligned}$$

in which $\mathbf{f} = \Sigma \mathbf{f}_i$ is the resultant force acting on the body. Therefore, in the absence of rotation, the work done on the rigid body is calculated in the same way as that done on a particle.

Next, consider a small rotation about the axis OO', as depicted in Figure 7-1*c*. We denote the angle of rotation (in radians) by $d\Phi$. Now, observe from the figure that the magnitude of the position change of point P_i may be expressed as

$$dr_i = \overline{OP_i} \sin \chi \, d\Phi$$

But this is also the magnitude of $\mathbf{u}_\Phi \times \mathbf{r}_i d\Phi$, where $\mathbf{u}_\Phi$ is a unit vector in the direction of the axis of rotation, and $\mathbf{r}_i$ is a vector from any point on the axis OO' to the point P_i. Finally, if we define the direction of rotation using the right-hand rule in conjunction with the direction of $\mathbf{u}_\Phi$, we can write the displacement increment as

$$d\mathbf{r}_i = (d\Phi \mathbf{u}_\Phi) \times \mathbf{r}_i$$

in which the direction and magnitude of the rotation of the body are completely characterized by the vector $d\Phi \mathbf{u}_\Phi$. Then the increment of work done by a set of forces acting on the body as it undergoes such an increment of rotation is given by

$$\begin{aligned} dW &= \Sigma(\mathbf{f}_i \cdot d\mathbf{r}_i) \\ &= \Sigma[\mathbf{f}_i \cdot (d\Phi \mathbf{u}_\Phi) \times \mathbf{r}_i] \\ &= \Sigma[(\mathbf{r}_i \times \mathbf{f}_i) \cdot (d\Phi \mathbf{u}_\Phi)] \\ &= (\Sigma \mathbf{r}_i \times \mathbf{f}_i) \cdot (d\Phi \mathbf{u}_\Phi) \\ &= \mathbf{M}_O \cdot \mathbf{u}_\Phi d\Phi \end{aligned}$$

in which $\mathbf{M}_O$ is the resultant moment of the forces about O. We note that the component of moment perpendicular to the axis of the rotation increment does not contribute to the work increment.

EXAMPLE

A torque of 120 in·lbf is applied to the handle of the screwdriver in order to drive the screw. If this torque does not vary significantly as the screw is driven, how much work will be done by the couple for each revolution of the screw?

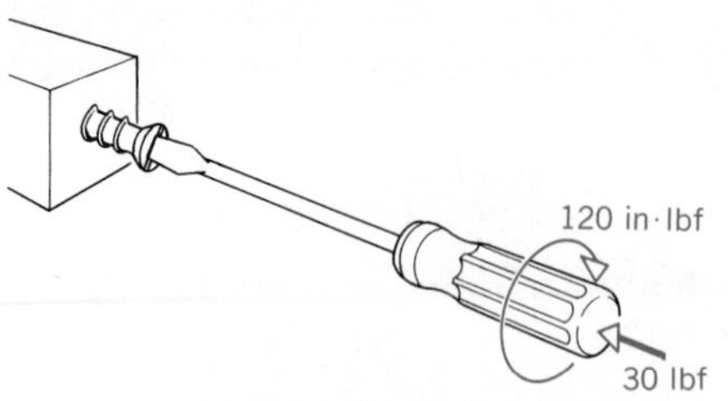

Since the moment and rotation increment vectors $\mathbf{M}_O$ and $\mathbf{u}_\Phi d\Phi$ are in the same direction, the work increment will be

$$dW = \mathbf{M}_O \cdot \mathbf{u}_\Phi d\Phi$$
$$= M_O d\Phi$$

The work for one revolution (2π rad) is then

$$W = \int_0^{2\pi} M_O d\Phi$$
$$= (120 \text{ in} \cdot \text{lbf}) \int_0^{2\pi} d\Phi$$
$$= 754 \text{ in} \cdot \text{lbf} = 85.2 \text{ J}$$

Note that this does not include the work done by the 30-lb axial force.

An *arbitrary* change in position of a rigid body can be expressed as a superposition of a translation equal to the displacement of any point O and a rotation about an axis through O. For a given repositioning, the translation depends on the point O selected, but the rotation $\mathbf{u}_\Phi d\Phi$ is independent of this selection. Far from obvious, these facts will be demonstrated in later chapters on kinematics. The increment of work done on a rigid body as it undergoes an arbitrary increment of change in position is, therefore,

$$dW = \sum \mathbf{f}_i \cdot (d\mathbf{R}_O + d\Phi \mathbf{u}_\Phi \times \mathbf{r}_i)$$
$$= \mathbf{f} \cdot d\mathbf{R}_O + \mathbf{M}_O \cdot \mathbf{u}_\Phi d\Phi$$

The work done as the rigid body undergoes a repositioning is then

$$\boxed{W_{1-2} = \int_1^2 [\mathbf{f} \cdot d\mathbf{R}_O + \mathbf{M}_O \cdot \mathbf{u}_\Phi d\Phi]} \qquad \textbf{(7-2)}$$

In general, **f**, $\mathbf{M}_O$, and $\mathbf{u}_\Phi$ can all vary as the body is moved a finite amount.

EXAMPLE

An upward force is applied to the end of the wrench handle, 528 mm from the elbow shown in Figure 2-4 (p. 35). Friction in the threads requires that a moment about the pipe axis of 79.2 N·m be applied to loosen it. What will be the work done by the force as the elbow is unscrewed 2°?

The component of moment about the axis of rotation is 79.2 N·m. Then, assuming this moment does not vary as the 2° rotation takes place, we have

$$W = \mathbf{M}_O \cdot \mathbf{u}_\Phi \int_0^{2^\circ} d\Phi$$

$$= (79.2\ \mathrm{N \cdot m})(2^\circ)(\pi\ \mathrm{rad}/180^\circ)$$

$$= 2.76\ \mathrm{J}$$

It is instructive to evaluate the applied force and linear travel at the end of the handle as the rotation takes place, and to compute the work using Equation 7-1.

PROBLEMS

7-10 Evaluate the work done on the winch handle in raising the car a height of 2 ft. The length of the winch handle is 2 ft and the spool onto which the line is wound is 6 in. in diameter.

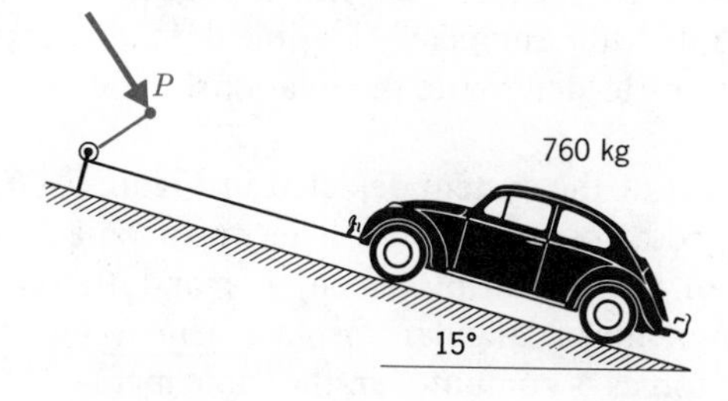

7-11 Evaluate the work done by the force T in raising the load to a height h, for each of the mechanisms of Problems 4-32 and 4-33.

7-12 Referring to the example above, compute the work of the force applied to the end of the wrench handle, using Equation 7-1.

7-13 If the pitch of the screw in the example on page 289 is $\frac{1}{16}$ inch per revolution, what fraction of the total work done by all forces on the handle is done by the 30-lbf axial force?

7-3 THE PRINCIPLE OF VIRTUAL WORK

The concept of work embodied in the above definitions is useful in the study of dynamics, because of its relationship to changes in certain quantities associated with motion. The closely related concept of *virtual work* is useful in both statics and dynamics.

Virtual work is defined as the work that would be done by all the forces acting on a system if the parts within the system were to undergo small increments of displacement. These increments of displacement, called the components of a *virtual displacement* of the system, are arbitrary; this means that we may consider as a virtual displacement any small change in configuration that suits the needs of the problem at hand, whether physically realizable or not.

Presently, we will show the equivalence between the previously used conditions for equilibrium and the *principle of virtual work*, which states: *A mechanical system is in static equilibrium if and only if the virtual work δW is zero for an arbitrary virtual displacement.* However, most readers will grasp the general idea more readily after first examining some examples.

EXAMPLE

The slider-crank mechanism shown in Figure 7-2a has a vertical force P applied to the piston and a couple with moment M applied to the crank. Assuming negligible friction, we want to determine the relationship between P and M for equilibrium.

Consider a virtual displacement of the system depicted in Figure 7-2b, in which the crank and connecting rod do not deform. The upper end of the connecting rod undergoes a small vertical displacement δx, and the crank undergoes a corresponding increment of angular displacement $-\delta\theta$. The corresponding virtual work of the forces is computed in the same manner as an increment of work on an actual set of displacements; with reference to the equation preceding (7-2), this virtual work may be written as

$$\delta W = -P\delta x + M(-\delta\theta)$$

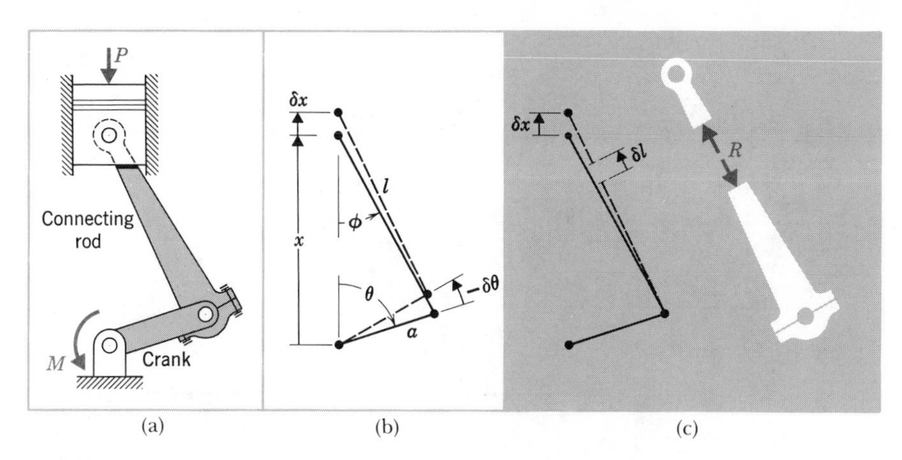

FIGURE 7-2

Before we can make use of this, we must determine the geometric relationship between δx and $\delta\theta$. To do this, we begin with two equations from the triangle formed by the crank, the rod, and the vertical:

$$x = a\cos\theta + l\cos\phi$$

$$a\sin\theta = l\sin\phi$$

Differentiation gives the first-order relationships among the variations in x, θ, and ϕ, as

$$\delta x = -a\sin\theta\,\delta\theta - l\sin\phi\,\delta\phi$$

$$a\cos\theta\,\delta\theta = l\cos\phi\,\delta\phi$$

Algebraic elimination of $\delta\phi$ from these two equations gives us the variation in x in terms of the variation in θ:

$$\delta x = -a(\sin\theta + \tan\phi\cos\theta)\delta\theta$$

using this result, we can write the virtual work expression as

$$\delta W = [Pa(\sin\theta + \tan\phi\cos\theta) - M]\delta\theta$$

Now, according to the principle of virtual work, if the system is in equilibrium, this is zero for an arbitrary value of $\delta\theta$. Therefore, we have

$$M = Pa(\sin\theta + \tan\phi\cos\theta)$$

If desired, a little trigonometry can be used to eliminate ϕ, with the result that

$$M = Pa\sin\theta\left(1 + \frac{(a/l)\cos\theta}{\sqrt{1 - (a^2/l^2)\sin^2\theta}}\right)$$

EXAMPLE

Determine the force in the connecting rod in Figure 7-2*a*.

For this problem, we consider a virtual displacement on which the desired force would do work. A variation in x with no variation in θ or deformation of the crank would require a variation in l, as shown in Figure 7-2*c*. This can be visualized as a separation somewhere along the rod, and we see that the internal surface force acting on the upper portion would do work equal to $R\,\delta l$. The relationship between δl and δx can be determined, as in the preceding example, by differentiating the equations for the triangle to obtain

$$\delta x = \cos\phi\,\delta l - l\sin\phi\,\delta\phi$$

$$0 = \sin\phi\,\delta l + l\cos\phi\,\delta\phi$$

and algebraically eliminating $\delta\phi$, with the result that

$$\delta l = \cos\phi\;\delta x$$

The virtual work is then

$$\begin{aligned}\delta W &= -P\delta x + R\delta l\\ &= (-P + R\cos\phi)\delta x\end{aligned}$$

and, because this must be zero for arbitrary δx,

$$R = P\sec\phi$$

The geometric analysis of the virtual displacements can usually be carried out more directly by sketching the small displacements, rather than resorting to formal differentiation as was done in the preceding examples. For example, the change in length of the rod in the last example can be determined as follows. Let the rod pivot a small amount about the crankpin, moving its upper end upward and to the right. Then let the rod lengthen until its upper end is at a point directly above its original position. The small triangle formed by the vertical and the path traced out by the upper end shows at once that

$$\delta l = \delta x \cos \phi$$

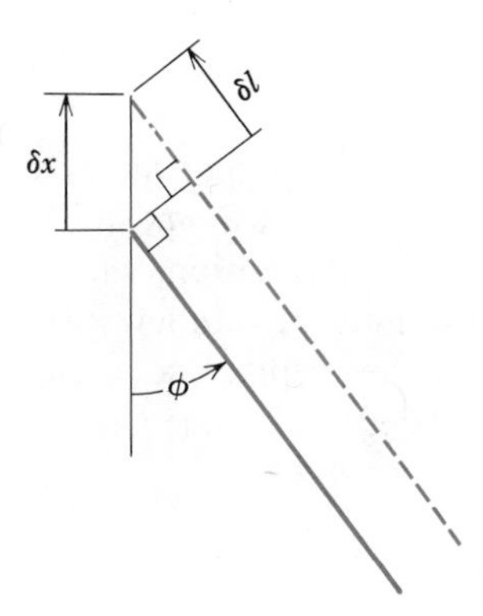

EXAMPLE

Figure 7-3a is a schematic of a platform balance scale. Determine the relationship among the dimensions so that the force ratio R/f_g is independent of

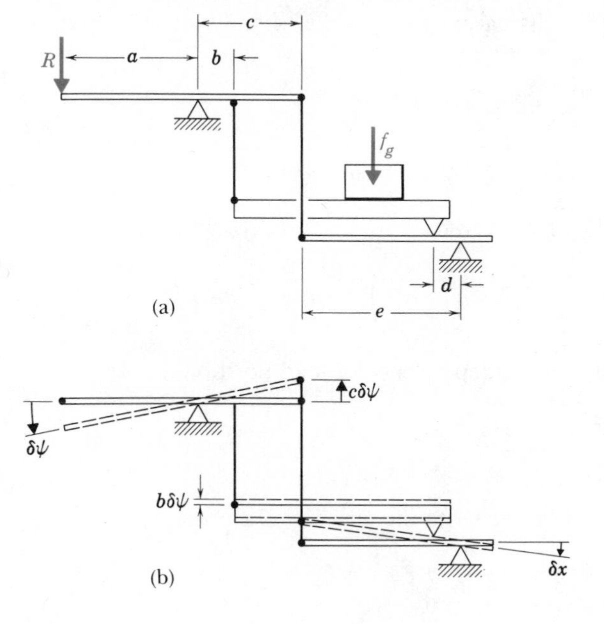

FIGURE 7-3

the position of the body on the platform. Also, determine the force ratio in terms of the dimensions.

Consider a virtual displacement of the system shown in Figure 7-3*b*, in which none of the members deforms. The virtual work principle will have the form

$$R\delta_R - f_g\delta_g = 0$$

in which δ_R and δ_g are the displacements of the points of application of the two forces. But if the value of R satisfying this is to be independent of the position of f_g along the platform, δ_g must also be independent of this position, which implies that during the virtual displacement the platform must remain level. In analyzing the geometry of the various components of the displacement, we observe that the vertical displacement δy of a point located a distance r from the pivot of a horizontal rod is related to the small angle of rotation $\delta\theta$ by

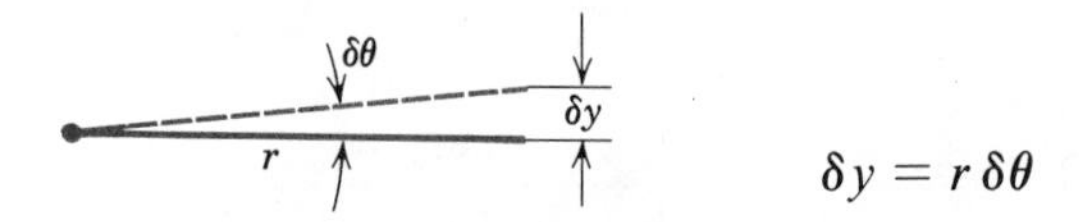

$$\delta y = r\,\delta\theta$$

Thus the right-hand end of the upper rod undergoes a displacement equal to $c\delta\psi$. Since this is equal to the displacement of the left-hand end of the lower rod,

$$e\,\delta\chi = c\,\delta\psi$$

We can express the requirement that the platform remain level as

$$b\,\delta\psi = d\,\delta\chi$$

$$= d\left(c\frac{\delta\psi}{e}\right)$$

which yields, as the condition for independence of load position,

$$\frac{b}{d} = \frac{c}{e}$$

To determine the force ratio we write the virtual work principle as

$$R(a\,\delta\psi) - f_g(b\,\delta\psi) = 0$$

which yields the relationship

$$f_g = \frac{a}{b} R$$

Those who doubt that the principle of virtual work holds advantages for some problems are urged to analyze this system by isolating free-bodies and writing corresponding equations of equilibrium.

Equivalence Between the Principle of Virtual Work and Vanishing Force Resultant

Consider a system comprised of a collection of particles, the resultant force acting on the ith particle denoted by $\mathbf{f}_i$. It will prove useful to distinguish between external forces (those of interaction with bodies outside the system) and internal forces (those of interactions among particles within the system). We denote the external force acting on the ith particle by $\mathbf{f}_{ie}$ and the force exerted on the ith particle by the jth particle by $\mathbf{f}_{i/j}$. Then the condition that the ith particle be in equilibrium may be written as

$$\mathbf{f}_i = \mathbf{f}_{ie} + \sum_j \mathbf{f}_{i/j} = \mathbf{0} \tag{7-3}$$

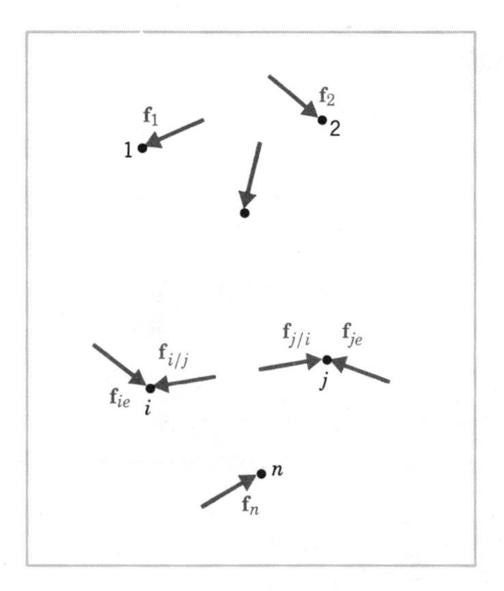

Now, let us dot multiply both sides of each such equation by a vector $\delta\mathbf{r}_i$ and add all the equations pertaining to the system:

$$\sum_i \mathbf{f}_{ie} \cdot \delta \mathbf{r}_i + \sum_i \sum_j \mathbf{f}_{i/j} \cdot \delta \mathbf{r}_i = 0$$

Newton's third law,

$$\mathbf{f}_{i/j} = -\mathbf{f}_{j/i}$$

permits simplification of the contribution to the double sum from the pair of terms

$$\mathbf{f}_{i/j} \cdot \delta \mathbf{r}_i + \mathbf{f}_{j/i} \cdot \delta \mathbf{r}_j = \mathbf{f}_{i/j} \cdot (\delta \mathbf{r}_i - \delta \mathbf{r}_j)$$

If the vectors $\delta \mathbf{r}_i$ and $\delta \mathbf{r}_j$ are small changes in position of the ith and jth particles, the vector $(\delta \mathbf{r}_i - \delta \mathbf{r}_j)$ is the change in the position vector $\mathbf{r}_{i/j}$ which locates the ith particle relative to the jth particle. Furthermore, we note that if $\delta \mathbf{r}$ is small compared to $\mathbf{r}$, the projection of $\delta \mathbf{r}$ onto $\mathbf{r}$ is equal to the change in the magnitude of $\mathbf{r}$; that is,

$$|\delta \mathbf{r}| \cos \measuredangle_{\mathbf{r}}^{\delta \mathbf{r}} = \delta |\mathbf{r}|$$

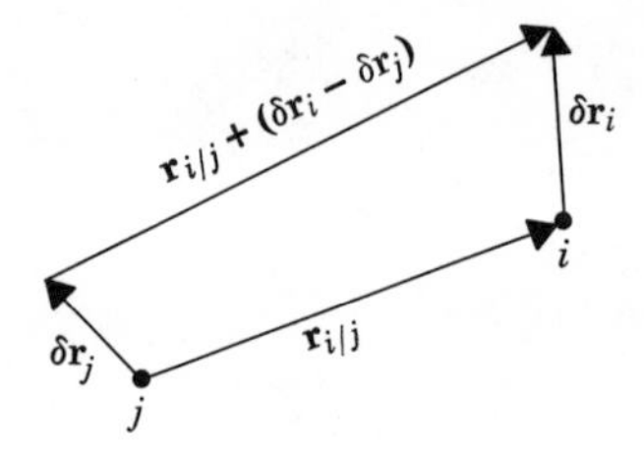

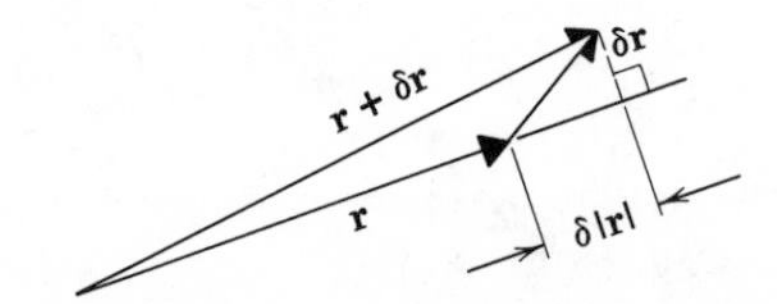

Then, because $\mathbf{f}_{i/j}$ is parallel to $\mathbf{r}_{i/j}$,

$$\begin{aligned} \mathbf{f}_{i/j} \cdot \delta \mathbf{r}_{i/j} &= |\mathbf{f}_{i/j}|\,|\delta \mathbf{r}_{i/j}| \cos \measuredangle_{\mathbf{r}_{i/j}}^{\delta \mathbf{r}_{i/j}} \\ &= -T_{ij} \delta |\mathbf{r}_{i/j}| \end{aligned}$$

where T_{ij} is defined as positive when the action-reaction is attractive and as negative when it is repulsive. With these results the equilibrium equation (7-3) can be rewritten as

$$\boxed{\delta W \equiv \sum_i \mathbf{f}_{ie} \cdot \delta \mathbf{r}_i - \sum_{i-j} T_{ij} \delta |\mathbf{r}_{i/j}| = 0} \tag{7-4}$$

in which the symbol Σ_{i-j} means sum over all pairs of particles, and $\delta\,|\,\mathbf{r}_{i/j}\,|$ is the increase in the distance between the ith and jth particles.

The left-hand side of Equation 7-4 is the virtual work of the forces acting on the system. We see from the above derivation that $\delta W = 0$ follows from $\mathbf{f}_i = \mathbf{0}$, so that if the system is in equilibrium the virtual work necessarily vanishes. Furthermore, if Equation 7-4 is satisfied for *arbitrary* $\delta \mathbf{r}_i$, the equation $\delta W = 0$ implies that $\mathbf{f}_i = \mathbf{0}$; therefore Equation 7-4 is also a sufficient condition for equilibrium.

The meanings of the terms in Equation 7-4 may be reinforced by a reexamination of the examples presented earlier. In determining the relationship between P and M for the slider crank mechanism, we considered a virtual displacement in which the distances between all pairs of particles that interact remain constant; therefore, the virtual work of the internal forces, $\Sigma_{i-j} T_{ij} \delta |\mathbf{r}_{i/j}|$, is zero in this case. Of the external forces, the reaction from the wall against the piston is perpendicular to the virtual displacement of the piston so that this $\mathbf{f}_{ie} \cdot \delta \mathbf{r}_i$ is zero. And because the $\delta \mathbf{r}_i$ for the reaction at the crank support is zero, the virtual work of this external force is also zero. This leaves us with

$$\begin{aligned} \delta W &= \sum_i \mathbf{f}_{ie} \cdot \delta \mathbf{r}_i \\ &= -P\delta x - M\delta\theta \end{aligned}$$

as written previously.

In determining the internal force in the connecting rod, we considered a different virtual displacement, this one in which the distance between the particles on either side of the separation undergoes a variation. That is, $\delta |\mathbf{r}_{i/j}| = \delta l$, and because these particles interact with a force R, the contribution to the virtual work there is

$$-\sum_{i-j} T_{ij} \delta |\mathbf{r}_{i/j}| = R\,\delta l$$

The only external force that would do work on this virtual displacement is P, so that

$$\sum_i \mathbf{r}_{ie} \cdot \delta \mathbf{r}_i = -P\,\delta x$$

In the analysis of the platform balance scale, the virtual displacement was chosen so that neither the internal forces nor the support reactions would do work. This illustrates the main strategy in using the principle of virtual work to best advantage: a virtual displacement that will involve as few unknown forces as possible in the equation of vanishing virtual work should be chosen; in this way one is spared the trouble of eliminating unknowns from a set of simultaneous equations.

FINAL EXAMPLE

Determine the bending moment at the section where the force is applied to the beam in Figure 7-4a.

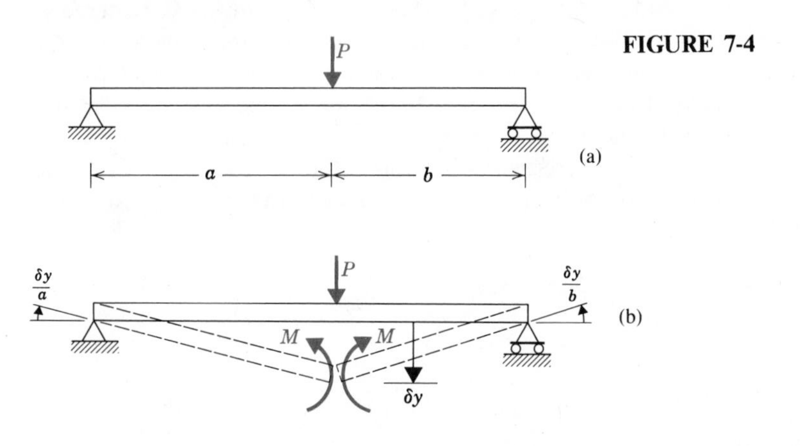

FIGURE 7-4

Consider the virtual displacement shown in Figure 7-4b. The virtual work that the bending moment would do on the left portion is equal to $-M\,\delta y/a$ and that which the corresponding reaction would do on the right portion is $-M\,\delta y/b$. The principle of virtual work then states

$$P\delta y - M\left(\frac{\delta y}{a} + \frac{\delta y}{b}\right) = 0$$

from which

$$M = \frac{abP}{a + b}$$

The previous analysis which led to this result required that at least one of the support reactions be determined first.

PROBLEMS

7-14 Work the example problem on page 293 without using the method of virtual work.

7-15 From sketches of small displacements, determine the relationship

$$-\delta x = a(\sin\theta + \tan\phi\cos\theta)\delta\theta$$

for the virtual displacement shown in Figure 7-2*b*.

7-16 As the two shafts of the gearbox turn, shaft *A* is observed to rotate faster than shaft *B*. A hand strength contest has been proposed, in which each of two contestants is to grip one shaft and try to cause the other shaft to rotate against the opponent's will. Which shaft would you select?

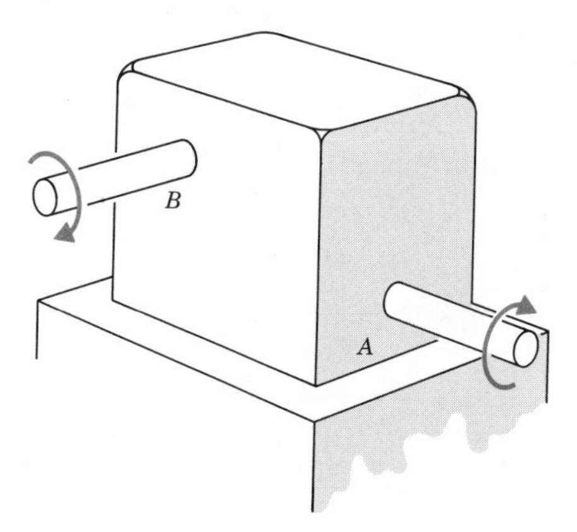

7-17 Referring to the gearbox of the previous problem, the directions of rotation of the two shafts is the same as viewed along the shaft toward the gearbox. Will the directions of the torques applied to the shafts be in the same or opposite directions for equilibrium?

7-18 Use the principle of virtual work to determine the bending moment at the section located by *x*.

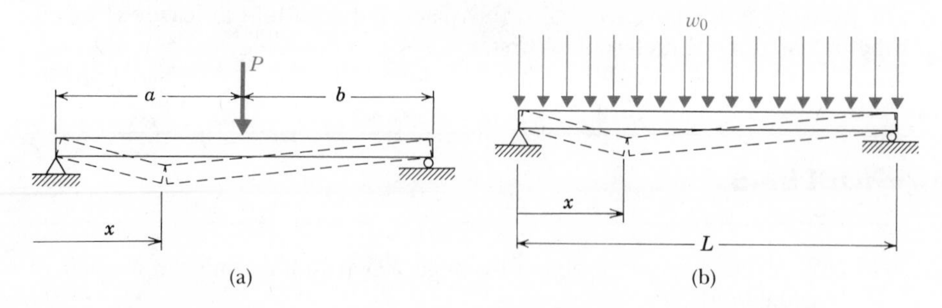

(a) (b)

7-19 Use the principle of virtual work to determine the force in member AB of the truss. Do *not* compute support reactions. All members are of equal length.

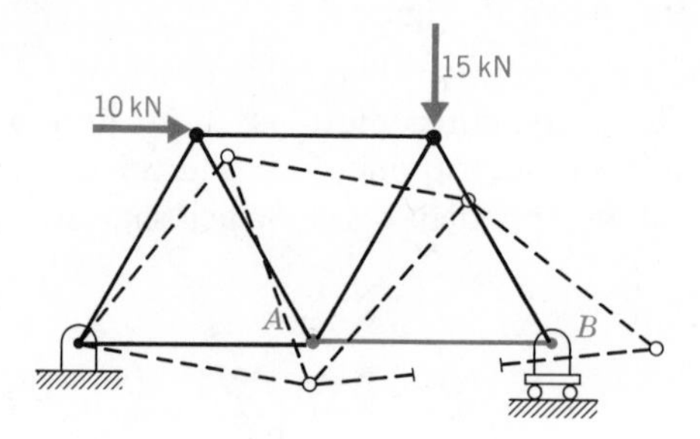

7-20 The mechanism shown is used to lift materials at a building site. What is the force in the cylinder in terms of the gravity force f_g and the angle θ? *Ans.* $4 f_g \tan \theta$.

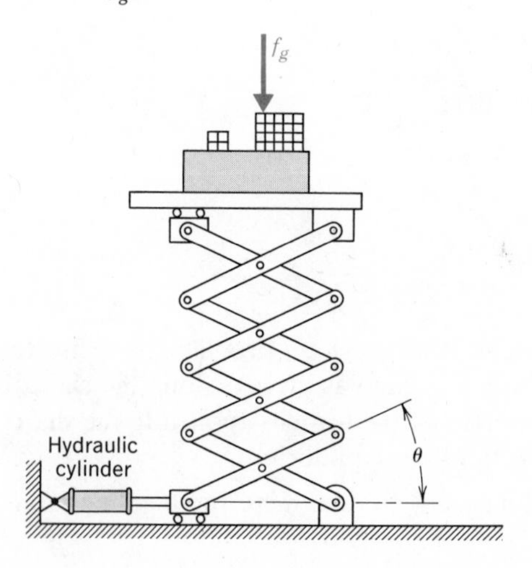

Use the method of virtual work to solve each of the following problems.

7-21 Problem 2-53 (p. 39).

7-22 Problem 2-57 (p. 41).

7-23 Problem 4-29 (p. 133).

7-24 Problem 4-31 (p. 133).

7-25 Problem 4-32 (p. 134).

7-26 Problem 4-33 (p. 134).

7-27 Problem 4-34 (p. 135).

7-28 Problem 4-35 (p. 135).

7-29 Problem 4-37 (p. 136).

7-30 Calculate the reaction at C of Problem 4-38(b) (p. 137).

7-31 Calculate the tension in the cable CB of Problem 4-39 (p. 137).

7-32 Problem 4-40 (p. 137).

7-33 Problem 4-41 (p. 138).

7-34 Problem 4-42 (p. 138).

7-35 Calculate the reaction at A of Problem 4-44 (p. 139).

7-36 Problem 4-49 (p. 141).

7-37 Problem 4-50 (p. 142). Note that the connecting rod moves as a rotation around point C', called the instantaneous center for the rod.

7-38 Calculate the moment M for Problem 4-51 (p. 142).

7-39 Calculate the reaction at C of Problem 4-54 (p. 143).

7-40 Problem 4-58 (p. 144).

7-41 Problem 4-61 (p. 145).

7-42 Problem 4-63 (p. 146).

7-43 Problem 4-82 (p. 153).

7-44 Problem 5-13 (p. 170).

7-45 Problem 5-14 (p. 171).

7-46 Calculate the forces in members AB and BC of Problem 5-27 (p. 174).

7-47 Problem 5-53 (p. 192).

7-48 Problem 5-54 (p. 193).

7-49 Evaluate $M(x)$ for Problem 6-2 (p. 215).

7-50 Problem 6-38 (p. 241).

7-51 Problem 6-39 (p. 242).

Appendix A

SOME USEFUL NUMERICAL VALUES

A-1 PHYSICAL CONSTANTS

Universal gravitational constant	6.672×10^{-11} N·m²/kg²
Mass of the earth	5.983×10^{24} kg
Speed of light	0.2998 Gm/s
Standard atmospheric pressure	101.325 kPa
Density of air (0°C, 1 atm)	1.29 kg/m³
Density of water (4°C, 1 atm)	1.0000 Mg/m³
Density of concrete	2.4 Mg/m³
Density of steel	7.7–7.9 Mg/m³

A-2 PREFIXES FOR SI UNITS

Exponents in higher order derived units apply to prefixes as well as units. For example, $km^2 = (10^3\ m)^2 = 10^6\ m^2$

Prefix	SI Symbol	Multiplication Factor
exa	E	10^{18}
peta	P	10^{15}
tera	T	10^{12}
giga	G	10^{9}
mega	M	10^{6}
kilo	k	10^{3}
hecto*	h*	10^{2}
deka*	da*	10^{1}
deci*	d*	10^{-1}
centi*	c*	10^{-2}
milli	m	10^{-3}
micro	μ	10^{-6}
nano	n	10^{-9}
pico	p	10^{-12}
femto	f	10^{-15}
atto	a	10^{-18}

*Use discouraged.

A-3 UNITS OF MEASUREMENT

Quantity	SI Unit	Other Units
length	meter (m)	inch = 25.400 mm foot = 0.304 800 m statute mile = 1.609 344 km nautical mile = 1.852 000 km fathom = 1.828 800 m
mass	kilogram (kg)	pound-mass = 0.453 592 37 kg slug = 14.593 903 kg grain = 64.798 910 mg
time	second (s)	minute = 60 s hour = 3600 s

Quantity	SI Unit	Other Units
temperature	kelvin (K)	degrees Fahrenheit $t_K = (t_F + 459.67)/1.8$ degrees Celsius $t_K = t_C + 273.15$
plane angle	radian (rad)	degree = $(\pi/180)$ rad minute = $(1/60)$ deg second = $(1/60)$ min
area	m^2	acre = 4046.956 m^2 hectare = 10^4 m^2
energy	joule (J) $J = N \cdot m$ $= kg \cdot m^2/s^2$	foot-pound-force = 1.355 818 J erg = 10^{-7} J calorie (mean) = 4.190 02 J British thermal unit = 1.055 kJ
force	newton (N) $N = kg \cdot m/s^2$	pound-force = 4.448 222 N kip = 4.448 222 kN poundal = 0.138 255 N dyne = 10^{-5} N kilogram-force = 9.806 650 N
power	watt (W) $W = J/s$ $= kg \cdot m^2/s^3$	horsepower (550 ft-lbf/s) = 745.7 W British thermal unit per hour = 0.293 071 W refrigeration ton = 3.517 kW
pressure	pascal (Pa) $Pa = N/m^2$ $= kg/m \cdot s^2$	bar = 100 kPa pound per square inch = 6.894 757 kPa centimeter of mercury (0°C) = 1.333 22 kPa centimeter of water (4°C) = 98.0638 Pa
velocity	m/s	kilometer per hour = 0.277 778 m/s mile per hour = 0.447 040 m/s knot = 0.514 444 m/s
volume	m^3	liter = $dm^3 = 10^{-3}$ m^3 fluid ounce = $2.957\ 353 \times 10^{-5}$ m^3 US liquid gallon = $3.785\ 412 \times 10^{-3}$ m^3 barrel = 0.158 987 m^3

Appendix B

PROPERTIES OF LINES, AREAS, AND VOLUMES

B-1 LINES

L = length $\quad c_i = \dfrac{1}{L}\int i\,dL \quad (i = x, y, z)$

(c_x, c_y, c_z) = coordinates of centroid

	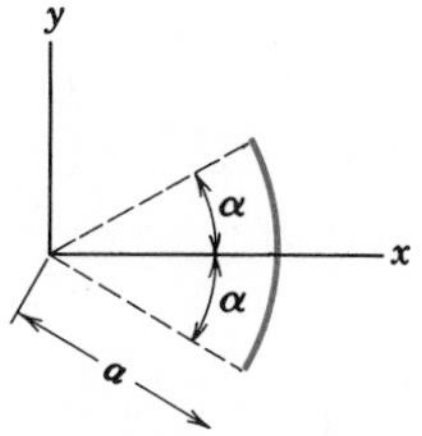 Segment of circle	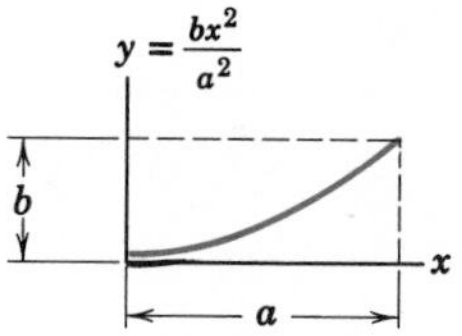 Parabola ($y = \frac{bx^2}{a^2}$)
$L =$	$2a\alpha$	$\frac{a}{2}\left[u + \frac{a}{2b}\log\left(u + \frac{2b}{a}\right)\right]$ $u = \sqrt{1 + (2b/a)^2}$
$c_x =$	$\frac{a \sin\alpha}{\alpha}$	$\frac{a^4}{12b^2L}(u^3 - 1)$ $u = \sqrt{1 + (2b/a)^2}$
$c_y =$	0	$\frac{a^3}{32bL}\left\{\left[1 + 2\left(\frac{2b}{a}\right)^2\right]u - \left(\frac{a}{2b}\right)\log\left(u + \frac{2b}{a}\right)\right\}$ $u = \sqrt{1 + (2b/a)^2}$

B-2 SURFACES

Plane Areas

A = Area $\quad c_i = \frac{1}{A}\int i\, dA \quad (i = x, y)$

$\mathcal{I}_{xx} = \int y^2\, dA \quad \mathcal{I}_{yy} = \int x^2\, dA \quad \mathcal{I}_{xy} = -\int xy\, dA$

	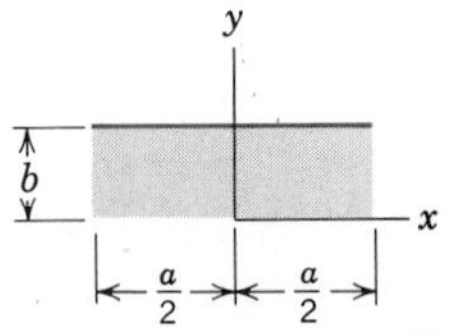 b; $\frac{a}{2}$, $\frac{a}{2}$	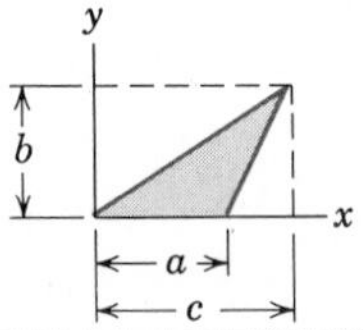 b; a; c	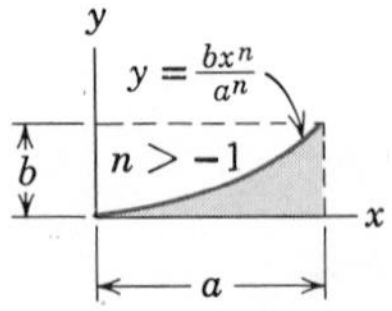 $y = \frac{bx^n}{a^n}$; $n > -1$; b; a
$A =$	ab	$\frac{ab}{2}$	$\frac{ab}{n+1}$
$c_x =$	0	$\frac{a+c}{3}$	$\frac{n+1}{n+2}a$
$c_y =$	$\frac{b}{2}$	$\frac{b}{3}$	$\frac{(n+1)b}{2(2n+1)}$
$\mathcal{I}_{xx} =$	$\frac{ab^3}{3}$	$\frac{ab^3}{12}$	$\frac{ab^3}{3(3n+1)}$
$\mathcal{I}_{yy} =$	$\frac{a^3b}{12}$	$\frac{ab}{12}(a^2 + ac + c^2)$	$\frac{a^3b}{n+3}$
$\mathcal{I}_{xy} =$	0	$-\frac{b^2a}{24}(2c + a)$	$-\frac{a^2b^2}{4(n+1)}$

	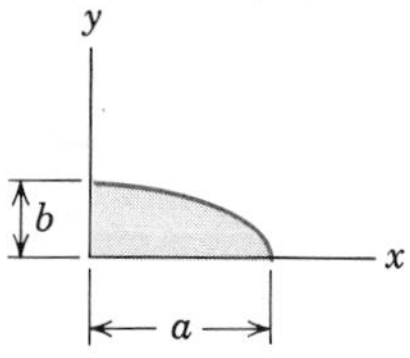 Quarter ellipse	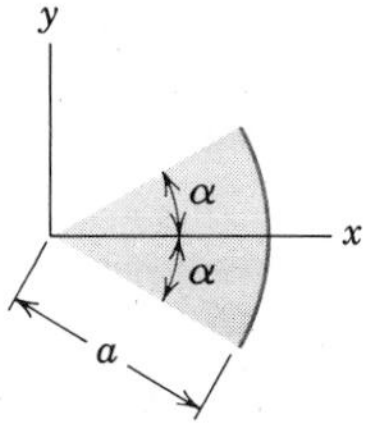 Sector of circle	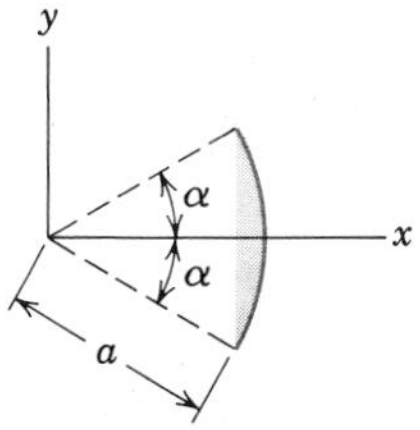 Segment of circle
$A =$	$\frac{\pi ab}{4}$	$a^2\,\alpha$	$a^2(\alpha - \frac{1}{2}\sin 2\alpha)$
$c_x =$	$\frac{4a}{3\pi}$	$\frac{2a \sin\alpha}{3\,\alpha}$	$\frac{4a}{3}\,\frac{\sin^3\alpha}{2\alpha - \sin 2\alpha}$
$c_y =$	$\frac{4b}{3\pi}$	0	0
$\mathcal{I}_{xx} =$	$\frac{\pi ab^3}{16}$	$\frac{a^4}{8}(2\alpha - \sin 2\alpha)$	$\frac{a^4}{8}(2\alpha - \frac{4}{3}\sin 2\alpha + \frac{1}{6}\sin 4\alpha)$
$\mathcal{I}_{yy} =$	$\frac{\pi a^3 b}{16}$	$\frac{a^4}{8}(2\alpha + \sin 2\alpha)$	$\frac{a^4}{8}(2\alpha - \frac{1}{2}\sin 4\alpha)$
$\mathcal{I}_{xy} =$	$-\frac{b^2a^2}{8}$	0	0

Sector of a Surface of Revolution

Area of the surface generated by revolving $\mathcal{C}$ about the axis through the angle θ:

$$A = \theta c_r L$$

For $\theta = 2\pi$, the formula becomes that for a complete surface of revolution and expresses the first Theorem of Pappus:

$$A = 2\pi c_r L$$

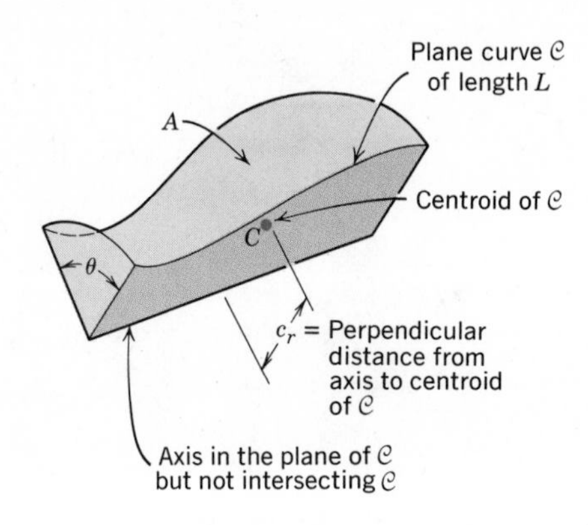

B-3 VOLUMES

V = volume $\quad c_i = \frac{1}{V}\int i\, dV \quad (i = x, y, z)$

(c_x, c_y, c_z) = coordinates of centroid

$h = a(1 - \cos\alpha)$

	Wedge	Segment of sphere
$V =$	$\dfrac{abc}{2}$	$\pi h^2 \left(a - \dfrac{h}{3}\right)$
$c_x =$	$\dfrac{2a}{3}$	$a \dfrac{(1 - h/2a)^2}{(1 - h/3a)}$
$c_y =$	$\dfrac{b}{3}$	0
$c_z =$	$\dfrac{c}{2}$	0

Cone

$V = \dfrac{Ah}{3}$

$c_x = \dfrac{3}{4} b_x$

$c_y = \dfrac{3}{4} b_y$

$c_z = \dfrac{1}{4} h$

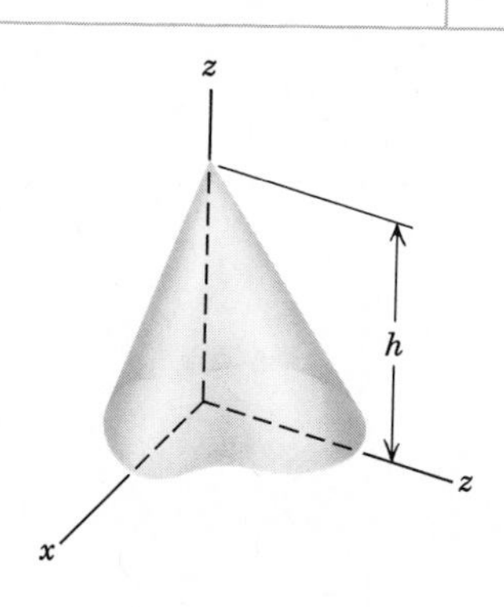

Area $= A$

$C(b_x, b_y)$

Base section of cone

Sector of a Body of Revolution

Volume of the sector generated by revolving $\mathcal{S}$ about the axis through the angle θ:

$$V = \theta c_r A$$

For $\theta = 2\pi$, the formula becomes that for a complete body of revolution and expresses the second Theorem of Pappus:

$$V = 2\pi c_r A$$

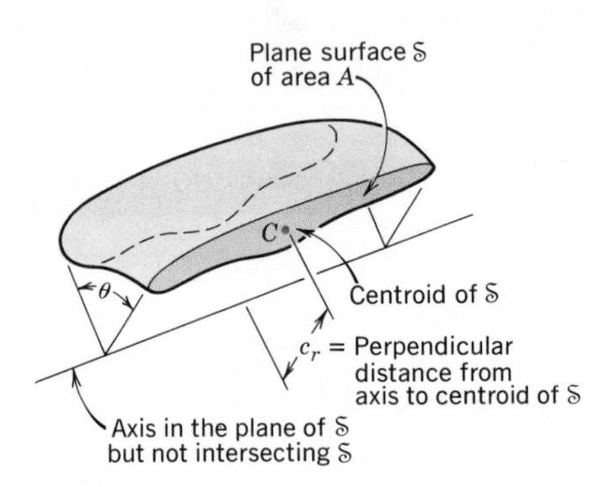

REFERENCES

IEEE Standard Metric Practice IEEE Std 268-1979 (The Institute of Electrical and Electronics Engineers, 1979)

Isaac Asimov, *Asimov's Biographical Encyclopedia of Science and Technology* (Doubleday and Company, 1972).

George W. Housner and D. E. Hudson, *Applied Mechanics Statics*, Second Edition (D. van Nostrand Company, 1961).

Edward J. Routh, *Analytical Statics*, *Vol I* (Cambridge at the University Press, 1896).

Murray R. Spiegel, *Vector Analysis* (Schaum's Outline Series, Schaum Publishing Company, 1959).

INDEX